Forensic Science

The Science Behind the Truth

Forensic Science

The Science Behind the Truth

By

Ian Shaw

University of Canterbury, New Zealand
Email: ian.shaw@canterbury.ac.nz

with

Anna Sandiford

The Forensic Science Group Ltd., New Zealand
Email: anna@theforensicgroup.co.nz

ROYAL SOCIETY
OF **CHEMISTRY**

Print ISBN: 978-1-83916-674-7
PDF ISBN: 978-1-83767-040-6
EPUB ISBN: 978-1-83767-041-3

A catalogue record for this book is available from the British Library

The Royal Society of Chemistry is a charity, registered in England and Wales, Number 207890, and a company incorporated in England by Royal Charter (Registered No. RC000524), registered office: Burlington House, Piccadilly, London W1J 0BA, UK, Telephone: +44 (0)20 7437 8656.

Visit our website at books.rsc.org

Printed in the United Kingdom by CPI Group (UK) Ltd, Croydon, CR0 4YY, UK

Dedication

To David and Pamela

With gratitude for support, smiles and love.

Forensic Science: The Science Behind the Truth
By Ian Shaw with Anna Sandiford

Published by the Royal Society of Chemistry, www.rsc.org

Preface

This book was inspired by a course in forensic science for criminal justice that IS created and runs at the University of Canterbury, New Zealand. The course has been altered, modified and tailored over the years to meet the needs of our ostensibly non-scientific students. We hope that our book meets the same needs.

The book is written for anyone interested in forensic science, particularly those in professions such as criminal justice and law, who may encounter forensic evidence in their work but lack scientific qualifications. We hope that our book will give them an insight into forensic investigations and the science that underpins them. We also hope that our book includes sufficient scientific detail to engage readers with scientific backgrounds.

For our non-scientist readers who might balk at the sight of a molecular structure, please consider the science as evidence of the complexity of forensic methodology and that there is a good scientific basis underpinning the techniques that forensic scientists use in their work – even if you do not understand the scientific minutiae.

We hope our book will also appeal to readers who want to understand the science behind forensic investigations; in your case, we hope that the underpinning scientific (particularly chemistry) detail will reinforce your appreciation and understanding of forensic techniques and their value as evidence.

Finally, for those of you who are 'just' interested in forensic science, we hope you enjoy our book and that at the end of it, you will appreciate how forensic science is used in criminal investigations beyond the glorified impression that might result from watching some popular TV series.

Forensic Science: The Science Behind the Truth
By Ian Shaw with Anna Sandiford

Published by the Royal Society of Chemistry, www.rsc.org

As authors, we have two very different roles: IS primarily authored the book, but since he is not a practicing forensic scientist, he needed input to ensure that the writing accurately reflects current forensic science practice – AS provided that all-important assurance.

The book is written in the first person singular to reflect its authorship and to create a more personal reading experience.

We hope you enjoy our book and, by the end, acquire a deeper understanding of the science behind the truth.

Ian Shaw and Anna Sandiford
Christchurch, New Zealand

Acknowledgements

So many friends and colleagues have helped in various ways during the writing of this book; we are grateful to them all.

Since the book covers a broad array of subject matter, we sought the expertise of colleagues in their specialist fields to fill the gaps in our knowledge and understanding.

IS thanks Professor Martin Kennedy from the University of Otago, NZ, who in his inimitably friendly way, tutored me through the complexities of the nucleic acids for Chapter 6. I am very grateful to Dr Helen Poulsen, a forensic toxicologist at ESR, NZ, for tactfully pointing out the innumerable mistakes in the first draft of Chapter 9 and for so willingly providing data and illustrations from forensic cases, and to Associate Professor Hilary Hamnet, Lincoln University, UK, for her invaluable expert advice on forensic toxicology case investigations. I thank Dr Heather Wolffram, University of Canterbury, NZ, for sharing her considerable expertise on the history of forensic science and for her invaluable comments on Chapter 2. Ian Harrison, Fingerprint Officer, Christchurch Central Police Station, NZ, willingly contributed his expertise to Chapter 6 for which I am very grateful. I thank Dr Andrew Miller, Senior Lecturer in Anatomical Pathology, University of Otago, NZ, for his guidance on *post-mortem* procedures and follow up investigations, as well as for reading and commenting on Chapter 8; and I thank The Royal College of Pathologists for facilitating my connection with Dr Rosa Miquel, Consultant Histopathologist at the Institute of Liver Studies, King's College Hospital, London, UK, who so willingly helped by providing important images for Chapter 8. Dr Jacinta Cording not only gave an excellent lecture in the University of Canterbury's Forensic Science for Criminal Justice

Forensic Science: The Science Behind the Truth
By Ian Shaw with Anna Sandiford

Published by the Royal Society of Chemistry, www.rsc.org

course upon which Chapter 10 is based, but she also willingly gave her time to review and correct the draft of Chapter 10. I thank Dr Chris Palenic from Microtrace LLC, USA for his help in explaining electrostatic detection and for providing the EDD image for Chapter 12. I am grateful to Dr François Bompart for translating and interpreting Bertillon's French which flummoxed an author who did not progress beyond school French!

AS is incredibly grateful to Paige McElhinney of The Forensic Group Ltd NZ for her advice and comments on the challenges of court attendance, DNA, probability and her extensive experience in historical case reinvestigations in Chapters 13, 14 and 15. Glynn Rigby of Zavest Investigations NZ proved equally invaluable in assisting with the writing of Chapter 11 and his always unending knowledge of firearms and firearms case investigation.

We wrote many letters and emails to copyright holders, requesting permission to reproduce images. Sadly, many did not even acknowledge our correspondence – a UK prison refused permission and even threatened us not to mention them in our book! But to those who gave us permission (acknowledged in Figure legends), we are very grateful: without illustrations, explaining forensic science would have been a far more difficult task. In particular, Brad Brown of Arrowhead Forensics, USA, gave us permission to reproduce several images – he was friendly, helpful and efficient, and Greg Callander (retired) of the New South Wales Police Force, Australia, granted permission within two hours of our request; once again, we are grateful.

Last, but by no means least, we thank IS's long-suffering partner, David Zehms, for reading every chapter and rectifying hundreds of grammatical glitches. This was a mammoth task for which we are deeply grateful.

Author Biographies

Ian Shaw is a Professor of Toxicology at the University of Canterbury in New Zealand. He began his career as a forensic scientist and progressed *via* forensic toxicology to general toxicology, accumulating over 40 years of experience in the field. He is a passionate science communicator and frequently appears on TV and radio and writes articles for newspapers and magazines. He won the New Zealand Association of Scientists' Science Communicator's Award in 2009 and was awarded a DSc by the University of Bath in 2019 for his work in toxicology and science communication. His greatest joy remains being honoured as the Science Lecturer of the Year by the University of Canterbury Students' Association in 2012, 2013, 2014, 2016, 2018 and 2020.

Anna Sandiford is a senior forensic science consultant originally trained in the UK and currently based in New Zealand. Initially educated in geology followed by forensic science, she went on to be the founder of New Zealand's only independent forensic science consultancy. She has many years of casework experience, ranging from drink driving through to multiple homicides and historical case reinvestigations. She is a passionate science educator for the legal profession, particularly for those who gave up science during school and now face the seemingly complicated mire of expert witness reports and statements. She is also an Adjunct Associate Professor at the University of Canterbury in New Zealand.

Forensic Science: The Science Behind the Truth
By Ian Shaw with Anna Sandiford

Published by the Royal Society of Chemistry, www.rsc.org

Units Used in This Book

This book is not intended only for a physical science literate readership. We welcome non-scientist readers and thought it might be helpful to explain the units used in our book.

Units used in this book and their meanings

Unit	In full	Meaning
Mass		
ng	nanogram	10^{-9} g One thousand millionth of a gram
μg	microgram	10^{-6} g One millionth of a gram
mg	milligram	10^{-3} g One thousandth of a gram
g	gram	As a guide, the mass of 1 mL water at 4 °C
kg	kilogram	10^{3} g One thousand grams
Volume		
μL	microlitre	10^{-6} L One millionth of a litre
mL	milliliter	10^{-3} L One thousandth of a litre
L		The volume of a 100 mm × 100 mm × 100 mm cube. Approx. 0.262 gallons
Length		
mm	millimeter	10^{-3} m One thousandth of a metre
cm	centimeter	10^{-2} m One hundredth of a metre
m	metre	The distance travelled by light in a vacuum over 1/299792458 s

Forensic Science: The Science Behind the Truth
By Ian Shaw with Anna Sandiford

Published by the Royal Society of Chemistry, www.rsc.org

(*Continued*)

Unit	In full	Meaning
km	kilometer	10^3 m One thousand metres; approx. 0.621 miles
Time		
s	second	The definition is complex but relates to the frequency of transfer between two ground states of the caesium-133 (^{133}Cs) atom. Frequency is measured in Hertz (Hz) which has the units 1/s or s^{-1}.
min	minute	60 s
h	hour	60 min, 3 600 s

Units of concentration

The Royal Society of Chemistry's general convention is to only allow the use of reciprocal units of concentration in the books it publishes. This means that in this book units of mass (m, *e.g.*, mg) per volume (v, *e.g.*, L) are expressed as $m\ v^{-1}$ (*e.g.*, mg L^{-1}). This is the same as milligrams per litre or mg/L.

Contents

Forensic Science: The Science Behind the Truth
By Ian Shaw with Anna Sandiford

Published by the Royal Society of Chemistry, www.rsc.org

1 The Whittle Case – 1976

Lesley Whittle was a young woman from a rich family in the UK Midlands. She was kidnapped by a man involved in a string of criminal activities that had escalated from petty crimes, through post office holdups to murder. The perpetrator of these crimes came to be known as the Black Panther. Had the ransom he demanded been paid, it would have made the Black Panther a rich man. He was elusive and devious; he left cryptic Dymo label maker messages at and around the scenes of his crimes. These messages often confused the police, but occasionally provided information. Clearly, the Black Panther was 'playing' with the police, pitting his wits against them to fulfil his warped sense of humour. The Whittle case is one of the most expensive murder hunts in the UK's crime history. The case relied on traditional forensic investigations and good old-fashioned policing. It is a perfect illustration of how forensic science was before DNA became the all-important, unequivocal evidence in so many modern cases. We use this case to highlight the value of piecing together the vast jigsaw of forensic, circumstantial and police evidence to reveal a picture leading to conviction.

1.1 Setting the Scene...

Take yourself back to 1976: think trendy teens wearing tight-fitting bell bottoms and platform shoes dancing to Abba's Dancing Queen at the disco. The first commercial Concorde flight had just taken off, the USA vetoed the UN resolution calling for a Palestinian state, the IRA bombed the West End of London, and the Apple Computer Company

Forensic Science: The Science Behind the Truth
By Ian Shaw with Anna Sandiford

Published by the Royal Society of Chemistry, www.rsc.org

was formed. To cap off the year, the first Punk Rock group blared its irreverence from transistor radios...

On 14th June 1976, the trial of the Black Panther (Donald Neilson; Figure 1.1) began in Oxford Crown Court. In July of the same year, he was convicted of Lesley Whittle's (Figure 1.2) murder and sentenced

Figure 1.1 Donald Neilson (left; reproduced from ref. 1 under Wikipedia Fair Use Guidelines) and Lesley Whittle (right; reproduced from ref. 2 under Wikipedia Fair Use Guidelines).

14th January, 1975
Lesley Whittle kidnapped
↓
7th March, 1975
Lesley's body found
↓
11th December, 1975
Donald Neilson arrested
↓
14th June 1976
Neilson's trial began
↓
1st July, 1976
Neilson sentenced to life in prison
↓
17th December, 2011
Donald Neilson died 'in prison'

Figure 1.2 Timeline of the Whittle case.

to life imprisonment by Justice Mars-Jones. Thirty-five years later, on 17th December 2011, he became ill (experiencing breathing complications associated with motor neurone disease) and was taken from Norwich prison to Norfolk and Norwich University Hospital, where he died (age 75 years) the next day.

The timeline of the Whittle case is shown in Figure 1.2.

1.2 The Whittle Family Feud

George Whittle, the owner of Whittle Coaches, had been living separately from his wife Salina for 30 years and was living with his common law wife Dorothy (the mother of George's two children Ronald and Lesley) in their impressive home at Highley (Figure 1.3) in

Figure 1.3 Map of Great Britain showing approximate locations of important places in the Lesley Whittle case. Created with Bing Image.

Shropshire. George died in 1970 (aged 65 years), but before he died, to avoid death duties, he gifted Ronald £107 000, Lesley £82 000 (put into a trust until she was older), and Dorothy three houses plus £70 000. He gave nothing to his estranged wife Selina and excluded her from his will – she relied on a paltry income of about £7.50 per week, consisting of a widow's pension and supplementary benefit. In 1972, Selina Whittle began legal proceedings to obtain money from her deceased husband's estate. She was granted £1500 a year from the Whittle estate for the rest of her life. This allayed her financial problems, and she thought that was the end of it.

However, the Whittle family feud was reported by the UK's Daily Express newspaper, and the resulting articles likely piqued the interest of the financially bereft Donald Neilson. Perhaps this is when Neilson formulated his plan to kidnap Lesley Whittle and demand a huge ransom for her safe return to her family. This would solve *his* financial woes. Problem solved!

Neilson's plan was to kidnap Lesley Whittle (there is some uncertainty about this, he might have intended to take another member of the family) and demand a ransom of £50 000 (approx. £577 000 in 2021) – he had deduced from their finances reported in the newspaper articles that the family could easily afford to pay this ransom for Lesley's safe return. Ex-army Neilson planned the kidnapping with military precision for the next three years.

1.3 The Kidnapping

On 14th January, 1975, Neilson kidnapped Lesley Whittle from her home in Highley. Dorothy had returned home from her friend's place at about 1.30 am; she saw that Lesley was fast asleep, took several sleeping pills and then went to bed herself.

Neilson cut the telephone wires to the house and entered through the garage. Dorothy heard nothing in her narcotic slumber as Lesley was bound and dragged out of the house, wearing only her dressing gown. She wore her mother's slippers as she was pushed out to Neilson's stolen Morris 1100 car, which was waiting on the road at the side of the house.

Neilson drove her to Bathpool Park, Kidsgrove (about 64 miles/100 km; $1\frac{1}{2}$ h drive), bound and gagged on the back seat of his car. He forced her down a drainage shaft where he left her steel wire-tethered by her neck to the balustrade of a 2-foot by 5-foot (0.6 m × 1.5 m) steel platform, naked and in the dark.

Neilson recorded Lesley's 'ransom' message in the drainage shaft. The chilling voice of Lesley heard above the sound of running water assured her mother: '*There is nothing to worry about, Mum. I am OK. I got a bit wet, but I am quite dry now and I am being treated very well. OK?*'

Neilson displayed an element of compassion (although this hardly balances the overall horror of the atrocity); he gave Lesley a sleeping bag and a foam rubber pad to lie on. He brought her food, magazines and even a bottle of brandy to keep her occupied and sustained during her 52-day (14th January–7th March, 1975) incarceration.

Dorothy was unaware of her daughter's abduction until she looked into Lesley's bedroom on the morning of 14th January – Lesley was not there. Dorothy searched the house, and drove to her son, Ronald's home (not far away) to see if Lesley was there. Lesley was nowhere to be found.

When Dorothy returned home with Ronald and his wife, Gaynor, they looked around the house one more time just in case Lesley had returned. Gaynor made a very sinister discovery in the sitting room: three coiled up Dymo labels on a candy box. They read:

1. *No police £50 000 ransom to be ready to deliver wait for telephone call at Swan shopping centre telephone box 6 pm to 1 pm if no call return following evening when you answer call give your name only and listen you must follow instructions without argument from the time you answer you are on a time limit if police or tricks death.*
2. *Swan shopping centre Kidderminster deliver £50 000 in a white suitcase.*
3. *50 000 all in old notes £25 000 in £1 notes and £25 000 in £5 notes there will be no exchange only after £50 000 has been cleared will victim be released.*

The police were called and this began one of the UK's biggest and most notorious hunts for the perpetrator of a crime.

1.4 The Time Between Lesley's Kidnapping and Finding Her Body

Lesley was kidnapped on 14th January, 1975 and her body was not found until 7th March, 1975 – what happened during this 2-month period?

There was, of course, a huge hunt (one of the biggest in the UK's history) for Lesley Whittle's kidnapper.

The intervening weeks between Lesley's disappearance and finding her body hanging in the drainage shaft were marked by poor organisation and botched attempts to track down and apprehend her kidnapper. Looking back at newspaper articles from that time and more recent summaries of the case, it is clear that the police were fumbling around not knowing quite what to do. And when they decided on a plan of attack, it more than often made the situation worse.

The Whittle family were, of course, worried beyond belief about Lesley. They simply wanted her home. They did not care how this was achieved, they just wanted her back. They decided to pay Neilson's ransom because they saw this as the sure way to get Lesley home. They were desperate.

The police were in a dilemma too: they were torn between two poles: wanting Lesley returned safe and sound to her family, and capturing her kidnapper. Ideally, they wanted both.

Over the past few years, there had been a steep rise in kidnappings in the USA and Italy. The police officer in charge of the Whittle case (Chief Superintendent Bob Booth, MBE) had studied the policing methods used, particularly in the USA. The formula was to secure the release of the kidnapped person first before pursuing the kidnapper. This is exactly what Chief Superintendent Booth attempted to do.

The family went to their bank to obtain the money in used notes and the denominations specified by Neilson in the Dymo message he left in their sitting room the day Lesley was kidnapped.

Arrangements were made for Lesley's brother, Ron, to deliver the ransom money to the telephone box at the Swan Shopping Centre as demanded in Neilson's Dymo message. The phone was tapped and the police hid nearby. Neilson caught on to the police's involvement and the plan failed.

The second attempt involved Ron taking the ransom money to Bathpool Park, where Neilson would be waiting. Neilson planned to use a flashing torch as a signal for Ron to leave the ransom money. The police equipped Ron with a transmitter so that they could track him and off he went to Bathpool Park. Ron got lost on the way and arrived 90 minutes late. Neilson had signalled repeatedly with his torch – apparently, he mistook a courting couple in a parked car for Ron! This organisational shambles became even worse when a police car arrived. The police car had nothing to do with the case – a police officer stepped out of his car and lit a cigarette; he was just taking a short break (but very much in the wrong place!). Neilson interpreted the police car as an ambush (albeit a very obvious one), and it is thought that he opened the man hole cover, descended the ladder to the platform to which Lesley was tethered, and pushed her off the platform to her death. Or was it an accident – did she simply lose her footing and fall?

Lesley's body was not found for 7 weeks.

1.5 Lesley's Death

Imagine the scene: Lesley was attached to the railing of a platform situated partway down the drainage shaft below its manhole access. A steel ladder led down from the manhole to the platform. Neilson regularly returned to the drainage shaft in Bathpool Park to bring Lesley food and perhaps books and magazines. So, Lesley was accustomed to his visits. His final visit was on 7th March, 1975...

In the trial, Neilson recalled that last visit and the circumstances of Lesley's death (as documented in the court proceedings quoted verbatim from S. Valentine, 1976, *The Black Panther Story* – see suggested reading at the end of this chapter):

> *As I took my foot off the ladder and while I was turning I came to face front and it was while I was doing this I looked around and she went over.*
>
> *The lantern was still lit. I grabbed this and stepped across to the other side of the landing. I put one foot down on the concrete ledge and went down as it were, in a squatting position in front of her.*
>
> *I had the lantern in my hand and as I went into the crouching position I held my left hand down towards her with the intention of pulling her back up.*
>
> *She was hanging, as it were, partly underneath, her head was lower than the gantry. There was some movement of her right hand and when I first got down there it was stretched out behind her, moving. Her other arm was bent at the elbow. The fist was clenched.*
>
> *There was nothing to grab hold of. The torch was pointing into her face. Her eyes flickered and stopped. Everything stopped. There was no movement. It was then I realised she was dead.*

An important conjecture in the trial was whether Neilson had killed Lesley (premeditated/first degree murder) or whether her death was an accident (manslaughter). The prosecution asserted that by tethering Lesley with a steel wire noose around her neck to a platform above a deep shaft, Neilson must have known that if she had fallen, she would die.

1.6 The *Post-mortem* Examination

The pathologist, Dr John Hunter Brown, reported that the body was of an emaciated young woman – she weighed only 44 kg (the average weight range for a 17-year-old female is 52–54 kg [2021 statistics]) – and that her stomach was empty, which meant she had not eaten for at least 24 h. Not having eaten meant that her blood sugar would be low, which might have led to a degree of unsteadiness which might have contributed to the reason she fell off the platform.

In court, Dr Hunter Brown recorded the cause of death as vagal inhibition, leading to the 'death' of her heart. Her brain would have taken longer to die, which means that if she had been recovered quickly and given the appropriate emergency treatment she might have survived.

Vagal inhibition occurs when the vagus nerve is compressed. The vagus nerve is the longest nerve in the body; it extends from the spinal cord, exiting through the vertebrae in the neck (Figure 1.4), to the heart, lungs and upper gastrointestinal tract; it controls (among other things) the heart. Pressure on the vagus nerve in the neck can stop

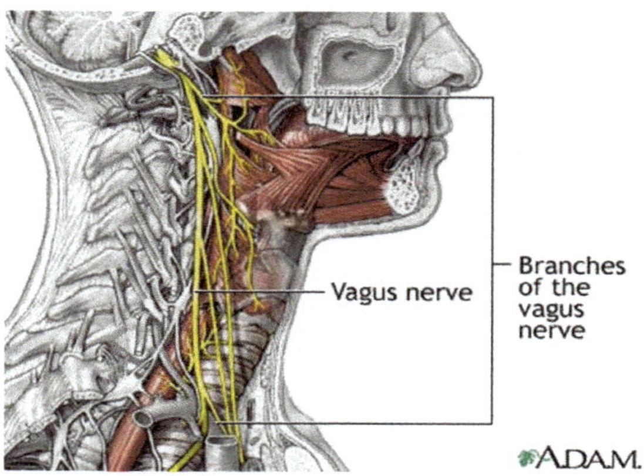

Figure 1.4 Position of the vagus nerve and its branches in the neck. Reproduced from ref. 3, https://skeptics.stackexchange.com/questions/3534/can-a-blow-to-the-base-of-the-neck-cause-unconsciousness-i-e-karate-chop, under the terms of the CC BY-SA 4.0 license https://creativecommons.org/licenses/by-sa/4.0/.

the heart – this is exactly what happened when Lesley fell off the platform with the steel wire noose around her neck. If the pressure had been released (*i.e.*, the noose was loosened) and the heart restarted to restore the blood supply to the vital organs (particularly the brain), Lesley would likely have survived. So, Neilson created the circumstances (putting the noose around her neck and not feeding her for at least 24 h, which would have led to low blood sugar and perhaps unsteadiness) that led to Lesley's death. Moreover, if he had exerted more effort to pull her back onto the platform, thereby reducing the noose's pressure on her neck, followed by cardiopulmonary resuscitation (CPR), she might well have survived the ordeal.

1.7 Neilson's Career of Crime

The Whittle murder marked the pinnacle of Neilson's significant career in crime. It all started when his fiancée, Irene Tate, persuaded him to leave the army to pursue a civilian career in Bradford in the north of England (Figure 1.3). He tried his hand at taxi driving, being a security guard, and carpentry; but all of them proved unsuccessful. In 1965, to supplement his income and support his now wife and daughter (Kathryn, born in 1960), he turned to a life of crime. He committed about 400 burglaries, escaping arrest due to his meticulous, military-like planning of each crime. These burglaries brought in a steady income, but Neilson desired more money (and perhaps more sophisticated crime). In 1967, he shifted his focus to the large sums of money held by post offices. Between 1967 and 1974 he raided 19 post offices in the north of England.

His post office raids were carried out in the dead of night when no one would be in the post office. In February 1972, he carried out a night raid of a sub-post office in Heywood, Lancashire (now in Greater Manchester). He cut the telephone line in the post office and wore a dark balaclava helmet (both signatures of his crimes) so that he could not be identified. His raid, however, did not go according to plan; the post master, Leslie Richardson, caught Neilson in the act. Neilson shot and injured Richardson. His crimes had taken a turn for the worse – he was prepared to shoot people who threatened his 'work'.

A couple of years later, during a robbery at a sub-post office in Harrogate, Yorkshire, he was disturbed by Donald Skepper, whom he shot dead. Neilson had now escalated to murder in pursuit of his crimes.

In the years between 1974 and the Whittle murder, Neilson killed three men – two were sub-post masters, the other was the husband of a sub-postmistress. They had all disturbed his nocturnal exploits.

The police floundered around; they simply did not have any idea about the perpetrator's identity, but they recognised the tell-tale insignia of the crimes which pointed to a single criminal, but who?

All descriptions of the criminal given to the police depicted a fit man wearing a balaclava helmet – the press christened him the Black Panther.

Among the tools of Neilson's trade was a squirty bottle of ammonia. When Neilson was eventually arrested in 1975 (more on this later) he was wearing a cartridge belt, carrying two knives, two razors, a piece of strangle cord, two balaclava helmets (one blue, one brown) and a bottle of ammonia–the tools of his trade.

About this time, one of the authors of this book (ICS) began an internship at the UK Forensic Science Service as part of his bachelor's degree studies. The balaclava helmets and ammonia occupied him for much of his time during his year as a trainee forensic scientist (more on this later as well).

1.8 Neilson's Arrest

On 11th December, 1975 (11 months after Lesley Whittle was kidnapped, and 9 months after her body was found), two police officers were sitting in their patrol car in a side road, keeping watch on the main road to Mansfield, 12 miles (19 km) north of Nottingham. They spotted a man carrying a holdall; this was not particularly noteworthy, until the man deliberately looked away from the policemen as he walked past – this piqued the police officers' attention. The police called the man over to talk to him: 'Good evening sir', said one of the police officers, 'At this time of night we like to check strangers out. Would you mind telling me where you have been this evening.' The man replied, I am just going home from work.' The police asked him for his name, and requested him to write it down. At this point, the man's demeanour changed. He said, 'Don't move!' and pulled a sawn-off shotgun out of his holdall.

Neilson forced one of the police officers into the back of the patrol car; Neilson got into the car and pushed his sawn-off shotgun into the police officer's armpit. He told the other police officer to drive to Rainworth, 3 miles (5 km) south east of Mansfield.

When they arrived at Rainworth, the police officers through a deft stroke, involving sudden braking, pulled the gun away from Neilson just outside a fish and chip shop; the gun went off, alerting the people in the queue for fish and chips, two of them ran to the police officers' assistance. Neilson was badly beaten into submission by the men, allowing the police to handcuff him. In fact, towards the end of the scuffle the police had to protect Neilson from the two adrenalin-driven would-be fish and chip consumers.

The police officers dragged Neilson to a nearby bus stop next to an iron railing fence. They handcuffed him to the fence until backup arrived. That was the end of the road for Donald Neilson, soon to be identified as the notorious Black Panther.

When Neilson was searched, he was found to have spare shotgun cartridges, razor blades, a 2-foot 6-inch (0.8 m) long strangle cord, two of his signature balaclava helmets ('panther hoods') and a bottle of ammonia solution. This hardly seemed like the equipment one would carry for a routine walk home from work!

1.9 Revelations in Neilson's Home

Once Neilson was safely ensconced in a police cell, the work on the case could begin. Of course, at that time the police did not know the man they had arrested was the Black Panther, but it did not take Neilson long to confess his alias and his string of crimes. He was proud of what he had 'achieved'.

The police went to his home in Grangefield Avenue, Thornaby-on-Tees (North Yorkshire), 97 miles (156 km) north of Mansfield – Neilson had lived in the house with his wife Irene (later convicted for cashing postal orders Neilson had stolen during his post office raids) and their daughter Kathryn for 15 years. They searched the house with the finest of fine-tooth combs. No item was considered too small to scrutinise. But not until they had exhausted their search without much luck, did they turn their attention to a locked attic room. There was no key anywhere to be found, so they broke the door down to reveal a military-style control room (referred to as the Panther's lair; Figure 1.5). The lair contained panther hoods, guns, ammunition, more ammonia solution, a Dymo label maker, gloves, rope, plastic core wire (similar to that used to secure Lesley Whittle to the balustrade of the drainage shaft in Bathpool Park). They had their man…

Figure 1.5 The Panther's lair. Photo from Alamy.com.

1.10 The Role of Forensic Science in the Case

If it had not been for Neilson's meticulous execution of his crimes, this would have been a textbook forensic case. There were so many forensic leads that abruptly reached dead ends because Neilson had covered his tracks so well. He pitted his wits against his enemy: the police. And he won, until his first wrong move in December 1975, which led to his arrest.

We will explore some of the forensic leads in the case and try to put ourselves in the place of frustrated police officers who continually drew a blank because of their opponent's cerebral superiority! We will also look at one or two lesser-known forensic investigations that added to the circumstantial evidence of the case – for this we will draw on the author's (IS) own work on dyes from Neilson's balaclava helmet(s) and the forensic investigation of the Dymo label messages.

1.10.1 Brown Shoes

A pair of brown training shoes was found in Bathpool Park. They were manufactured by a major company and so the chances of tracking down a single pair out of the 30 000 pairs manufactured was not very likely. However, through meticulous collaboration with the manufacturers, the shoes were identified as one of 42 pairs made between

1971 and 1975, and better still, this batch of trainers had been sold to three outlets in Coventry, Leicester and Leeds. I imagine a tingle of excitement ensued at this point, but this was soon dashed when none of the salespeople in the stores could remember to whom they had sold the shoes.

1.10.2 Binoculars

A pair of binoculars was found by police right next to the manhole leading down to the drainage shaft in which Lesley's body was hanging. The manufacturers traced the binoculars back *via* their serial number to a camera chain store, and they had been sold in their Manchester branch on 27th October, 1972. Excitement was building! Then came the crucial piece of evidence: the man who purchased the binoculars had completed the guarantee form, and what's more, they had a copy. The purchaser was identified as Mr Turner of Wilmslow, Cheshire. Could he be Lesley's murderer? The police visited Turner's address only to find that it was not a private house, but a company. And there was no Mr Turner. Neilson had won again.

1.10.3 Steel Noose Wire

Then the steel wire used as the noose by Neilson to tie Lesley to the platform in the drainage shaft was investigated. It had been manufactured by a company in Redditch in the UK's Midlands. The wire had an unusual yellow polyvinylchloride (PVC) core, and the manufacturers were certain that it was part of an unsuccessful small experimental batch – only ten 30 ft (9 m) lengths were ever made. The manufacturers sold the wire off to a shop in Walsall, Staffordshire. The shop owner told the police that all of the wire was sold between 1st September and 14th October, 1973. The police even tracked down the sales assistants who had sold the wire – this had to be the lead they needed, surely – but the assistants had no recollection whatsoever of who had bought the wire. Another dead end. Three nil to Neilson!

1.10.4 Lesley's Sleeping Bag

Then came a potentially significant breakthrough. Lesley's sleeping bag from the drainage shaft still had its £6.00 price tag attached. Interestingly, the price was handwritten, and the zeroes were underlined and were smaller than the number 6. This is a characteristic of continental European writing. The brand of sleeping bag was sold by several retail camping shops in the north Midlands. The police could find no one in the stores whose handwriting matched that on the price tag. Four nil to Neilson!

There is a gaggle of similar dead-end investigations. Either Neilson was very careful or very lucky.

Following Neilson's arrest, his signature balaclava helmets were recovered from his lair in the attic of his home – police found a cornucopia of tools, firearms, a Dymo label maker, binoculars, boots, a bottle of ammonia solution and much, much more there.

1.10.5 Fingerprints

Several partly obliterated fingerprints were found in the drainage shaft. They did not tally with fingerprints in the police database, which suggested that the kidnapper was not already 'known' to the police. When Neilson was arrested, his fingerprints were compared with those lifted from the drainage shaft: they were considered a match. The police had scored a point against Neilson!

1.10.6 Spots of Blue Dye on the Post Office Floor and Neilson's Signature Balaclava Helmet

Some spots of blue dye were found on the floor of one of the post offices that Neilson raided. The colour was similar to that of one of Neilson's balaclava helmets, and so it was decided to investigate further. Careful scraping of the blue spot from the floor followed by dissolving it in a solvent gave an intense blue coloured solution. This was applied to the bottom left-hand corner of a thin layer chromatography (TLC) plate and developed in a specific mixture of solvents in one direction. When the solvent reached the top of the TLC plate, the plate was dried with a warm air blower (like a hair dryer); then the plate was revolved 90° anticlockwise and developed in a different solvent system until the solvent reached the top of the plate. The TLC plate was removed from the solvent and dried in warm air. The first development separated the components of the dye into their different colours; the second development further separated some of the already separated colours, producing a pattern of coloured spots on the TLC plate (this is termed two-dimensional TLC). The pattern of spots and their colours are characteristic of a particular dye (Figure 1.6; see Chapter 4 *Forensic Chemistry* for a more detailed explanation of TLC).

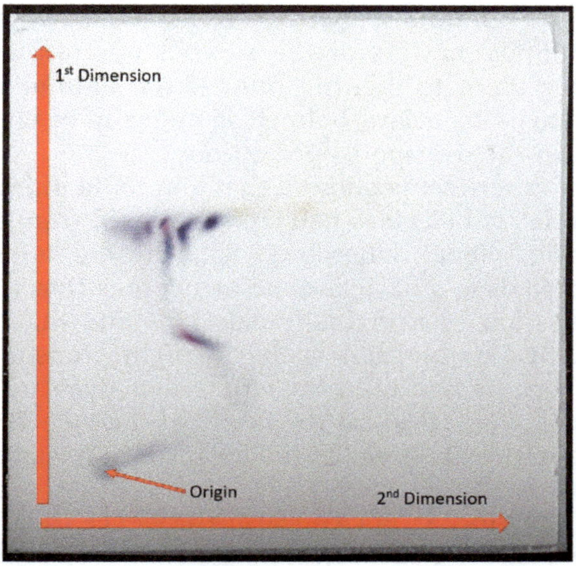

Figure 1.6 Two-dimensional chromatography of a very similar dye to the one used to dye Neilson's balaclava helmet. The dye extract was applied to the origin of a silica TLC plate and run in the 1st dimension in butan-1-ol-water-ammonia (aq. specific gravity = 0.88) (5:1:2:5). This gave three blue spots. It was then run in the 2nd dimension in the organic layer from ammonia (aq. specific gravity = 0.88)-trichloromethane-water-methanol (2:11:1:7), further separating the 3 blue spots to give a complex pattern of separated dye components. The pattern is characteristic of the dye. For an explanation of TLC, see Chapter 4 *Forensic Chemistry*. For details of the method, see ref. 4. Photograph taken by Matt Walters.

Dye was solvent extracted from a small sample of fibres from the balaclava helmet. It was analysed by two-dimensional TLC using the same two solvent mixtures that were used to analyse the dye taken from the post office floor. The dye separation patterns and colours from the 'floor' dye and balaclava dye were compared. They were identical. This indicated (but did not prove) that the dye on the post office floor might have originated from the balaclava helmet. This suggested that the balaclava helmet had been in the post office.

But why would the dye drip from the balaclava helmet? Further work showed that the dye could be easily stripped from Neilson's balaclava helmet by ammonia solution. This was interesting because a bottle of ammonia was found in Neilson's control room. In addition, it was clear that during the post office raids ammonia had been used as a weapon. Ammonia is a powerful eye irritant that causes temporary blindness; perhaps Neilson used ammonia spray to disable sub-postmasters or the sub-postmistress and make it impossible for them to identify him. If the ammonia had been directed on to his balaclava helmet, it might have eluted the dye and resulted in the dye spots on the floor.

Still further experiments showed that ammonia did not elute the dye from a series of balaclava helmets purchased from a store close to the forensic science laboratory. So, why did ammonia elute the dye from Neilson's balaclava? It turned out that the balaclava helmet in question was originally pale blue and that it had been dyed to make it dark and less visible at night. Neilson was clearly not a dye expert; he had used the wrong dye (dyes are specific for wool, cotton or synthetics) for the fabric of the balaclava and so it did not bond to the fibres well. This means that it was easily eluted by ammonia solution.

While the dye story is not conclusive evidence, it provided additional circumstantial evidence, linking Neilson (or rather his balaclava helmet) to the scene of one of his crimes.

1.10.7 Dymo Messages

Next, there was an investigation of the all-important Dymo label maker and the Dymo tape messages (Figure 1.7) Neilson had left in the Whittle's sitting room on the day Lesley was kidnapped. He left many more Dymo messages in and around his crime scenes. The messages were sometimes riddles intended to confuse the police or divert them in the wrong direction. Neilson was clearly having fun.

Like typewriters, Dymo label makers have idiosyncrasies of printing. The letter D might be skewed or perhaps at an angle or the gap between one letter and its adjacent letter might be larger than that between two other letters. These foibles help in identifying a Dymo printed tape as having originated from a particular labelmaker. There were very specific idiosyncrasies in the label messages left by Neilson, and these idiosyncrasies were common to all of the messages with particular letters or sequences of letters. Then the jackpot: the Dymo label maker found in the Panther's lair exhibited the same idiosyncrasies as those in his messages.

This sounds more conclusive than it might be. It is possible (if not likely) that individual Dymo label makers from a manufacturing

Figure 1.7 A Dymo label maker (top), a reconstruction of part of one of the messages Neilson left in the Whittle's sitting room on the day Lesley was kidnapped, showing inconsistencies in printing (middle), and the geometry of the letters ING on a Dymo tape, showing that the I and G deviate from the horizontal line created by the N, and that the separation I → N and N → G is different (bottom) [none of these examples are from the Whittle case, but are shown here to illustrate the principles].

batch all exhibit the same (or very similar) idiosyncrasies. So, it is not possible to say conclusively that the label maker found in Neilson's home was the same machine that produced the messages left at the crime scenes. Nevertheless, this adds to the circumstantial evidence pointing to Neilson.

1.11 The Trial

On 14th June, 1976, Oxford Crown Court was heavily guarded, the public gallery was full and the press benches were at full capacity. Anticipation pervaded the court's atmosphere. This was the day Donald Neilson was brought to trial in front of Mr Justice Mars-Jones for the killing of two sub-postmasters, the husband of a sub-postmistress and Lesley Whittle.

The case was opened: the prosecution sought a conviction for murder in the first degree (premeditated murder); the defence (George Gray, QC) sought manslaughter on the grounds that Neilson had not intended to kill his victims.

The long and convoluted story of Neilson's descent into a life of crime after leaving the army, facing failures in business attempts, stealing weapons to facilitate his sub-post office raids, escalating to murder in pursuit of his crimes, and finally engaging in kidnapping and murder. He was examined by a psychiatrist who attested that he was not insane, which meant that he must take responsibility for his actions.

The jury found Neilson guilty of the murder of all four victims for which Mr Justice Mars-Jones sentenced him to 4 life-sentences. He was also sentenced to a further 21 years for kidnapping Lesley Whittle, 10 years for blackmailing Dorothy Whittle, 30 years in total for a series of arms burglaries and possession of a sawn-off shotgun intended to endanger life – in total he was handed down 4 life-sentences plus 61 years, to be served concurrently. It was recommended that he serve the whole life tariff (*i.e.*, no opportunity for parole).

He was sent to Norwich Prison to serve his sentence.

Neilson appealed to the High Court in 2008, seeking to have his minimum term reduced to 30 years instead of life. The appeal was heard by Mr Justice Teare who upheld the whole life tariff saying in his decision speech:

> *This is a case where the gravity of the applicant's offences justifies a whole life order. The manner in which the young girl was killed demonstrates that it too involved a substantial degree of premeditation or planning. It also involved the abduction of the young girl. The location and manner of Lesley Whittle's death indicates that she must have been subjected by the applicant to a dreadful and horrific ordeal.*

Neilson remained in Norwich Prison until his death at the age of 75 years on 17th December, 2011.

Further Reading

1. S. Valentine, *The Black Panther Story*, New English Library Ltd, UK, 1976.

References

1. File: Donald Neilson mugshot.jpg [Wiki]. Available from: https://en.wikipedia.org/w/index.php?curid=19271616 [accessed 4th September 2023].
2. File: Lesley Whittle02.jpg [Wiki]. Available from: https://en.wikipedia.org/wiki/File:Lesley_Whittle02.jpg [accessed 18th April 2024].
3. Can a blow to the base of the neck cause unconsciousness? (*i.e.*, "Karate...CHOP!") [Internet], Skeptics Stack Exchange. Available from: https://skeptics.stackexchange.com/questions/3534/can-a-blow-to-the-base-of-the-neck-cause-unconsciousness-i-e-karate-chop, cited 4th September 2023.
4. I. C. Shaw, *Analyst*, 1980, **105**, 729–730.

2 The History of Forensic Science

History is important in our understanding of the past. The triumphs and mistakes of the past guide the development and success of the future. History is not about setting precedents like case law in court; it is far more discriminating. The errors of the past must prevent us from carrying them into the future. The triumphs of the past should be magnified and embraced. The development of forensic science serves as a perfect example of history demanding our attention and urging us to address its deficiencies – its biases, and enhance its reliability and relevance. After all, forensic science must stand as an honest, undisputable science that underpins the truth.

When we look back at the history of forensic 'science' from its historical foundations in the Roman and Greek empires to Dr Edmond Locard's (1877–1966) application of the firm scientific principles of transfer of evidence, we see many examples of malfunction and misuse of this apparently robust science. Sir Bernard Spilsbury (1877–1947), although lauded as a great man in forensic science was, perhaps, no more than the greatest personality in forensic science. His word was 'fact', but should it have been questioned? Who dared to question Spilsbury? How many people were falsely convicted based on Spilsbury's forensic deliberations? How many innocent individuals faced the gallows?

Fast forward to today: the science underpinning criminal investigations is robust, quality assured, unquestionable – or is it? Forensic scientists no longer dabble at crime scenes. They carry out highly controlled searches, collecting items and samples and transferring

Forensic Science: The Science Behind the Truth
By Ian Shaw with Anna Sandiford

Published by the Royal Society of Chemistry, www.rsc.org

them in sealed and labelled bags to their labs for investigation under pristine conditions to minimise cross-contamination.

And as time progresses, the multidisciplinary forensic approach is being replaced by DNA technicians who focus on the probability of a DNA profile originating from a particular person. Is this genuine progress or should we question the forensic science history being shaped today?

2.1 When Did Forensic Science Begin?

It all began with the desire to understand, to know why someone had died. In the beginning the goal was not to find out the cause of death in a crime setting, but rather to understand why someone was no longer living. But, before this, humans were dissected by other humans to find out what was inside, to look at the body's contents and to view this as art or anatomy – or both. Someone had to take that first step; that first cut into a fellow human. There is no record of the first person to dissect a human, but only medical doctors were allowed to explore cadavers internally as part of their profession until the 1500s.

2.1.1 Leonardo da Vinci and Andreas Vesalius Laid the Foundations

Leonardo da Vinci (1452–1519) is famous for his wonderful anatomical sketches (Figure 2.1) which had to be kept secret because he was not a medical doctor and so he was not permitted to dissect another human. Despite this, he obtained bodies *via* graverobbers and later a hospital director provided cadavers for da Vinci's work. The catholic church eventually permitted his activities, which led to hundreds of detailed anatomical drawings that presented a new insight into the human body and its organisation. Da Vinci's anatomically precise and artistically stunning drawings illustrated the Renaissance's resistance to polarising science and the arts. Arguably da Vinci's ideas spawned Andreas Vesalius' (1514–1564) publication of *De Humani Corporis Fabrica Libri Septem* (On the Fabric of the Human Body in Seven Books). The book (which exploited recent advances in printing) recorded the detailed anatomy of the human body and gave new insights into the body organs and their interconnections. Once the taboo of dissection had been overcome, it was a relatively small step (in a historical timescale sense) to understand normal anatomy and organ structure in contrast to the anatomy and underlying

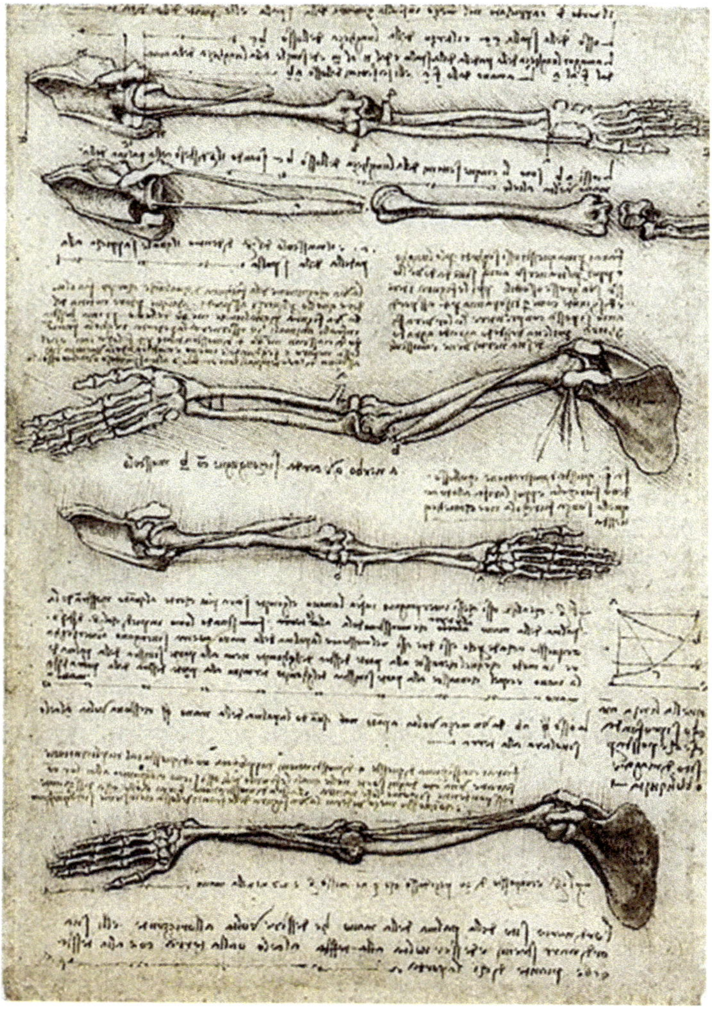

Figure 2.1 Leonardo da Vinci's anatomical sketches of the human arm. Their art is beautiful and the level of detail astounding. Figure is in the Public Domain.

mechanisms of disease (pathology) – this will be discussed in more detail in Chapter 8 *Forensic Pathology*.

2.1.2 Was Imhotep an Ancient Egyptian Forensic Pathologist?

While da Vinci and Vesalius were among the first to record the detailed anatomy of the human body, the idea of using dissection to determine the cause of death dates back much, much further. Imhotep (2650–2600 BC) is arguably the first forensic pathologist – he served as both a Chief Justice and personal physician to pharaoh Djoser (reigned 2686–2648 BC), thus bringing together medicine and law. Ancient Egyptian law required dissection (*post-mortem* examination) to be carried out following the unexpected death of someone from the higher echelons of their society. It is unlikely that Imhotep and those like him knew the specifics of determining the cause of death (unless the cause was obvious like a stab wound to the heart), but the principle of determining why someone had died was established.

2.1.3 Caesar's Death Initiated a Murder Investigation

Two and a half thousand years after Imhotep, Julius Caesar (100–44 BC; Figure 2.2) was murdered. His death and the ensuing investigation is an excellent example of forensic pathology at work. Caesar's *post-mortem*

Figure 2.2 Julius Caesar (100–44 BC), his murder by Marcus Junius Brutus led to what might be considered one of the earliest forensic *post-mortem* examinations to determine the cause of death. There were 23 stab wounds, one was deemed fatal because it penetrated Caesar's heart. Figure is in the Public Domain.

examination revealed 23 stab wounds of which one was deemed fatal, and so the great man had been murdered. There were many men who might have wanted Caesar dead; they became the prime suspects in the case. Simple 'police' reasoning narrowed down the possibilities: the senators aspiring for Caesar's position were very unlikely candidates for committing the murder, lacking sufficient knowledge of daggers and their effective use. But one eminent military man well understood the efficient use of weaponry: his name was Marcus Junius Brutus – and the rest, as they say, is history:

> CINNA: O Cæsar, —
> CÆSAR: Hence! Wilt though lift up Olympus?
> DECIUS: Great Caesar, —
> CÆSAR: Doth not Brutus bootless kneel?
> CASCA: Speak, hands, for me!
> CÆSAR: *Et tu Brutè!* Then fall, Cæsar! [Dies]
> CINNA: Liberty! Freedom! Tyranny is dead! Run hence, proclaim, cry it about the streets!
>
> Shakespeare: Julius Caesar, Act 3, Scene 1

2.1.4 The Carolina Code of Criminal Procedures

In 1532, the *Constitutio Criminalis Carolina* (Carolina Code of Criminal Procedure) was published in Germany. This marked a sea change in thinking in relation to criminal investigations. It introduced the inquisitorial approach to criminal investigation, relying on evidence that was impartially assessed without being influenced by the interests of those involved in the case. The judgement was, therefore, more likely than ever before to be fair and evidence-based. Moreover, expert opinions might be sought to help assess the case – this could involve circulating case files to, for example, medical experts in universities.

So, by the early 1600s, there existed both a detailed knowledge of the anatomical organisation of the human body and the concept that cause of death might be determined by examining changes in the body, and there was a well-established inquisitorial process for trying a case in the courts. Progress indeed!

2.1.5 Zacchia's Quæstiones Medico-Legales

The Italian physician, Paolo Zacchia's (1584–1659) *Quæstiones Medico-Legales* was published in 1557. As its title implies, Zacchia's book brought medicine and the law together in a context of causal determination. Despite its very promising title, there is still a liberal smattering of magic, witches and demons – all deeply ingrained in 16th century thinking. But the idea that medicine could be used in criminal investigation settings was a move in the right direction.

2.1.6 Setting the Scientific Scene for Modern-day Forensic Science

At this stage, diagnosis (both clinical and in the context of the cause of death in a medicolegal setting) was purely structural – determining whether the 'art' of the body had been disrupted. Were there stab wounds that entered the heart or some other vital organ? The idea that physiological or biochemical changes (*e.g.*, by poisoning) might lead to death was not even a twinkle in the eye of the medicolegal practitioners (if we can call them that) of the time. The term 'physiology' as we understand it today (*i.e.*, the function of the body and bodily parts) was first used in 1564. However, Hippocrates (460–370 BC) introduced the idea of physiology by describing the four humours – black bile, phlegm, blood, and yellow bile – which he said controlled bodily function, but this was of no use whatsoever for determining the cause of death in a forensic context! The word 'physiology' was introduced by the French physician Jean Fernel (1497–1558) towards the end of his life. In the late 1800s, the science of physiology, as we understand it today, developed. There was a huge accumulation of physiological knowledge in the mid- to late-1800s, culminating in the German botanist Matthias Schleider's and physician Theodor Schwann's cell theory – a breakthrough that revolutionised physiological understanding, and was arguably the beginning of what we now know as biochemistry. But it was not until the German physiologist Felix Hoppe-Seyler (1825–1895) coined the term *biochimie* to describe physiological chemistry that a real understanding of how the body functions began.

This is not a book about the history of our understanding of the human body! Nevertheless, a cursory glimpse at the development of the science of the human body and the understanding of its complex chemistry is important for appreciating how scientists in the late 1800s might have 'thought' based on their knowledge. This, in conjunction with significant developments in the way cases were tried in the courts, opened the door for the development of modern-day forensic science.

2.2 Modern-day Forensic Science

The stage was set for a revolution in forensic investigations. Nineteenth-century scientists had a good understanding of physiology and biochemistry, as well as the anatomy of the physical (*e.g.*, stabbing) causes of death. They knew if someone had died in suspicious circumstances, and they could ask the question, 'Had they been murdered?' But it was not until the Spanish toxicologist (much of his career was spent in France) Mathieu Orfila (Figure 2.3) – the grandfather of toxicology – came along that such investigations started utilising chemical analyses to determine the cause of death in poisoning cases.

2.2.1 Murder by Poisoning

Murder by poisoning is rare nowadays, but in Orfila's time, it was far more common. This was likely due to the ready availability of poisons such as arsenic. In the late 1800s, arsenic salts (usually arsenic

Figure 2.3 Mathieu Orfila – the father of forensic toxicology. Lithograph by Alexandre Collette (1814–1876). Figure is in the Public Domain.

trioxide – As_2O_3) were mixed with chalk (calcium carbonate) and vinegar (acetic acid) to make a paste that was consumed by women to enhance their complexion! At the other end of its use-spectrum, arsenic was used as a rat poison and an antifungal in wallpaper paste. This multifarious use of arsenic meant that it could be purchased from chemists' shops (pharmacies) and was likely to be found under the kitchen sink or in the medicines cabinets of many 19th century homes.

The penchant for the use of arsenic as a murder weapon grew in Victorian Britain – there are many classic cases involving the poisoning of one's 'nearest and dearest' with the help of a few doses of arsenic trioxide in a nice glass of milk at bedtime over a few days or weeks.

Arsenic was a favourite conduit to inheritance. All you had to do was poison your elderly aunt who had left you a tidy sum in her will, and your life would be all the better for it. It is for this reason that arsenic was nicknamed the inheritor's powder in Victorian times.

2.2.2 The George Bodle Case 1832

A classic case involving the inheritor's powder occurred in Britain in 1832. John Bodle was a well-off tenant farmer on his father George's farm in Plumstead, not far from London. He lived with his wife, Catherine, their sons John and George (26), their skivvy, Mary Higgins and Catherine's nephew, Henry.

The poisoning saga begins with George Bodle and his wife, Ann, suffering a severe gastric upset. Since food-related illness was common in those days, no one thought much of it, but it persisted and the doctor was called. The doctor decided (just from the symptoms) that the cause was likely to be white arsenic (arsenic trioxide).

Shortly after George's arsenic experience, he was persuaded by his son-in-law, Samuel Baxter, to change his will. The family did not know what George's original will specified, but they suspected that his considerable wealth had been left to his son, John. Clearly, this is not what Samuel wanted. The new will made Samuel a significant benefactor. The plot is now well and truly set!

George's demise was slow (a characteristic of arsenic poisoning – more on this later) and suspicions about who administered the poison began to emerge. The case was handed over to the magistrate to sort out.

The courts of that time were beginning to insist on substantial evidence in poisoning cases – no longer relying solely on a medic's hunch, as had been the case previously. This led to more in-depth forensic investigation and analysis. Introducing John Gordon Smith (1792–1833) the University of London's recently appointed Professor of Medical Jurisprudence – the first professor in England in this field. He brought the 'medic's hunch' with a modicum of analytical chemistry to courts. Importantly, Professor Smith recognised the importance of ensuring that samples were not tampered with, thus making the evidence more robust than the courts might have been used to.

The hunch that arsenic trioxide was the cause of death was investigated further, following a *post-mortem* examination of George Bodle's decaying corpse on his kitchen table. Samples (including stomach contents and a sample of the morning coffee) were taken and accompanied a police constable on a pub crawl before reaching the chemistry lab!

At the time, it was important to actually demonstrate scientific results in the courtroom for them to be considered credible. It simply was not good enough for a scientist to describe their results; they had to be seen to be believed.

Professor Gordon Smith (for the prosecution) secured the help of a young analytical chemist, James Marsh (1794–1846; Figure 2.4), an ordinance chemist at the Royal Arsenal, Woolwich. Marsh also worked with Michael Faraday (later Professor of Chemistry at the Royal Institution, London, who later became one of the world's most eminent chemists) at the Royal Military Academy. Marsh carried out the standard (at the time) test for arsenic which involved heating the test sample with concentrated hydrochloric acid (HCl) and hydrogen sulphide (H_2S), producing a characteristic yellow precipitate of arsenic trisulphide (As_2S_3) (Figure 2.5). The arsenic trisulphide precipitate was weighed – the weight is proportional to the amount of arsenic in the original sample.

Vogel describes the 'Determination of arsenic as trisulphide' in the 1939 edition of his book, *A Textbook of Quantitative Inorganic Analysis Theory and Practice*:[1]

> *Pass a rapid stream of washed hydrogen sulphide through a solution of the arsenious arsenic in 9N hydrochloric acid at 15° to 20°C. Allow to stand for an hour or two, and filter through a weighed filter crucible (Gooch, sintered glass or porous porcelain). Wash the precipitate with 8N hydrochloric acid saturated with hydrogen sulphide, then successively with alcohol, carbon disulphide (to remove any free sulphur which may be present) and alcohol. Dry at 105°C to constant weight, and weigh as* As_2S_3.

Figure 2.4 James Marsh (1794–1846), the young chemist who invented a reliable, quantitative test for arsenic. The Marsh test became the mainstay of arsenic analysis until the 1970s. Figure is in the Public Domain.

$$As_2O_3 + 3H_2S \rightarrow As_2S_3 + 3H_2O$$

Figure 2.5 Reaction of arsenic trioxide (As_2O_3) with hydrogen sulphide (H_2S) (in concentrated hydrochloric acid), producing a yellow precipitate of arsenic trisulphide (As_2S_3). If a test sample treated in this way forms a yellow precipitate, the sample contains arsenic.

The problem was that by the time the case came to court, Marsh's yellow precipitate, proving the presence of arsenic in George Bodle's morning coffee, had degraded (it slowly re-forms arsenic trioxide in air) and was no longer visible for the judge and jury to see.

Not to be defeated, Marsh set about devising a new test that would be more suitable for the judge's and jury's need to see the evidence themselves. He mixed arsenic trioxide with zinc powder in sulphuric acid in a flask. The flask was heated in a water bath and arsine gas (AsH_3) was generated from the arsenic trioxide. The arsine gas was passed through a drying tube (containing anhydrous calcium chloride) and then into a glass tube, which was gently heated at one point. The arsine decomposed to form arsenic at the point where the tube was heated, resulting in the formation of an arsenic mirror (Figures 2.6 and 2.7). This mirror is diagnostic of arsenic, and the mass of arsenic deposited on the glass tube was proportional to the amount of arsenic in the flask.

A sample suspected of containing arsenic trioxide (*e.g.*, Bodle's coffee) was put into the flask with zinc and sulphuric acid and allowed to react to form arsine gas. If an arsenic mirror formed on the

Step 1

$$As_2O_3 + 6\,Zn + 6\,H_2SO_4 \rightarrow 2\,AsH_3 + 6\,ZnSO_4 + 3\,H_2O$$

Step 2

$$2\,AsH_3 \rightarrow 3\,H_2 + 2\,As$$

Figure 2.6 The chemistry of the Marsh test. Step 1: arsenic trioxide (As_2O_3) is heated with zinc (Zn) powder and sulphuric acid (H_2SO_4) to form arsine gas (AsH_3). Step 2: the arsine is burned to form an arsenic (As) mirror on a cold glass surface. When the test is carried out on a test sample (*e.g.*, the coffee in the Bodle case), the presence of an arsenic mirror indicates that the sample contained arsenic.

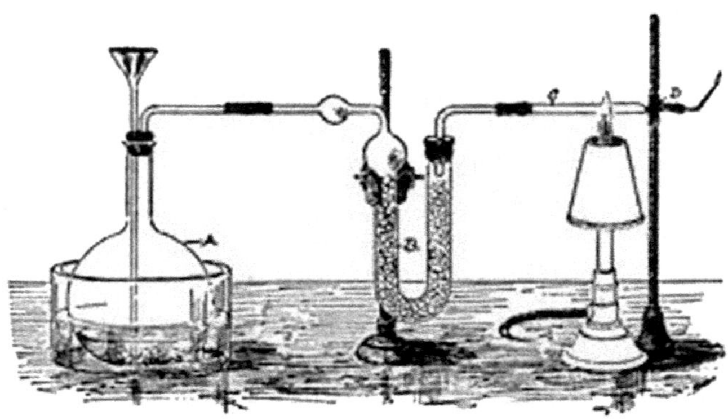

Figure 2.7 Equipment used for the Marsh test. Left: flask containing zinc, sulphuric and the test sample - arsine (AsH_3) gas is liberated if the sample contains arsenic (see step 1, Figure 2.6). Middle: the arsine gas is dried in a U-tube containing anhydrous calcium chloride. Right: the arsine gas is heated in a glass tube using a Bunsen burner; the arsine gas is broken down to form arsenic and hydrogen (Figure 2.6, step 2); arsenic forms a mirror-like deposit on the inside of the glass tube just beyond the Bunsen burner flame (*i.e.*, where the tube is cooler). If arsenic is deposited, it indicates that the test sample contained arsenic. Figure is in the Public Domain.

inside of the tube, the sample contained arsenic. The weight of the 'mirror' was used to calculate the amount of arsenic in the test sample.

The arsenic mirror was stable and could be shown in court as proof of the presence of arsenic in Bodle's coffee. The Marsh test is very sensitive; it can detect as little as 0.02 mg arsenic in a test sample. Marsh published his new test in the *Edinburgh Philosophical Journal* in 1836. In his paper, Marsh described the formation of the arsenic mirror:[2]

> *... a glass tube, from a quarter to half an inch in diameter (or according to the size of the jet of the flame), and eight or ten inches in length, is to be held vertically over the burning jet of gas, in such a manner that the gas may undergo perfect combustion, and the arsenic combined with it may become sufficiently oxydized; the tube will thus, with proper care, become lined with arsenious acid* in proportion to the quantity originally contained in the mixture.*
>
> *... three very good indications of the presence of arsenic may be obtained at one operation;* viz., *metallic arsenic will be found deposited in the tube at the part nearest where the flame impinges, — white arsenic† or arsenious acid* at a short distance from it ...*
>
> *$As(OH)_3$
> †Arsenic trioxide (As_2O_3)

The Marsh test remained the standard test for arsenic until the 1970s.

Enough arsenic was found in the coffee grounds from George Bodle's morning brew to have caused his death. George Bodle's grandson, John, was arrested for the murder.

In the Magistrates' court, a complex salacious interrelationship between family and servants emerged, befitting the best Victorian drama and the case was referred to the Crown Court. Surprisingly John Bodle was acquitted of his grandfather's murder.

But there was a sting-in-the-tail of the Bogle case: John Bogle, under a false name, turned up again in the Crown Court for extorting money from Lord Abbingdon's butler. This time, he was found guilty and was sentenced to 20 years' transportation. Under the safety of the double jeopardy rule, he confessed to his grandfather's murder. Case closed.

2.3 Arsenic Poisoning

From the focus on arsenic in poisoning cases above, and the efforts analytical chemists made to refine their analytical methods for courtroom applicability, it is evident that arsenic was a prime instrument for murder.

Understanding both arsenic's mechanism of toxicity and the symptoms of its toxicity is important in diagnosing arsenic as the culprit in a murder case. At the time of the Bogle case, little was known about arsenic's mechanism of toxicity, except that it was highly toxic.

2.3.1 How Much Arsenic Is Fatal?

A single high oral dose of arsenic can be fatal (the lethal dose of arsenic in humans is 0.6 mg per kg body weight; *i.e.*, a single dose of about 50 mg), but in murder cases, arsenic was usually administered at low dose over a few days or weeks – its effects are cumulative. This meant that the victim slowly became ill, which diminished the suspicion of poisoning. This chronic toxicity approach led to irreversible changes that cascaded to cause sufficient biochemical and physiological impacts to result in death.

2.3.2 Symptoms of Arsenic Poisoning

The symptoms of acute arsenic poisoning include red or swollen skin, abdominal pain, nausea, vomiting and diarrhoea, muscle cramps, and abnormal heart rhythm. While these symptoms are serious, they are fairly general and could be due to a number of clinical issues, which means that arsenic might not be the first potential cause that comes to an investigating physician's mind. The symptoms of chronic arsenic poisoning are even more general, including constant sore throat and persistent digestive issues, later accompanied by an unusual and highly diagnostic darkening of the skin.

2.3.3 Why Is Arsenic Toxic?

Surprisingly, we still do not fully understand the biochemical mechanism underpinning arsenic's extreme toxicity. This is probably because arsenic interferes with a multitude of cellular processes, rather than targeting one pathway. This, in turn, leads to the generalised

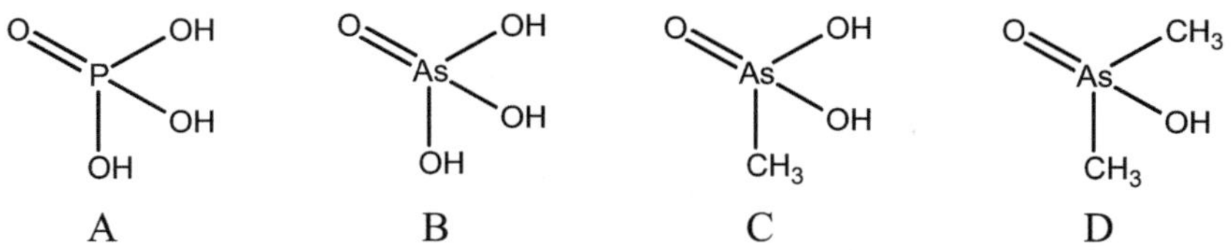

Figure 2.8 Structure of phosphate (A), arsenate (B) and the two main metabolites of arsenic trioxide, methylarsonic acid (C) and dimethylarsinic acid (D) shown in their non-ionised forms. It is clear that they have structural analogies, which might mean that arsenic compounds interfere with phosphate-dependent cellular processes, leading to their broad toxicological profiles.

symptoms of arsenic poisoning rather than a specific symptom related to a defined cellular target. One theory is that some arsenic ions (*e.g.*, arsenate – AsO_4^{3-}) and arsenic metabolites (*e.g.*, methyl metabolites of arsenic trioxide) mimic phosphate and can therefore interfere with phosphate-dependent cellular processes, hence contributing to arsenic's general toxicological profile (Figure 2.8).

And finally for completeness, arsenic is carcinogenic following long-term exposure to low doses (*e.g.*, environmental exposure), but this is, of course, not relevant to murder using arsenic – unless the murderer is prepared to wait for a very long time!

2.4 Orfila and the Lefarge Case – 1840

It is widely believed that Orfila's investigation of the Marie Lefarge case was the first application of the Marsh test for arsenic to an arsenic murder investigation. You will see from our discussion above that the Bodle case (1832) pipped him to the post. The Lefarge case was, however, the first conviction based on arsenic analysis using the Marsh test. The case broke new ground because it was the first case to be successfully prosecuted on forensic toxicology grounds. It marked a new era – the advent of forensic toxicology.

Marie-Fortunée Lefarge (née Capelle) (1816–1852) (Figure 2.9) had still not found a husband when she turned 23 years old. Her father presumably did not like the idea of being financially responsible for his daughter for the rest of his life, prompting him to enlist the services of a dating agency (then, rather formally termed 'marriage brokers') without her knowledge.

The sole candidate who met M. Capelle's condition that 'no marriage contract should be made with a man whose only income is his

Figure 2.9 Marie-Fortunée Lefarge (1816–1852) poisoned her husband Charles with arsenic because her marriage was not quite what she expected.[4] Figure is in the Public Domain.

salary as a subprefect' [a French government administrative official] was Charles Pouch-Lefarge, a 28-year-old, rather large and coarse man – different from what Marie had envisaged! To add to their woes, Charles Lefarge had advertised himself with the marriage broker as wealthy with property worth Fr200 000 and an annual income of Fr30 000 (both very significant sums at the time). To say the least, he had inflated his financial standing: in fact, early in 1839, he had been declared bankrupt. He saw a 'good marriage' as his only way out of debt. M. Capelle presented Charles Lefarge as a friend of Marie's aunt so that Marie did not suspect the marriage broker's jiggery-pokery behind the scenes. Charles and Marie were married on 10th August, 1839. They moved to Charles' estate (Le Glandier) to begin their married life.

From the beginning, Marie was disillusioned; the house was in disrepair and rat infested, and her in-laws were bawdy peasants – not quite what she had expected! She tried to make the best of her far from ideal marriage; she even wrote enthusiastic letters about her marriage and home to her friends back in Paris. She tried to help Charles to get a job by writing letters of recommendation. But all of this failed – Marie had to face reality; she was desperately unhappy.

In December 1839, Charles went to Paris on business, but before he left Marie she made a will bequeathing her inheritance to her husband. She did this with the understanding that Charles would reciprocate *via* a will leaving all he owned to her. He did as he was told and off he went to Paris. Marie did not know that shortly after making his will, Charles changed it, leaving the Le Glandier house to his mother, which meant that if he died Marie would get very little...

While Charles was in Paris, his 'devoted' wife sent him letters, her picture and a Christmas cake. Charles became very ill after eating just one slice of the Christmas cake. Cholera was rife in Paris at that time and Charles thought that he might have contracted the disease because the cake his wife sent had been contaminated *en route*. He threw the cake away.

When he returned home, he was still unwell; his 'devoted' wife sent him to bed where she pampered him with venison and truffles to stimulate his appetite. He became even more sick. He developed severe leg cramps and dehydration (probably due to diarrhoea) (see Section 2.3.2 *Symptoms of Arsenic Poisoning*). The doctor was called – he diagnosed cholera and prescribed egg nog (a cream, raw egg and brandy drink).

A friend, Anna Brun, was staying with the Lefarges at that time. She noticed Marie putting a white powder into Charles' egg nog and later into his soup. Charles became even more ill. The family was very concerned. Anna told the family and Charles' doctor what she had seen. Later, the gardener told Anna that he had bought arsenic for Marie; Marie insisted that this was to poison the rats that infested the house. They believed her.

Charles' condition worsened and a second doctor was called. He suspected poisoning, but it was too late. Charles died on 13th January, 1840.

Charles' death was investigated by the magistrates. 'Rat poison' was found all over the house – Marie asserted that this proved why she had bought arsenic. The pharmacist from whom Marie had purchased arsenic had records of sale (this was required for poisons), confirming that Marie had bought arsenic the day before she sent the Christmas cake to Charles in Paris, and then again the day before he returned home.

The Marsh test was used to analyse Charles' stomach contents and the 'rat poison' from around the house. The 'rat poison' contained no arsenic; it turned out to be a paste of flour, soda and water. The analysis of the stomach contents was unreliable; despite this, the doctors claimed that arsenic had been found. Perhaps the most compelling evidence was a significant quantity of arsenic found in a little malachite box where Marie kept her personal things; and Anna had seen her taking a white powder from this box and putting it into Charles' egg nog and soup.

2.4.1 Marie Lefarge's Trial

The case came to court on 3rd September, 1840. Word had spread about the case throughout Europe. The court was packed. Marie entered wearing mourning clothes and carrying a bottle of smelling salts – the paragon of a wrongly accused woman devastated at the loss of her beloved husband.

The case hinged on whether Charles' body harboured arsenic. Only the stomach had been retained prior to burial. It was negative for arsenic. The great toxicologist Mathieu Orfila was consulted and he sent an affidavit declaring that if the arsenic had been absorbed, it would not be found in the stomach, but might be elsewhere in Charles' body. The body was exhumed, but the court decided not to commission Orfila to carry out the Marsh test on the body organs. Instead, three local chemists were given the task. They found no arsenic in the organs. That was the end of the prosecution case.

Because of her suspicions, Anna had sequestered samples of the food, including the egg nog, that Marie had given her husband. The court ordered that these samples be analysed using the Marsh test. They commissioned the same three chemists to carry out the analyses. This time they tested positive for arsenic; the chemists declared in court that there was enough arsenic in the egg nog to kill three men.

There was still considerable uncertainty about the case. Marie had purchased arsenic at just the right time, Charles' food contained arsenic and Marie's little malachite box also contained arsenic; but there was no arsenic in Charles' body; surely there would be if it was the cause of his death. Orfila was called to settle the conjecture.

Orfila took the three chemists into an anteroom in the courthouse and re-analysed the body organs using the Marsh test. This time arsenic was found. There was now a connection between Marie, Charles' food and arsenic in his body.

On 19th September, 1840, Marie-Fortunée Lefarge was convicted of the murder of her husband and sentenced to life imprisonment.

After 17 years of her sentence, Marie contracted tuberculosis, and in June 1857, she was released from prison on compassionate ground by Napoleon III. She lived in Ussat, south-west France, until her death on 7th November of the same year.

2.5 The Advent of the Celebrity Forensic Pathologist

The Lefarge case established forensic toxicology as a crucial component in murder investigations (remember poisoning was a murder weapon of choice in the 19th century) and promoted Orfila to international status. Forensic toxicology retains a key place in crime investigations to this day, even though murder by poisoning is now rather uncommon (see Chapter 9 *Forensic Toxicology*).

By the beginning of the 20th century, forensic medicine was firmly ingrained in the medical curriculum. J. Dixon Mann's textbook, *Forensic Medicine and Toxicology* published in 1902, is among the first comprehensive guides for students of medicine. Its chapter titles set the scene for the huge advances in the application of science and medicine to criminal cases over the previous 10 years. Its chapters include, Examination of the Dead Body, Modes of Dying, Blood and Other Stains, Rape and Unnatural Offences, a series of chapters on death by asphyxia, electrocution, starvation and burns, Mechanical Injuries and Wounds, Poisons in their General Aspect, followed by a series of chapters on specific classes of poisons.

The international hype *via* the press led to celebrity status for scientists and medics involved in key criminal cases. This began our love affair with forensic science, which continues today with a plethora of somewhat dubious depictions of the glamour of forensic science and forensic scientists – isn't it interesting that according to TV all forensic scientists are handsome or beautiful and they always solve their cases? This is not like the forensic science *we* know!

Forensic pathology in the 1920s had star appeal too. It was guaranteed to grace the newspapers, especially if the infamous Sir Bernard Spilsbury was leading the case.

2.5.1 Sir Bernard Spilsbury

Sir Bernard Spilsbury (1877–1947; Figure 2.10) was a British forensic pathologist. He is a legend in his field. His court appearances could

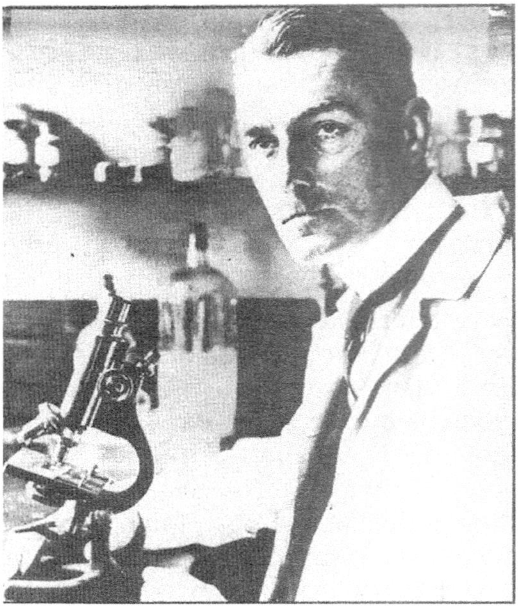

Figure 2.10 Sir Bernard Spilsbury (1877–1947) the legendary forensic pathologist whose celebrity status might have influenced the value of his forensic evidence. Figure is in the Public Domain.

fill the public gallery with his effortless, overwhelmingly dominant style. He simply was never wrong! Or at least it seemed that way.

He served as the forensic pathologist for a number of very high-profile cases, including the Crippen and Seddon cases, the Major Armstrong poisonings, the 'Brides in the Bath' murders, and the Crumbles murders.

His approach came under significant criticism in the 1940s, not only his dominant style, but also the validity of his scientific methodology. The eminent forensic pathologist, Professor Sir Sydney Smith (1883–1969; see Section 2.9 *Professor Sir Sydney Smith (1883–1969) – The Internationalisation of Forensic Science*) remarked about Spilsbury that he is '*very brilliant and very famous, but fallible...and very, very obstinate.*' And another eminent forensic pathologist, Professor Keith Simpson (1907–1985) said that Spilsbury's '*positive evidence had doubtless led to conviction at trials that might have ended with sufficient doubt for acquittal.*' A damning indictment indeed. But most damning of all is the comment by Manchester University's science historians Professor Ian Burnley and Dr Neil Pemberton in their *British Medical Journal* article published in 2010 (see *Further Reading* at the end of this chapter); commenting on Spilsbury's style they wrote, this '*lent him an aura of infallibility that for many raised concerns that it was his celebrity rather than his science that persuaded juries to credit his evidence over all others.*'

On balance, Spilsbury was detrimental to the development of forensic science in Britain. He refused to collaborate, take part in research projects or attend scientific meetings. He was an opinionated loner. It is probably because of Spilsbury that the development of the team approach to crime scene investigation was slow in the UK, while it moved apace in continental Europe (but more of this later...). To be fair though, as Spilsbury's biographer, Andrew Rose points out, '*He must not be judged with the wisdom of hindsight.*' In his era, Spilsbury was a truly great man.

Spilsbury took his own life by gassing himself in his laboratory at University College London on 17th December, 1947.

2.5.2 The Crippen Case

Hawley Harvey Crippen (Figure 2.11) was a homeopath from Michigan, USA. He moved to the UK with his second wife in 1897 – his first wife had died of a stroke. Despite his inability to practice medicine in the UK, he was still known as Dr Crippen (a name that now sends a chill down the spine!). Crippen's wife was Corrine (Cora) Turner, a music hall performer, whose stage name was Belle Elmore.

Figure 2.11 Dr Crippen and his wife Cora (also known by her stage name, Belle Elmore). Figure is in the Public Domain.

Crippen had worked for Dr Munyon's, a homeopathic pharmaceutical company; he continued to distribute patent medicines for Munyon's even after relocating to London. Munyon's sacked him in 1899 for spending too much of his time organising his wife's stage appearances. He moved to Drouet's Institution for the Deaf as a manger. He hired a secretary, Ethel de Neve, with whom he soon began an affair.

In 1905, the Crippens moved to 39 Hilldrop Crescent off the Camden Road in London, where they began taking in lodgers to make some extra cash. Cora soon began an affair with one of the lodgers. In response, Crippen entered into a relationship with Ethel de Neve as his mistress. This polyamorous relationship sets the scene for Cora Crippen's fate.

The Crippens held a party at their home on 31st January, 1910: Cora was never seen again. Crippen claimed that Cora had returned to the USA, and that she had died there and been cremated in California. Ethel de Neve moved in with Crippen and they began living as man and wife – Ethel was often seen around town wearing Cora's clothes and jewellery.

The police were alerted about Cora's disappearance by a music hall friend of Cora's (a strongwoman known as The Vulkan) and indirect acquaintance of Scotland Yard's (London's central police department) Superintendent Frank Froest. Crippen was interviewed by the police; he admitted that he had fabricated the story about Cora's return to America and her subsequent death – his excuse was that he was embarrassed because Cora had left him and run off to America with another music hall star. The police believed Crippen's story.

The police interest in Cora's disappearance spooked Crippen and de Neve and they fled to Belgium and onwards to Canada where they thought they would be far enough away to escape further police scrutiny.

Police suspicion was aroused by Crippen and his lover's sudden disappearance, prompting them to search 39 Hilldrop Crescent again. They had the most gruesome of finds under the basement floor: a human torso.

The Home Office senior forensic analyst Sir William Willcox was called in to investigate – he found the sedative scopolamine in the cadaver's tissues. Clearly, the victim had been drugged to facilitate its (the sex of the torso could not be determined) murder with minimum struggle and noise. Since the head and limbs were nowhere to be found (they were never found), identification of the torso was impossible. Enter Sir Bernard Spilsbury...

Spilsbury's *post-mortem* examination of the torso revealed an abdominal scar (Figure 2.12) which he claimed matched with Cora's

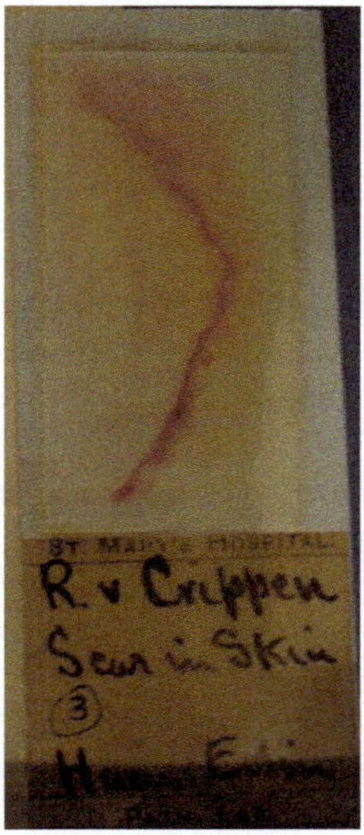

Figure 2.12 Cora Crippen's scar tissue, stained with haematoxylin and eosin and mounted on a microscope slide, was a key exhibit at Dr Crippen's trial. Figure is in the Public Domain.

medical history. During the trial, there was significant conjecture about the scar tissue. The defence pointed out that the scar had hair follicles, which is not a characteristic of scar tissue; Spilsbury retorted that the hair follicles were at the ends of the scar and not at its centre, which meant that they could be part of the surrounding 'normal' tissue and not the scar *per se*. Also, bleached blond hair, matching with Cora's, was found on the torso, along with a fragment of material that matched a pair of pyjamas given by Cora to Crippen.

Crippen had purchased scopolamine a few days before Cora's disappearance – the prosecution linked Crippen's purchase with the scopolamine found in the torso. The defence suggested that Crippen had given scopolamine to Cora as an aphrodisiac and that he had accidentally administered a fatal overdose. When Cora died, Crippen panicked and concealed her body under the cellar floor. Therefore, Cora's death was a terrible accident! This begs the question why had he severed the corpse's legs, arms and head? Surely, one might expect that in his panic, he would have hidden the entire body.

On 22nd October, 1910, Hawley Harvey Crippen was found guilty of the murder of Corrine Crippen. The presiding judge, Lord Alverstone, donned the traditional black 'cap' and sentenced Crippen to death.

Crippen was hanged at Pentonville Prison, London, at 9.00 am on 23rd November, 1910.

2.5.3 Reappraisal of the Crippen Case a Hundred Years On

With our modern forensic eye, it is clear that so many facets of the Crippen case simply would not stand up in court today. The hairs found on the torso were long and bleached, but just because Cora had long and bleached hair does not mean that the hair belonged to her. And was the scar that Spilsbury used as a basis for his decisions really a scar? Or was it simply a fold in a fragment of tissue?

Much of the evidence presented in court back in 1910 would have little credence in court today. This thinking prompted scientist to re-evaluate the Crippen case in 2010 using the most up to date techniques available (see *Further Reading* at the end of this chapter).

A microscope slide of tissue from the torso (Figure 2.12) still exists, and it was possible, using DNA technology (see Chapter 6 *DNA*) to determine that the cell nuclei had Y chromosomes; therefore, the torso belonged to a male and could not have been Cora's. In addition, scientists tracked down several of Cora's female descendants and using mitochondrial DNA (inherited down the female line – see Chapter 6 *DNA*) analysis, it was shown that the torso could not be traced back to any of the female antecedents.

The hair samples found on the torso and Cora's hairbrush (with some hairs attached) have been kept by the Metropolitan Police (London). Scientists have requested the hairs for DNA analysis to see if the hairs from the torso match the hairs from the hairbrush. The police have refused to release the samples for analysis.

So, a hundred years after Crippen's execution, questions remain about the validity of the forensic investigation. It has even been suggested that the police planted the torso under the cellar floor of 39 Hilldrop Crescent, but who knows? One thing appears to be certain; the torso was not Cora's.

2.6 Systematic Crime Scene Investigation – The Birth of CSI

The hallmark of the Spilsbury era was individualism; one man investigating along a path dictated by the case – an individualistic, non-systematic approach. There was no or very little team work, no organised, systematic approach, and perhaps little attempt to prevent cross-contamination. The situation in continental Europe was far more advanced in all these aspects. Arguably, Spilsbury's dogmatism inhibited the European approach, crossing the English Channel and the North Sea.

Reform began in the UK in the 1930s, whereas in continental Europe, it had been underway since the late 1800s. The Austrian criminologist, Hans Gross (1847–1915; Figure 2.13) introduced a systematic investigative approach based on scientific methods, and emphasised the importance of professional ethics in criminal investigation. His book *Handbuch für Untersuchungsrichter als System der Kriminalistik* (Criminal Investigations, a Practical Textbook) was published in 1893; it marked the modernisation of forensic science.

2.6.1 Modernisation of Crime Scene Investigations – Hans Gross

Hans Gross (Figure 2.13) initiated a revolution in the way crime scenes were investigated. His ideas and processes form the foundation of

Figure 2.13 Hans Gross (1847–1915), the grandfather of crime scene investigation.[3] Figure is in the Public Domain.

today's highly systematic approach to crime scene investigation (CSI). He is the grandfather of CSI.

Gross' advice to crime scene investigators included:

- Maintain absolute calm.
- Exclude everything that happened after the moment of the crime.
- Resist the urge to touch or move anything.
- Treat the crime scene like an open-ended puzzle.

He explained the need to make careful sketches of the crime scene and to note accurately exactly where items were by using a grid system (he used squared paper to sketch on) (Figure 2.14). He described his grid approach in *Handbuch für Untersuchungsrichter als System der Kriminalistik* [translated]:

> *Suppose that a certain number of drops of blood which have dried upon a plank have to be depicted. The large sketch represents the actual*

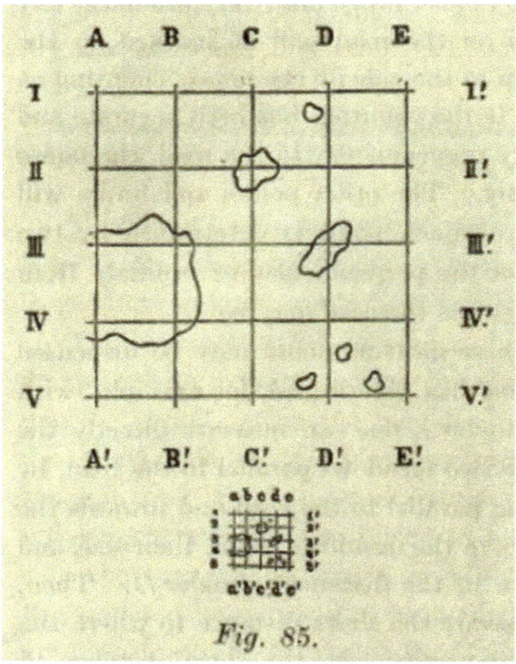

Figure 2.14 Hans Gross' small and enlarged grid sketch of blood spots on a plank at the scene of a crime. He used number (roman numerals) and letter notations for the horizontal and vertical grid lines to make it easy to reference the position of an object (in this case a bloodspot).[3] Figure is in the Public Domain.

> *subject-matter and the small one the finished drawing. First examine the proportion of the plank to be drawn and its dimension. This is then delimited by means of a set-square setting down the lines A A′ & I I′ at right angles. These are then divided into a certain number of equal parts, the more there are of them the more accurate the result will be: parallels are then drawn so as to obtain a number of squares of equal size.*

Gross extended his thinking from accurate depictions of crime scenes to realistic observations and recollections of what might be important at the scene. He cautioned against the human trait of exaggeration and 'seeing' things that might not be there. He points out that human nature tends to be selective, but that for crime investigation, accuracy is crucial:[3]

> *It is only in conformity with human nature to stop the more willingly at what is more interesting than at what belongs to everyday life. We like to discover romantic features where they do not exist and we even prefer the recital of monstrosities and horrors to that of common every-day facts. This is implanted in the nature of every one, and though in some to a greater, in some to a lesser extent, still there it is. A hundred proofs, exemplified by what we read most, by what we listen to most willingly, by what sort of news spreads the fastest, show that the majority of men have received at birth the tendency to exaggeration.*

Gross set the ground rules for forensic science as we know it today – the *fin-de-siècle* CSI!

2.7 Advent of the Crime Lab – Dr Edmond Locard, 1912

The French criminologist Dr Edmond Locard (1877–1966; Figure 2.15) is considered the pioneer of modern forensic science. In 1910, Locard persuaded the Lyon (France) police department to give him an attic room and two scientific assistants – this became the first police lab in the world, and Locard became the first forensic scientist.

2.7.1 Locard's Exchange Principle

Locard's goal was to introduce hard science and logic into his crime laboratory investigations. As a result of his extensive practical crime

Figure 2.15 Dr Edmond Locard (1877–1966) created the first police lab in the world and said 'Any action of an individual or a criminal and any kind of violation cannot take place without leaving any trace.' This became Locard's Exchange Principle, the foundation of modern forensic science. Figure is in the Public Domain.

work infused with a degree of logic, he made the most important statement in forensic science:

> *Toute action de l'homme, et a fortiori, l'action violent qu'est un crime, ne peut pas se dérouler sans laisser quelque marque. [Any action of an individual or a criminal and any kind of violation cannot take place without leaving any trace].*

This is Locard's Exchange Principle – Every Contact Leaves a Trace. It is the fundamental tenet of modern forensic science.

In addition, Locard created the 12-point fingerprint identification method, which is used even today (see Chapter 7 *Fingerprints*), established the International Academy of Criminalistics in Lausanne, Switzerland (destroyed in World War II), and between 1931 and 1935, he published 7 volumes of his *Traité de Criminalistique* (Treaty of Criminalistics). Locard and forensic science are synonymous.

Locard's Exchange Principle might involve the transfer of clothing fibres from a rape victim to the rapist, hairs transferred from a murder victim to the murderer or perhaps plant debris on the shoes of a burglar that might match plants in the garden of a burgled house. To investigate the hairs, fibres and plant debris, a microscope is necessary. The microscope became the centrepiece of forensic science at the beginning of the 20th century.

Marie Latelle's murder in Lyon in 1912 tested Locard's Exchange Principle and his technical skills to their limits. His investigation was a perfect example of the new approach to forensic investigation – the innovative Locard used the transfer of Marie's makeup to her murderer to solve the case.

2.7.2 The Marie Latelle Murder Case – 1912

Marie Latelle was found strangled to death in her parent's parlour, her boyfriend, Emile Gourbin was the prime suspect because he and Marie had been together the same day. However, Gourbin had an excellent alibi: he had been playing cards and drinking with a group of friends at his home the whole evening and his friends had not left (rather drunk) until 1.00 am. This time period covered the estimated time of Marie's death, which meant that Gourbin could not have murdered her.

Dr Locard was brought in to investigate the case. He scraped under Gourbin's finger nails and when he looked at the scrapings under the microscope, he identified skin tissue – just what might occur if he had strangled Marie. At that time, it was not possible to determine whether the cells were human or from whom they might have originated. For these reasons, this line of investigation was dropped.

Locard looked at the cells once more under his microscope and noticed that they were coated with a pink powder. What could this be?

Separation of the powder from the nail scrapings made it possible for Locard to identify it as rice starch using a chemical test (iodine solution turns blue on coming into contact with starch). Further chemical analysis showed the presence of bismuth, magnesium stearate, zinc oxide and Venetian red (a red brown iron oxide-containing pigment). Dr Locard analysed Marie's face powder using the same methods he had used for the powder he isolated from under Gourbin's nails: the chemical compositions of the powder from the Gourbin's nail scrapings and Marie's face powder were identical. Gourbin had Marie's makeup under his nails!

When confronted with the evidence Gourbin confessed to Marie's murder. During his confession, he told the police that he had adjusted his clock to show a later time, making his friends believe that they had left at 1.00 am, when in reality they left at about 11.00 pm. This gave Gourbin just enough time to murder Marie.

2.8 Professor John Glaister (1856–1932)

John Glaister (Figure 2.16) was promoted to the chair of forensic medicine and public health at the Glasgow Royal Infirmary Medical School in 1888 – he was only 32 years old. He occupied the chair until 1931, a year before his death. His most notable contribution to forensic science was his textbook: *A Textbook of Medical Jurisprudence, Toxicology and Public Health* published in 1902. Glaister's son, John Jr., also became a forensic practitioner and succeeded his father as a Professor of Forensic Medicine at Glasgow University. Professor John Glaister Jr. had a particular interest in hairs (both human and animal) in the context of Locard's Exchange Principle; his work

Figure 2.16 Professor John Glaister (1856–1932) made a significant contribution to forensic science. He wrote *A Textbook of Medical Jurisprudence, Toxicology and Public Health* in 1902. Figure is in the Public Domain.

significantly increased the robustness of the use of hairs in forensic casework. In 1931 Professor Glaister Jr. made the important point:

> *The avenue of medico-legal investigation demands an exhaustive study of animal and human hairs... For the purpose of identification, it is essential that the examiner should have at hand a comprehensive collection of known hairs for comparison.*

Interestingly, Professor Glaister Jr.'s nephew, the television producer Gerard Glaister, was responsible for one of the first TV series about forensic science – *The Expert* (BBC, 1968–1971) – loosely based on his uncle's work. The series attracted large audiences (including one of this book's authors, IS) and kicked off the forensic science TV era (and IS's interest in forensic science!).

2.9 Professor Sir Sydney Smith (1883–1969) – The Internationalisation of Forensic Science

The growth and development of forensic science had focussed on Europe from those first inklings of a 'real' scientific approach to criminal justice starting with Professor John Gordon Smith's use of John Marsh's analytical method for arsenic in the Bodle case in 1832 (see Section 2.2.2 *The George Bodle Case 1832*), Hans Gross' revolution in the way crime scenes were investigated in the late 1890s, followed by Dr Edmond Locard's transfer principle and establishment of his first crime lab in 1912. The scene was well and truly set for the scientific approach to criminal investigation we know today.

Sydney Smith (1883–1969; Figure 2.17), a young New Zealand pathologist, introduced an international approach to forensic science. He graduated in medicine from Victoria College of Wellington (now Victoria University of Wellington) in 1912, worked as a GP for a short while, and then went to the Department of Forensic Medicine at Edinburgh University to do research for his doctorate. He moved back to New Zealand in 1914 for a couple of years, and then he was appointed as a medico-legal advisor to the Egyptian government and a senior lecturer in forensic medicine at the Cairo School of Medicine. In 1925, he published his *Textbook of Forensic Medicine*, and in 1928, he was appointed to the Regius Chair of Forensic Medicine at the University of Edinburgh. He became a world authority on ballistics and firearms (see Chapter 11 *Firearms*). Arguably his international experience was instrumental in

Figure 2.17 Professor Sir Sydney Smith (1883–1969) introduced an international approach to forensic science. He began as a young pathologist in New Zealand, lectured in forensic medicine at the Cairo School of Medicine and ultimately held the position of Regius Professor of Forensic Medicine at the University of Edinburgh. Courtesy of Royal College of Physicians of Edinburgh.

the development of his expertise – he was the first truly international forensic scientist.

He worked on many high-profile cases, many involving firearms. However, his work in forensic anthropology opened a new field in the identification of murder victims. This is well-illustrated by the Buck Ruxton 'Savage Surgeon' case in 1935.

2.9.1 Buck Ruxton – The 'Savage Surgeon' 1935

Ruxton (1899–1936; Figure 2.18) was an Indian-born doctor working in Lancaster, England. He murdered his wife and their housemaid on 15th September 1935, dismembered their bodies, and dumped them in a river near Moffat in Scotland – they were found on 29th September 1935 by a passer-by, who noticed a roughly wrapped parcel lodged against a boulder in the fast-flowing river under a bridge – there was an arm protruding from the parcel...

Professor John Glaister (see Section 2.8 *Professor John Glaister (1856–1932)*) conducted the *post-mortem* examination of the maggot-ridden

Figure 2.18 Buck Rushton, the 'savage surgeon' was found guilty of the murder of his wife and housemaid, dismembering their bodies and dumping them in a Scottish river. He was hanged in Manchester, England, on 12th May, 1936. Figure is in the Public Domain.

body parts; he called in an entomologist (Alexander Mearns) to determine the larval stage of the maggots.

Blowflies lay their eggs on a corpse in the open air very soon after it is killed or dumped; identifying the larval stage and knowing the time it takes for the maggots to develop from the egg to the particular larval stage helps to estimate the time of death (or in this case, dumping the body parts) (see Chapter 5 *Forensic Biology*).

Mearns identified the blowfly pupae on the body parts as *Calliphora vicina*, but noted that the maggots were of a different species. This suggested that the body parts had been initially in one place where the *C. vicina* eggs had been laid (and they had fully developed to the pupa stage), and then moved to another place later where the eggs of the maggots were laid. Furthermore, the body parts must have been in the place where they were found for 14–17 days (based on the larval stage of the maggots). This was one of the first instances where forensic entomology was used as evidence in court.

The forensic pathologists pieced together the body parts to reassemble the corpse. They discovered that there were in fact two corpses! The press christened the case "The Jigsaw Murders". They called in Professor Sydney Smith from the University of Edinburgh to help identify the bodies. By this time, it was suspected that they might be the bodies of two missing women from Lancaster. Smith was the first to use forensic anthropology to identify a corpse: he used X-rays to examine the remains of the heads of the two corpses and superimposed photographs of the faces of Ruxton's wife and housemaid onto the X-rays. The face shapes, including the jaw structures, matched. The corpses were identified! The ensuing investigations strongly implicated Ruxton as the murderer...

The case began at the Manchester High Court of Justice on 2nd March 1936, under the jurisdiction of Mr Justice Singleton. Ruxton was found guilty of the murder of his wife, Isabella Ruxton, and their housemaid, Mary Jane Rogerson – he was sentenced to death. The case was appealed but the appeal failed. Ruxton was hanged at Manchester prison on 12th May, 1936.

It is clear from this case that collaboration, bringing together ideas and techniques, and significant innovation underpinned a good forensic investigation. Just imagine if Sir Bernard Spilsbury had been the forensic pathologist in charge. His hunch would have been 'correct' and accepted without question by the court without the rigours of team work and the innovation that a group of clever minds can facilitate. Forensic science had come a long way in just a few years.

2.10 The Meteoric Rise of Forensic Science: The 1940s, 50s, 60s and Beyond...

This section almost requires a book of its own. There was so much that had changed so quickly once the techniques available to forensic scientists developed and were perfected. Validation and probability were considered crucial for determining the weight of a particular piece of forensic evidence in the context of the case as a whole. Did the hair found on the overalls of an accused murderer come from the murder victim just because the colour matched the victim's hair? Or could it have come from someone else who had nothing to do with the case? If it was a brown hair, the probability of it being from the brown-haired murder victim is much lower (because brown is a very common hair colour) than if it was half brown and half blond (suggesting that it had been bleached and that the bleaching was growing out) and the murder victim had bleached brown hair. This way of thinking – of interrogating data – came to the fore in the 1940s as forensic science matured from a hunch to hard scientific evidence.

This, coupled with the amazing speed of development of analytical techniques, made forensic science a very powerful tool in the cause of criminal justice.

2.11 Dr Harry Söderman (1902–1956)

Harry Söderman (Figure 2.19) was a Swedish forensic scientist who significantly influenced the development of forensic science worldwide. His book *Modern Criminal Investigation* (1935; co-authored with John O'Connell, Deputy Chief Inspector, New York police) was the standard text for decades. The book was the first 'modern' forensic science text – reading the book today does not give a feeling of 'out of date' methodology and thinking. The only things missing from *Modern Criminal Investigation* are the sophisticated techniques that forensic scientists have at their fingertips nowadays. For example, Söderman and O'Connell's chapter on arson mentions nothing about identifying accelerant residues using headspace gas liquid chromatography (GLC) (see Chapter 4 *Forensic Chemistry*) as a text written today would. There is a good reason for this; GLC was not described until 1952 (see *Further Reading* at the end of the chapter), twenty years after *Modern Criminal Investigation* was published.

Söderman had a world view of criminology and applied this to his forensic science thinking. This perspective emerged because after he graduated from the chemical college in Malmö, Sweden, he went to Altenburg, Germany for further studies (in forestry!) and then on a

Figure 2.19 Dr Harry Söderman about 1950. His book *Modern Criminal Investigation* (1935; co-authored with John O'Connell) was the standard text for decades. Figure is in the Public Domain.

world tour; during this period, he was commissioned by a police magazine to write travelling letters; he focussed on how the Police in different countries solved crimes. This piqued his interest in criminology and gave him a perspective beyond his predecessors. He went on to study with Dr Edmond Locard (see Section 2.7 *Advent of the Crime Lab – Dr Edmond Locard, 1912*) and graduated with a PhD on ballistics in 1926 from the University of Lyon (France). Dr Söderman went on to become one of the world's most influential forensic scientists of modern times.

2.12 Latter-day Forensic Scientists Embrace Scientific Robustness

After the 1950s, there was a significant growth in forensic science with a concomitant explosion of forensic scientists. Forensic science institutes and societies were created, international scientific meetings devoted to forensic science were regular features of the academic calendar, forensic science course became popular in universities worldwide, and more recently, degrees in forensic science were created. Forensic science became a household term. Journalists wrote about its use to solve crimes and newspaper readers loved the (often gory) detail, and TV programmes abounded and became prime-time favourites. This was the Renaissance period of forensic science.

There are too many individual forensic scientists, each of whom did their bit towards the development of the science, to mention individually. I will take a closer look at one or two who made relatively small contributions *per se*, but their myriad small contributions added up to a huge overall advance worldwide.

Dr H. J. Walls (1907–1988) was the director of the Metropolitan Police Lab, London; he continued the lineage of eminent forensic scientists who were appointed to direct the UK's first forensic laboratory. His contribution involved overseeing the work of the laboratory and authoring books largely intended for a general readership; for example, *Expert Witness* published in 1972, which covers the cases Walls was involved in across the spectrum of his broad ranging lab's work.

In 1970, Walls co-authored (with Alistair Brownlie) *Drink, Drugs & Driving* (2nd edition 1985), which became the standard text on the subject. In many ways the book was ahead of its time because it tackled the now accepted, but then just becoming appreciated, issues of blood alcohol concentrations and driving impairment (drunk driving), setting legal limits, roadside testing and its legal ramifications. At about the same time when *Drink, Drugs & Driving* was published, Dr Tom Jones OBE (1935–2013) had invented the first electronic breathalyser test device (Jones received the Queen's Award for Technological Achievement in 1980), which became the mainstay for drink driving prosecutions.

The University of London's professor of chemical toxicology, E. G. C. Clarke (1905–1978), brought together analytical methods to measure drugs in *post-mortem* (and other clinical) materials in his book *Isolation and Identification of Drugs* published in 1969. This book was, and to some extent remains, the bible of forensic toxicologists. The book has chapters authored by the big names of 1960s and 70s forensic science, including Dr Alan Curry (who became the Director of the British Forensic Science Service) and Dr H M Stevens from the Home Office Central Research Establishment.

Dr Alan Curry (1925–2007) pioneered GLC blood alcohol analysis (see *Further Reading* at the end of the chapter) which has led to many an over-the-limit driver's conviction and safer roads for the rest of us. Curry was not only a forensic science pioneer but he also had a keen sense of humour as illustrated by the 'constitution' (or lack of) of his brainchild, the Crippen Club:

> *The club shall have no Constitution or Mission Statement.*
>
> *It shall be a dining club for forensic toxicologists who are outstanding in their professional achievements.*
>
> *At the dinner, a Member shall give a ten-minute talk which shall be aimed at distillation rather than evaporation.*
>
> *A toast will be drunk at each dinner to the memory of the late Professor E G C Clarke.*

Dr Stuart Kind (1925–2003) was the Director of the Wetherby Forensic Science Laboratory and founded the Forensic Science Society in 1959 (now the Chartered Society of Forensic Sciences). He pioneered and applied geographical profiling to the Yorkshire Ripper case which led to the serial killer Peter Sutcliffe's arrest on January 2nd, 1981.

Dr Margaret Pereira (1928–2016) was a prominent figure in forensic science, ascending from the bottom to the very top of Britain's Forensic Science Service – she replaced Curry as its controller in 1982 and, as the Guardian obituary put it, '*She broke every glass ceiling on the way*'. On her journey through the forensic science ranks she developed blood grouping techniques for minute bloodstains (see *Further Reading* at the end of the chapter) – a significant advance at the time. As the controller, she oversaw the introduction of DNA technology, which fundamentally changed practical forensic science worldwide forever.

Dr H. M. Stevens (1926–1988) was a great, rather unsung, hero of forensic science. He beavered away on his pet subject, the stability of drugs and poisons in putrefying human tissues (see *Further Reading* at the end of the chapter). This was important work because it identified putrefaction breakdown products in human poisoning cases as well as assessing the residence time of poisoning agents in rotting corpses. In addition, it fuelled many dinner party conversations (I know, I was at several of them).

2.13 Sir Alec Jeffreys (Born 1950)

But there is one scientist, Sir Alec Jeffreys (Figure 2.20), who revolutionised forensic science through his introduction of DNA technology in criminal identification. This changed the face of forensic

Figure 2.20 Sir Alec Jeffreys is the architect of arguably the most important revolution in forensic science: DNA profiling. Reproduced from ref. 6, https://doi.org/10.1371/journal.pgen.1000765, under the terms of the CC BY 4.0 license https://creativecommons.org/licenses/by/4.0/.

science as we knew it forever (see Chapter 6, *DNA*). It arguably led to the demise of forensic laboratories and the redundancies of many forensic scientists as police and the courts' dependence on the ostensibly definitive DNA evidence grew, and grew, and...

Further Reading

1. S. Hempel, *The Inheritor's Powder*, Weidenfeld & Nicolson, London, 2013, ISBN 978 0 297 86742 5.
2. M. Hughes, Arsenic toxicity and potential mechanisms of action, *Toxicol. Lett.*, 2002, **133**, 1–16.
3. I. Burney and N. Pemberton, The rise and fall of celebrity pathology, *Br. Med. J.*, 2010, **341**, 1319–1321.
4. D. R. Foran, B. E. Willis, B. M. Kiley, C. B. Jackson and J. H. Trestrail, The conviction of Dr Crippen: new forensic findings in a century old murder, *J. Forensic Sci.*, 2010, doi: 10.1111/j.1556-4029.2010.01532.x
5. I. Burney and N. Pemberton, Making space for criminalistics: Hans Gross and *fin-de-siècle* CSI, *Stud. Hist. Philos. Biol. Biomed. Sci.*, 2013, **44**, 16–25.
6. P. Gill, A. J. Jeffreys and D. J. Werrett, Forensic application of DNA 'fingerprints', *Nature*, 1985, **318**, 577–579.
7. H. M. Stevens, *J. Forensic Sci. Soc.*, 1984, **24**, 577–589.
8. M. Pereira *et al.*, *Med., Sci. Law*, 1969, **9**, 116–121.
9. A. S. Curry *et al.*, *Analyst*, 1966, **91**, 742–743.
10. A. T. James and A. J. P. Martin, *Biochem. J.*, 1952, **50**, 697–690.

References

1. A. I. Vogel, *A Textbook of Quantitative Inorganic Analysis Theory and Practice*, Longmans, Green and Co, London, New York, Toronto, 1939, p. 516.
2. J Marsh, Account of a method of separating small quantities of arsenic from substances with which it may be mixed, *Edinburgh New Philos. J.*, 1836, **21**, 229–236.
3. H. Gross, *Handbuch für Untersuchungsrichter als System der Kriminalistik*, 1906, p. 468.
4. Memoirs of Madam Lefarge, Carey & Hart, Philadelphia, 1841.
5. T. Rothwell, Margaret Pereira obituary. *The Guardian*, 2017 Jan 26, Available from: https://www.theguardian.com/science/2017/jan/26/margaret-pereira-obituary.
6. J Gitschier, The Eureka Moment: An Interview with Sir Alec Jeffreys, *PLoS Genet.*, 2009, **5**(12), e1000765.

3 Crime Scene Investigation

Locard's Transfer Principle (Chapter 2 *The History of Forensic Science*) has had a huge and positive impact on forensic investigations: the transfer of fibres, blood, hairs, pollen and DNA serves as valuable evidence, indicating that something has been in contact with something else, some place or somebody. However, Locard's Principle can wreak havoc on crime scene findings if the scene is not searched systematically and logically, with the absolute minimum chance (ideally zero, of course) that cross-contamination of evidence might occur.

Söderman (Chapter 2 *The History of Forensic Science*) realised the importance of minimising transfer and so he introduced systematic approaches into crime scene investigation and evidence collection to minimise cross-contamination. As forensic science has progressed over the years, particularly with the advent of DNA profiling, the need to be ultra-fastidious to prevent even the transfer of a few DNA molecules from one item to another is of paramount importance. Modern day crime scene investigators should wear anti-static overalls (to prevent particles and fibres from adhering to them), gloves, boots, hair coverings and masks to ensure that transfer of miniscule particles of evidence is as close to zero as possible. Any breech of these fastidious protocols may significantly reduce the admissibility of evidence in court.

Furthermore, the bagging and transfer of samples from the crime scene to the laboratory should be carried out in such a way that every step in the sample's journey is recorded – including documenting every person who has handled the packaged sample on its way to the lab. When the sample reaches the laboratory, the rules are equally strict to prevent the transfer of material on the lab bench. This is

Forensic Science: The Science Behind the Truth
By Ian Shaw with Anna Sandiford

Published by the Royal Society of Chemistry, www.rsc.org

crucially important because lab staff will likely handle samples from both the suspect and their victim(s) or complainants. The need to separate items and samples from different areas of a case is paramount.

And throughout this incredibly well-tracked process, we have to be cognizant of the remote possibility that an unscrupulous police officer or forensic scientist might interfere with a sample to alter the science behind the truth...

3.1 What Is a Crime Scene?

Before we start discussing how a crime scene is investigated, it is important to know what a crime scene is. This might sound obvious, but crime scenes vary very considerably. The US Department of Justice defines a crime scene as '*any location that may be associated with a committed crime.*' Wikipedia extends this in its excellent, all-encompassing definition:

> *The location of a crime scene can be the place where the crime took place or can be any area that contains evidence from the crime itself. Scenes are not only limited to a location, but can be any person, place, or object associated with the criminal behaviours that occurred.*

The most important aspect of the crime scene is to secure it quickly, prevent any tampering and note any item of interest in its original location (*i.e.*, observe and record) before it is collected for further investigation. Our understanding of the importance of the crime scene and the potential evidence that it includes has evolved significantly over time. This evolution aligns with major advances in forensic science; for example, Gross' 'refined' view of the scene of a crime and Locard's Exchange Principle underpin today's belt and braces approach to crime scene investigation.

3.2 Today's Approach to Crime Scene Investigation

Hans Gross (1847–1915) is the grandfather of crime scene investigation (see Chapter 2 *The History of Forensic Science*); in the early 1900s, he recognised the need to '*exclude everything that happened after the moment of the crime*', '*resist the urge to touch or move anything*' and '*treat the crime scene like an open-ended puzzle*'. Before Gross came along, the crime scene was akin to a smash and grab event. Exhibits were picked up, examined, taken away, left behind, moved, and contaminated...not conducive to a good scientific investigation of a complex evidential jigsaw!

Gross recognised the importance of making careful notes of the crime scene and creating accurate sketches to facilitate the reconstruction of the scene, even if only in the minds of the investigating scientists at a later date.

Gross realised that Locard's Exchange Principle (see Chapter 2 *The History of Forensic Science*) worked both ways: primarily it was important in showing that someone (*e.g.*, a suspect) had been in a particular place, but it could also result in wrongful accusations if fragments of material (*e.g.*, fibres) were transferred by investigation police or scientists from one place to another. For these reasons, modern forensic scientists and officers working at the crime scene should wear special clothing (including non-static overalls, overshoes, gloves and masks) to minimise the risk of transferring material in the crime scene setting (Figure 3.1).

Figure 3.1 A mock crime scene showing the investigating forensic scientist wearing overalls (non-static), a mask and gloves to reduce the risk of material transfer or contamination.

3.3 The Process of Crime Scene Investigation

The primary goal of crime scene investigations is to uncover material that can be pieced together to identify what happened at the scene and who was involved. The process of achieving this goal focuses very strictly on preventing contamination and the transfer of material; from the perspective of the forensic science team, it involves

1. A visual 'investigation' of the scene – to get a general idea of what has happened and where the major areas of interest lie (*e.g.*, a body, weapons, or a broken window). [This is not part of the forensic investigation, but if there are injuries, medical assistance is sought, and if someone is fleeing from the scene the police might make an arrest; preservation of life takes priority].
2. Secure the scene – the scene is cordoned off, usually with 'crime scene tape' and an officer is given responsibility for controlling access to the scene.
3. Controlled access to the scene. Everyone who enters the scene is noted (being police or forensic employees, their fingerprints and DNA profiles will be on record). This allows their elimination from the investigations.
4. Noting the conditions – this includes the weather, time of day, *etc.*
5. Reporting fully and accurately during a detailed investigation of the scene.
6. Releasing the scene when the investigation is complete – once the scene is released, there is no going back!

Point 5 is the responsibility of the forensic investigation team; it involves a very carefully orchestrated approach to noting, recording and collecting items and samples for further investigation in the laboratory.

Before delving into the details of the forensic investigation of the scene, it is important to understand who might be working at the crime scene and what their responsibilities are.

3.4 Protecting the Crime Scene

If a crime scene is outdoors or otherwise exposed to the weather it might be necessary to protect it to minimise the impact of the weather on anything of relevance within the scene. Clearly, if the scene comprises a large area outdoors, it is unlikely that it will be possible to protect all of it. In such cases, important areas or specific items of the

Figure 3.2 A tent erected at a crime scene to protect key information (in this case blood patterns) from the weather at an outdoor crime scene. Courtesy of Arrowhead Forensics, Lenexa, Kansas, USA.

scene are often protected separately (*e.g.*, a body in a murder case). The protection is primarily intended to safeguard the potential evidence, but in the case of a body at the scene, it also serves to preserve the dignity of the deceased.

The most common way to protect evidence at the scene from the weather is to erect a tent (Figure 3.2); this is a common sight in news reports of serious outdoor crimes.

3.5 Crime Scene Officers

There are key groups of people who might be working at the crime scene. The police staff working at the scene are referred to by various terms such as scenes of crimes officers (SOCOs), forensic scene investigators (FSIs), crime scene investigators (CSIs) or crime scene examiners (CSEs), depending on your location in the world. The roles and ranks of these scene investigators also vary from country to country. In some countries they are sworn police officers, in others they are employed by the police service, but are not sworn police officers or they might be employees of a country's forensic science service. Despite the specifics of their titles and employment contracts, their roles are similar; they are trained in the investigation and collection of information from the scene of a crime. I will refer to them as scene officers to avoid any bias or confusion.

3.6 Searching, Recording and Collecting Evidence from the Crime Scene

The scene officer(s) conduct their business at the crime scene in a very specific and methodical way. They may follow the following approach:

1. Survey the scene and note any 'obvious' potential evidence.
2. Search the scene.
3. Document the scene and potential evidence (using sketches, photographs, videos, and recordings).
4. Collect and preserve samples, material and items (*e.g.*, to be sent to the laboratory for further investigation).
5. Release of the scene.

I will deal with points 2, 3 and 4 in detail.

Back in the early 1900s, Hans Gross proposed a grid approach for accurately searching and recording the crime scene (see Chapter 2 *The History of Forensic Science* and Figure 3.3); his general approach is still used by scene officers today with a few variations.

Gross used a grid recording system. Scene officers today might prefer a spiral or, perhaps, a line search method (Figure 3.3). The important point is that the way the scene officer moves around the scene to record it and collect information reduces the likelihood of missing anything, and that during the process, they are unlikely to disturb anything before they have noted or marked it on their scene plan.

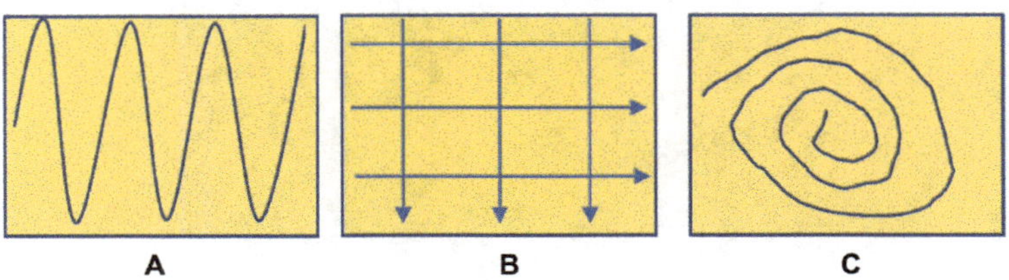

Figure 3.3 The three main crime scene search methods are: strip method (A), grid method (B) and spiral method (C). The crime scene (*e.g.*, a room) is represented by the orange square; the blue line shows the path through which the scene officer moves through the crime scene, noting and collecting potentially relevant information without disturbing anything before recording its presence and position.

As the scene officer moves through the crime scene observing and collecting information, he or she should make a detailed record of everything in the scene. This can take the form of a detailed sketch with measurements (Figure 3.5), a sketch on grid-ruled paper (Chapter 2 *The History of Forensic Science*; Figure 2.15) to help accurately pinpoint from where the item (*e.g.*, a blood stain) was recovered or noted, photographs, or a recording of the scene officer describing the crime scene or even better a combination of methods. Usually, items of interest found at the scene are identified with a number label often with an adjacent ruler to indicate size and photographed (Figure 3.4). The number markers may be included on the sketch of the crime scene. This should make it possible to pinpoint a particular item or sample on the scene sketch or plan, see a photograph of it and aid its identification in the lab.

The physical forms of scene documentation (*e.g.*, sketches) should be dated and signed so that the person taking the notes can be identified. Any recordings should also be assigned a unique filename if they are recorded electronically, or verbally identified on the recording if they are not. Identification of the scene records is of paramount importance.

The crime scene might include larger structures that require separate searching at a separate location (*e.g.*, a motor vehicle at a fatal crash scene then transferred to a tow yard for further examination). Continuing with the example of a motor vehicle; it may be divided into zones (Figure 3.6) to aid in searching and recording the findings.

Figure 3.4 Photograph of a number marker at a crime scene with an adjacent ruler to indicate the size of the relevant item. Courtesy of Arrowhead Forensics, Lenexa, Kansas, USA.

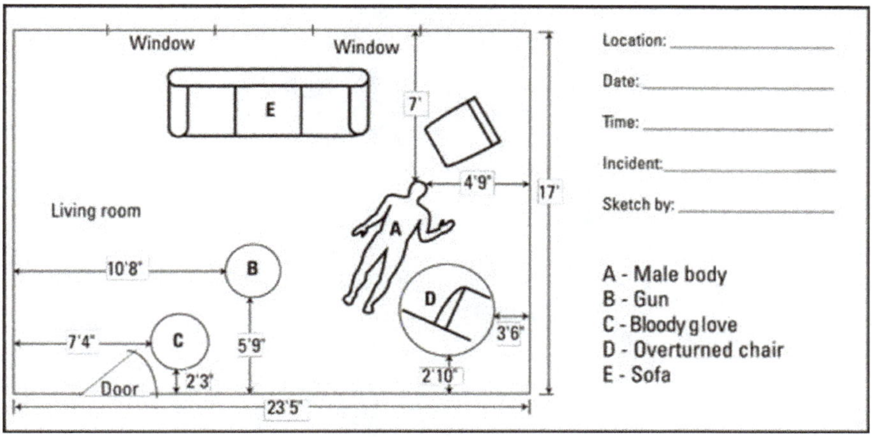

Figure 3.5 A clearly labelled sketch, including dimensions, of a crime scene. The pro forma on which the sketch was drawn (or electronically created) has specific places to record the scene officer's or examiner's name, date, time, location and description of the incident. Reproduced from ref. 1 with permission from John & Wiley & Sons, Copyright 2019 John Wiley and Sons.

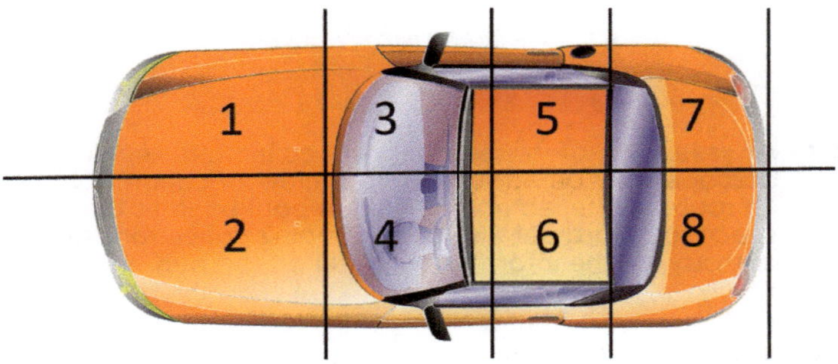

Figure 3.6 Use of zones in a motor vehicle to aid searching and recording the position of any potential evidence found. Zone 1 – left bonnet (hood), zone 2 – right bonnet (hood), zone 3 – left front seat, zone 4 – right front seat, zone 5 – left back seat, zone 6 – right back seat, zone 7 left boot (trunk), and zone 8 – right boot (trunk). Created with Bing Image.

3.7 Collecting Samples and Items as Potential Evidence

During the search of the crime scene, the scene officer or forensic scientists may collect samples to take back to, or submit to, the lab for further investigation. This involves removing the sample from its position at the scene after it has been carefully and clearly

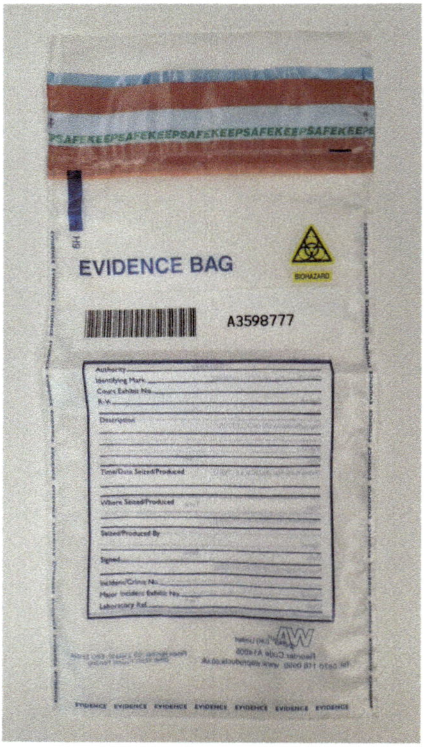

Figure 3.7 An example of a sample/item collection (evidence) bag: a barcode may be added to aid its identification during transfer en route to the lab and when it is booked in at the lab. There is space to record details of the sample's collection and transfer of custody on the information panel on the bag.

documented as described above. It may be placed into a specially designed plastic or paper bag (other containers may be used for specific samples; *e.g.*, tightly sealed containers for arson samples – see Chapter 4 *Forensic Chemistry*) that should be tamper-evident sealed and permanently labelled.

When bagged, the sample should be labelled with some form of case identification information, the name or initials of the person collecting the sample and the date and time of collection.

Every time the sample is transferred from one person or one place to another, the transfer should be recorded, either on the sample bag or using another tracking method (*e.g.*, a barcode). The person to whom the sample is transferred should be recorded along with the date and time of transfer. If properly completed, this allows for all of the sample's movements (chain of custody) to be retraced if anything goes wrong or there are questions about the security of the sample. This process is of paramount importance to prevent sample tampering or contamination – or at least discover it in the unlikely event that it happened. In addition, nowadays sample bags often have barcodes pre-printed or added to aid their identification during transfer and when received by the lab (Figure 3.7).

In addition to the details recorded on the sample bag label to allow tracing of the chain of custody, modern bags may also have tamper-evident seals. Such seals show clearly if an attempt has been made to open the bag before it reaches the lab. This further minimises the risk of tampering. It also allows assessment of whether the bag was properly sealed.

I am currently involved in investigating a suspected horse doping that is increasingly resembling a sample tampering case because of inconsistencies in the blood sample tubes and indications of tampering with the sample bags. The importance of the integrity of sample tubes and bags is illustrated well by this case.

3.8 A Case of Horse Doping or Was It Sample Tampering?

Race horses can be worth a very great deal of money if they win races regularly. In addition, the prize money for winning a race can be significant. All of this can lead unscrupulous trainers or owners to dose their horses with substances that will enhance their performance, thereby increasing their chances of securing the winner's pot of money. Equally, unscrupulous opponents might attempt to discredit an owner or trainer by making it look as though he or she has doped their horse to enhance its performance.

The case involves a very experienced trainer who had never been accused of horse doping before. Before an important race, horses are often blood sampled to check for doping. In this case, testosterone cypionate (TC) was found in the horse's blood.

3.8.1 Testosterone Cypionate

TC is a drug illegally used to enhance race horse performance. Interestingly, it is used legally in human medicine in transgender female to male transitions to enhance testosterone levels and the consequent development of male characteristics. Administered by intramuscular injection, it operates by slowly breaking down to release the male hormone, testosterone (Figure 3.8); the testosterone enters the bloodstream *via* capillaries from the muscle. In the context of racehorses, enhanced testosterone increases masculinity, including increased muscle bulk and perhaps enhancing the animal's vigour.

The finding that TC was present in the horse's blood unequivocally proved that it had been doped. The trainer emphatically denied that he had administered the drug.

3.8.2 Uncertainties About Blood Sample Collection

On closer investigation, it was found that the person who took the blood sample before the race had used the wrong type of blood tube to send the blood sample to the lab for analysis. In addition, the blood tube had not been taken from a sealed packet immediately

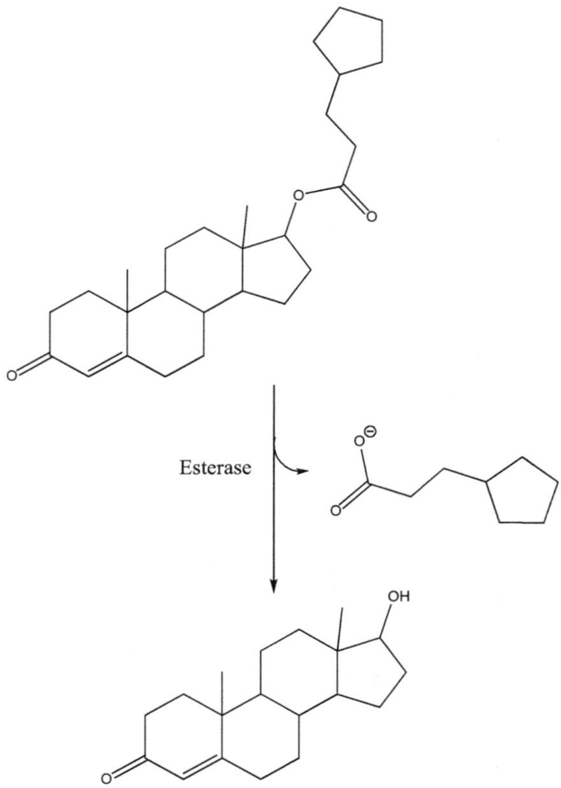

Figure 3.8 Conversion of testosterone cypionate (top) to testosterone (bottom) catalysed by an esterase (a group of enzymes found throughout the body).

before receiving the blood sample. Furthermore, the bag used to transport the blood sample tube to the lab was allegedly opened *en route*.

Because of the suspicion that someone might have added TC to the blood tube before it was used for the blood sample, or perhaps added it to the blood sample en route to the lab, it was decided to take a mane hair sample and test it for TC. Hair analysis for drugs and toxins will be discussed in Chapter 9 *Forensic Toxicology*: in essence, some drugs and toxins are laid down in hair as it grows; therefore, if the hair sample is cut into segments and analysed for the substance, in this case TC – its position along the hair will indicate an approximate timeline of when the substance was first administered and for how long (based on the hair's growth rate). In this case, no TC was detected in the mane hair.

This posed a dilemma: TC was present in the horse's blood sample, but not in its hair; the blood tube was not taken from a sealed packet and there was alleged tampering with the sample bag. Could it be that someone added TC to the blood tube before it was used to collect the blood sample? Or might someone have added TC to the blood sample en route to the lab? Or did the trainer dope his horse with TC? If the latter is the case, why was TC not found in the horse's hair sample?

This case was sub judice (permission was given to use the outline details of the case) at the time of writing this book and so the outcome is not known, but it illustrates well the importance of chain of custody records and tamper-proof bags when transporting samples to the lab.

3.9 What Happens When the Sample Arrives at the Laboratory?

The arrival of a sample at the forensic science laboratory marks a continuation of the chain of custody. However, at this point the recording process may shift from being recorded on the sample bag (Figure 3.7) to transferring the sample to the lab's receipt and recording system. Nowadays most forensic systems use bar codes which are read by the person receiving the sample on behalf of the lab and used throughout the passage of the sample through the lab's investigation process. If a bar code system is not used, the sample should be assigned a lab number which should accompany it during its time in the lab.

3.10 Fingerprints and Other Primary Investigations at the Scene

So far, I have discussed the three key groups of people (different scenes require different people in attendance, this is just an example) present at the crime scene, namely, the police officer(s) controlling entry to the scene, the scene officer(s) and the forensic scientist(s). There will, of course, be others who need access to the scene, including investigating police officers, a pathologist (*e.g.*, in cases involving death), other medical personnel (*e.g.*, in cases involving injury), fingerprint officers and photographers.

3.10.1 Fingerprint Officers

If items are small enough to be removed from the scene and can be removed and bagged without damaging any latent prints, they may be taken directly to the fingerprint lab for further investigation or submitted to the lab later (see Chapter 7 *Fingerprints*). However, there might be prints on large items (*e.g.*, furniture) or parts of the fabric of the building (*e.g.*, a door) that are more easily studied *in situ*; for this reason, fingerprint officers may be present at the scene examining it for prints or other marks (see Chapter 7 *Fingerprints*).

3.10.2 Pathologists

If a death is involved and the body is at the scene, a pathologist may conduct an examination before the body is removed and taken to the mortuary for *post-mortem* examination (see Chapter 10 *Forensic Pathology*). A scene examination allows the pathologist to see the body *in situ*, and to see anything on or near to the body that might be important in determining the cause of death. For example, vomit on the floor near to the mouth of the body may suggest that the person vomited and died in the position in which the body was found, or the vomit might have nothing directly to do with the case.

When the body is removed from a scene and taken to the mortuary for *post-mortem* examination, the *post-mortem* might produce samples for laboratory analysis (*e.g.*, a liver sample for drug analysis; see Chapter 9 *Forensic Toxicology*) or trace material examination (DNA, body fluids, fibres, hairs, projectiles, *etc.*). These samples are usually sent to the lab in labelled (*e.g.*, case number and nature of sample) glass jars or plastic containers inside a tamper-evident bag on which the chain of custody may be recorded.

3.10.3 Other Medical Personnel

A scene might include a person who has, for example, been injured or sexually assaulted. Preservation of life and the wellbeing of the person are of paramount importance and should (of course) be given top priority. For this reason, medical personnel (*e.g.*, a doctor, paramedics) may be given access to the scene. The injured person may be a key witness to the incident and so it is important to consider involving them in the investigations when their immediate wellbeing is assured. For example, in alleged rape cases, it is important to conduct a specialized medical examination in order to collect relevant samples from intimate and non-intimate locations. Such samples can be examined for trace materials and/or screened for body fluids and potentially sent for DNA analysis (see Chapter 6 *DNA*). This is just one example of the need for medical examination; there are, of course, many more.

3.10.4 Photographers

This needs no explanation, except that here I am referring to specialist photographers rather than the scene officer taking his or her own photographs as a record of the scene. For important pieces of information (*e.g.*, a body at the crime scene), specialist photographers might be necessary to capture high quality photographs to record the situation accurately.

3.11 Inteference with Evidence: The Crewe Double Murder 1970

This is a very complex case. Its immense detail is not necessary to appreciate the importance of evidence tampering and how this became clearer as the case progressed through the courts and beyond. However, in order to appreciate sample tampering in the context of the Crewe murders, it is important to appreciate the key features of the case.

3.11.1 Outline of the Case

On 22nd June 1970, the police were called to a house in Pukekawa (60 km south of Auckland), New Zealand, to find a blood-spattered room and a crying baby. No one else, dead or alive, was anywhere to be found. The house was the family home of Jeannette and Harvey

Crewe and their 18-month-old daughter. The Crewes had farmed there since shortly after their wedding in 1966. Despite an extensive local search, no bodies were found.

On 16th August 1970, Jeannette's body was found in the Waikato River, not far from their home. There was a 0.22 calibre bullet with '8' stamped on its base (identified as a CAC Industries Pattern 8 bullet) lodged in her brain – this bullet became the most important piece of evidence in the case. The police conducted an intensive search for the cartridge from the bullet (see Chapter 11 *Firearms* for technical details about firearms and bullets) in the house and garden. They found nothing. The fact that nothing was found is also important in the later investigation of the case.

The police extended their search to a 5 mile radius of the house. This search involved firing guns owned by residents and looking at the bullet and its cartridge, hoping to find a match to the bullet recovered from Jeanette's brain (see Chapter 11 *Firearms*). Just two guns were found that could have fired the bullet that killed Jeannette; they belonged to Arthur Allan Thomas and to the Eyre family. A test fire of Thomas' gun resulted in a score mark on the bullet which was not present on the bullet from Jeannette's brain. This likely ruled out Thomas' gun as the murder weapon.

Harvey Crewe's body was found in the Waikato River, a little upstream of Jeannette's, on 16th September 1970. The body was weighted down with part of an axel from a car. Forensic examination of the axel identified the car as belonging to Thomas' father.

Despite the test fire of Thomas' gun suggesting it was not the murder weapon, the police turned their attention from their primary suspect, Jeannette's father, to Thomas. On 27th October 1970, another police search of the Crewe's garden found a cartridge case. And what is more, it could have been fired from Thomas' gun: Thomas was arrested and charged with the murder of Jeannette and Harvey Crewe.

3.11.2 Arthur Allan Thomas' Trial and Appeals

Thomas' trial began on 15th February 1971 and he was found guilty and sentenced to life in prison on 2nd March. It all seemed like a cut-and-dried case with a very successful outcome for the police.

Thomas appealed his conviction, but it was dismissed on 18th June, 1971. He then petitioned the Governor General of New Zealand in late 1971 and a second trial was ordered on 26th February; the trial began one month later. The key forensic science witness for the defence in this trial was Dr James Sprott, who questioned whether the cartridge case found by police in the Crewe's garden could have come from the cartridge containing the bullet that killed Jeannette (see Chapter 11 *Firearms* for the detailed anatomy of a cartridge). Despite this, Thomas was found guilty again on 16th April, 1971.

Thomas further petitioned the Governor General of New Zealand in 1974 – the case was referred to the Court of Appeal again. The appeal was declined on 8th January 1975 on the grounds that the cartridge found in the garden could not have fired a CAC Industries Pattern 8 bullet. This was the end of the road for Thomas' petitions of innocence.

3.11.3 Petitions of Innocence Beyond the Courts

Dr Sprott felt so strongly about the case that he co-authored (with Pat Booth) a book, *ABC of Injustice* – an exposé of the Thomas case. This caught the attention of British author, David Yallop who published *Beyond Reasonable Doubt?* in 1978. The book presented a beautifully researched, detailed and compelling case for Thomas' innocence.

3.11.4 Arthur Allan Thomas' Pardon and the Royal Commission

Arthur Allan Thomas was pardoned in December 1979 and a Royal Commission was established to explore the facts of the case and determine Thomas' compensation for false imprisonment. The Commission's investigation was long and detailed and only its key aspects are summarised here:

- the CAC Industries Pattern 8 bullet from Jeannette's brain was made between 1948–1963 (*i.e.*, their manufacture ceased before Jeannette was shot in 1970).
- Old Pattern 8 cartridges were found in Thomas' garage.
- The bullet found in Jeannette's brain could not have come from the cartridge shell found in the garden.
- The prosecution forensic scientist showed 'a disturbing lack of neutrality' and made significant errors in his investigation of the bullet.
- The bullet from Jeannette's brain could not have been fired from Thomas' gun.
- The cartridge case was only found after a second search. The first intensive search found nothing.

- The police disposed of the cartridge case found in the Crewe's garden even though it was important evidence in the case.
- Thomas had a good alibi (at home with his family and their lodger) for the night of the murder.

3.11.5 The Police Planted Evidence

The Royal Commission found that the cartridge case had been planted by Detective Inspector Hutton and Detective Sergeant Johnson.

Arthur Allan Thomas was awarded NZ$1 087 450 and 38c: a large and very precise sum of money.

This finding is an excellent example that the safeguards in place to protect the security of items collected at the crime scene were surmounted by police officers who, by virtue of their jobs, were deemed to be upright and honest. Today's safeguards (*e.g.*, tamper-proof bags and chain of custody records) would make it much more difficult for DI Hutton and DS Johnson to plant their evidence. But if they were involved in the scene search, they could have 'collected' and bagged their planted evidence without anyone noticing. However, in the end, inconsistencies in the science raised questions, but even then, it took the tenacity of a few people to persuade the 'system' that something was badly awry.

The case remains unsolved to this day.

Further Reading

1. D. A. Yallop, *Beyond Reasonable Doubt?*, Hodder & Stoughton, UK, 1979 [This is the 1st edition; there are several later editions, including e-books and Kindle versions].
2. R. H. Fox and C. L. Cunningham, *Crime Scene Search and Physical Evidence Handbook*, Paladin Press, USA, 1987.
3. R. L. Taylor, J. B. Gordon and A. H. Johnston, *Report of the Royal Commission to Inquire into the Circumstances of the Convictions of Arthur Allan Thomas for the Murders of David Harvey Crewe and Jeanette Lenore Crewe*, P. D. Hasselberg, Government Printer, Wellington, New Zealand, 1980. Available at https://web.archive.org/web/20110609174212/http:/homepages.paradise.net.nz/r.christie/thomas_royal_commission_1980.pdf.

Reference

1. D. P. Lyle, *Forensics For Dummies*, John Wiley and Sons, 2nd edn, 2019.

4 Forensic Chemistry

In its broadest sense, forensic chemistry is the application of chemistry in a legal setting. This includes the chemical methods used for investigations at the crime scene (*e.g.*, visualising otherwise invisible bloodstains: Chapter 5 *Forensic Biology*) and the methods used in the laboratory to continue the investigations on the samples collected from the crime scene (*e.g.*, chromatography to compare dyes extracted from fibres). Forensic chemistry encompasses a wide range of chemical analyses, including forensic toxicology (*e.g.*, chemical analysis of *post-mortem* samples for drugs), fingerprints (chemical methods used to visualise latent fingerprints) and some firearms investigations (*e.g.*, the chemical composition of gunpowder residues). These applications of forensic chemistry will be discussed separately in their respective chapters: Chapter 7 *Fingerprints*, Chapter 9 *Forensic Toxicology* and Chapter 11 *Firearms*.

The purpose of this chapter is to facilitate an understanding of the chemical methods that underpin many forensic investigations (*e.g.*, analysing accelerant residues in arson cases) and how they are applied to samples taken from the crime scene. Understanding the methods is important in appreciating their validity, robustness and appropriateness – in short, are they fit for the intended purpose and are they reliable?

There are three basic 'types' of methods that will be introduced in this chapter: chromatography – separation of often very similar molecules (*e.g.*, separating volatile accelerant components in arson cases); mass spectrometry – definitive identification of molecules (*e.g.*, identifying a suspected illicit drug seized by customs); and

Forensic Science: The Science Behind the Truth
By Ian Shaw with Anna Sandiford

Published by the Royal Society of Chemistry, www.rsc.org

other spectroscopic methods used to detect and identify substances (*e.g.*, *infrared spectroscopy*).

4.1 How Does Chemistry Fit into a Forensic Investigation?

Recall the Whittle Case (Chapter 1 *The Whittle Case – 1976*): the Black Panther's balaclava helmet had been dyed with an inappropriate dye that dissolved in the ammonia solution squirted by a post office worker in defence. The dried drops of purple dye on the post office floor were scraped off and compared with the dye from the balaclava helmet – thin layer chromatography (TLC) was used for the dye analysis (more on this later).

TLC is a form of chromatography, but the underpinning chemistry is the same for all types of chromatography. So, I will begin with a discussion of the basic principles of chromatography and then explore the specific types of chromatography and how they are used in forensic casework.

4.2 The Basic Principles of Chromatography

Chromatography works on the principle that molecules exhibit varying affinities to the surfaces of solids or differing solubilities in liquids. This is illustrated well by the partition of a chemical between two liquids: if I put hexane ($CH_3(CH_2)_4CH_3$) and water in a container together, they will not mix – the hexane will float on top of the water. If I add a small amount of the red dye, Sudan IV (Table 4.1) to the hexane/water mixture and shake the mixture vigorously and then allow the two layers to separate, the dye will end up in the upper hexane layer. This is because Sudan IV is soluble in hexane and pretty well insoluble in water. If I repeat the experiment, but add the blue dye, methylene blue (Table 4.1), and Sudan IV to the hexane/water mixture and shake it as before, when the two layers separate, the top (hexane) layer will be red and the lower (water) layer will be blue. This is because Sudan IV is soluble in hexane but insoluble in water, and methylene blue is soluble in water but insoluble in hexane. Sudan IV has partitioned into the hexane layer while methylene blue has partitioned into the water layer. We have separated a mixture of the two dyes through simple partition. This is exactly how chromatography works.

Table 4.1 Structures and water and octanol solubilities of Sudan IV and methylene blue.

Substance	Structure	Solubility
Sudan IV	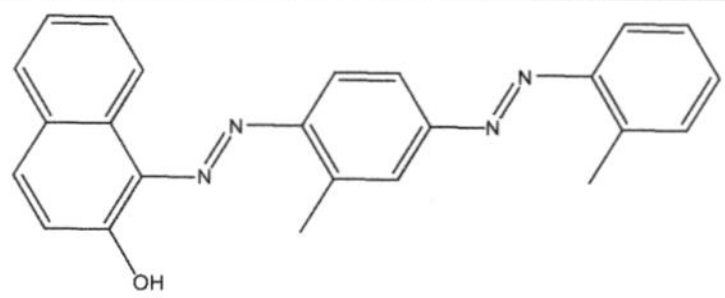	$\log P_{OW}$[a] $= 5.5$[b] Very low water solubility[b]
Methylene blue	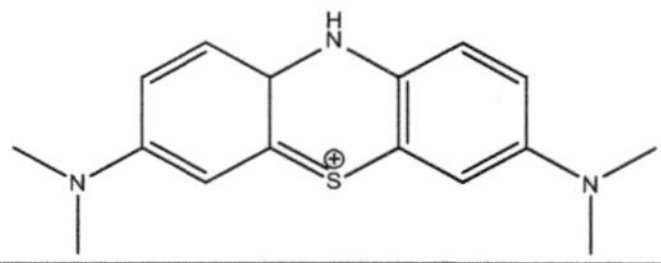	$\log P_{OW} = 0.75$[c] Water solubility $= 43\ g\ L^{-1}$ (25 °C)

[a] $\log P_{OW}$ is the log of the partition of a substance between water (hydrophilic) and octanol (hydrophobic).
[b] A value of 5.5 means that the substance is $10^{5.5}$ (*i.e.*, 316 228) times more soluble in octanol than water (*i.e.*, it is practically insoluble in water).
[c] A value of 0.75 means that the substance is $10^{0.75}$ (*i.e.*, 5.6) times more soluble in octanol than water (*i.e.*, soluble in both water and octanol, but more soluble in octanol).

There are four types of chromatography – paper chromatography (rarely used nowadays), TLC, high performance liquid chromatography (HPLC) and gas liquid chromatography (GLC) – they all work on the principle of partition explained above.

4.2.1 Paper Chromatography

Even though paper chromatography is rarely, if ever, used in forensic investigations nowadays, I will include it here because it effectively shows the partition principle in a chromatography setting.

Paper chromatography is carried out by applying small (this is important) spots of samples for analysis (*e.g.*, inks from different pens) on the baseline of a piece of chromatography paper (this is pure, standardised, composition-defined paper much like blotting paper). The paper is then hung in a glass chromatography tank with its bottom edge (below the test spots) dipped into a solvent (usually a mixture of solvents); the chromatography tank is covered with a glass plate (to saturate the atmosphere with the solvent and prevent it from evaporating off the paper) and the solvent is allowed to move up the paper (by capillary attraction). As the solvent moves up the paper, the components of the test spot separate through partition between the solvent and the paper (Figure 4.1). This time, however, the attraction to the paper is not solubility in the paper, but an attraction to the

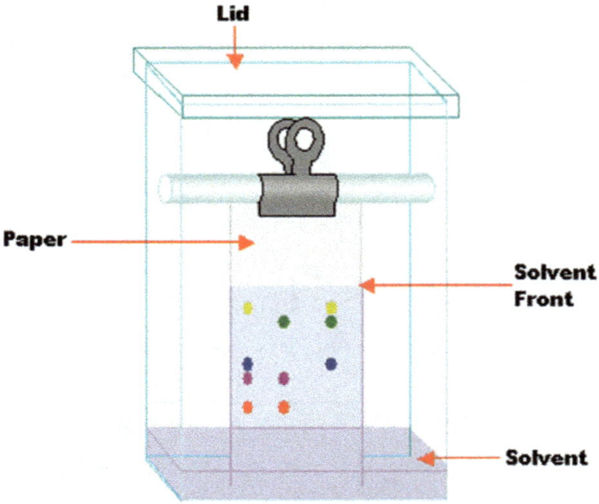

Figure 4.1 A paper chromatography setup. Reproduced from ref. 1, https://commons.wikimedia.org/wiki/File:Chromatography_tank.png, under the terms of the CC BY-SA 3.0 license https://creativecommons.org/licenses/by-sa/3.0/.

surface chemistry of the paper (adsorption) – *i.e.*, partition towards the paper occurs if components of the test spot mixture have chemical affinity for the paper (I will explain this later); partition towards the solvent as it moves up the paper is solely based on differential solubility. The more soluble a component of the test spot is in the solvent, the more likely it is to be carried by the solvent as it moves up the paper. The partition between adsorption to the paper and solubility in the solvent is what separates the components of the test spot.

If a test spot, for example, consists of a green pen ink it will separate into the components of the ink that make up the green colour – this might include blue inks and yellow inks (*i.e.*, blue + yellow = green). There might be nuances in the colours of particular inks that are created using different and complex component inks. So, for example, a dark green ink might have blue, yellow and black components to create a particular shade. Since different green inks might be manufactured using complex combinations of different dyes to give just the right colour, it might be possible to distinguish between these inks through the different separation patterns achieved by paper chromatography. This, of course, is useful for a forensic scientist examining inks extracted from a forged cheque; for example, it might show that a different green pen was used to 'adjust' the monetary value of the cheque (see Chapter 12 *Forensic Document Examination*).

4.2.2 A Closer Look at the Partition Chemistry of Paper Chromatography

Paper is basically cellulose, a polymer of glucose (Figure 4.2). The polymer has a plethora of hydroxyl groups (OH) which can bond (*via* hydrogen bonds) to molecules with specific chemical groups or be attracted to molecules carrying the opposite charge (electrostatic interaction; hydroxyl groups have a partial negative charge denoted as $\delta-$). If a component of the mixture (*e.g.*, the pen ink discussed above) to be separated by paper chromatography is capable of bonding to or is attracted to the hydroxyl groups of the 'paper molecules', it will adsorb to the paper and be retarded as the solvent moves up. On the other hand, the solvent (this is termed the mobile phase because it is moving up the paper) is usually a mixture of solvents (*e.g.*, butan-1-ol; $CH_3CH_2CH_2CH_2OH$ plus acetic acid; CH_3COOH) that is predominantly hydrophobic (*i.e.*, water repelling) – not all solvent systems are hydrophobic, but most are. This means that if a molecule in the mixture to be separated is also hydrophobic – *i.e.*, it has parts of its molecule that are capable of interacting with hydrophobic solvents; these groups are often aliphatic chains like $-CH_2CH_2CH_2CH_2CH_2CH_3$ (hexyl) or phenyl groups (Figure 4.2 and Table 4.2); it will partition into the

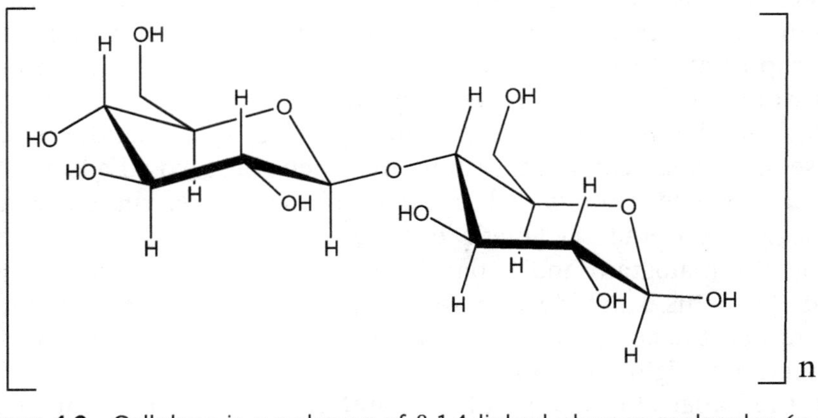

Figure 4.2 Cellulose is a polymer of β-1,4-linked glucose molecules (*n* denotes many repeats). Its many hydroxyl (OH) groups allow hydrogen bonding to molecules or electrostatic interactions between positively charged molecules and the partially charged ($\delta-$) oxygen atoms of the OH groups, leading to adsorption to the stationary phase in paper chromatography.

Table 4.2 Examples of two hydrophobic molecules, one is an aliphatic chain (hexane) and one comprises two phenyl groups (biphenyl) showing their very low water solubilities and, for comparison, highly hydrophilic acetic acid (in water it exists as an equilibrium between charged and uncharged forms).

Substance	Structure	Water solubility mg L^{-1} at 25 °C
Hexane		9.5
Biphenyl	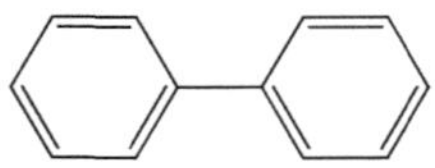	4.5
Acetic acid	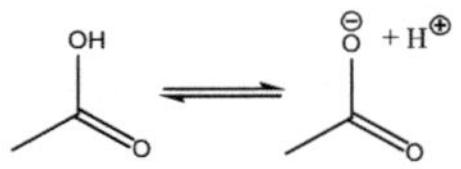	Miscible[a]

[a] Infinitely soluble.

mobile phase instead of adsorbing onto the paper (termed the stationary phase). This means that the molecule will be carried up by the solvent, whereas the paper-adsorbing molecule will move much slower, if at all. This will separate the two molecules. Differential partition between the mobile and stationary phases leads to the separation of components of a mixture by paper chromatography (or any other type of chromatography as we will see later).

Returning to our original liquid/liquid partition example where Sudan IV (hydrophobic) was separated from methylene blue (hydrophilic) (Table 4.1) by shaking with hexane (hydrophobic) and water (hydrophilic); if we subjected a mixture of Sudan IV and methylene blue to paper chromatography using hexane as the mobile phase, we would achieve separation: methylene blue would adsorb and Sudan IV would move with the mobile phase. The separation would probably be extreme – *i.e.*, methylene blue would stay close to the origin of the chromatogram and Sudan IV would move up with the solvent front. Since this is not ideal, we would tinker with the solvent system to ensure that the components of our mixture would end up somewhere between the origin and solvent front.

In the paper chromatography separation of Sudan IV from methylene blue, there is no liquid/liquid partition. In this case, the solvent (hexane) is hydrophobic and the paper (because of the cellulose OH groups) is hydrophilic, and so the partition occurs between attraction to the OH groups and solubility in hexane.

4.3 Thin Layer Chromatography

The principles underpinning TLC are akin to those for paper chromatography (*i.e.*, partition between the thin layer stationary phase and the solvent); however, the stationary phase is a thin layer (hence the name of the method) of a finely powdered stationary phase, which is often silica (silicon dioxide; SiO_2). Just like cellulose in paper chromatography, silica's oxygens carry $\delta-$ charges and thus will attract and adsorb positively charged molecules like methylene blue (Figure 4.3). So, if we separated our Sudan IV/methylene blue mixture by silica TLC with hexane as the solvent, methylene blue would adsorb onto the stationary phase and Sudan IV would move with the solvent and they would be separated.

You might be wondering why bother with TLC when it operates similarly to paper chromatography? The answer is that TLC is far more efficient and can separate molecules that are very similar (which paper chromatography might not achieve). This is because the surface area of the stationary phase determines the separation efficiency and since the particles of silica on a TLC plate are very small (10–14 µm), they provide a very large total surface area and thus a high separation capability (this is termed the plate number).

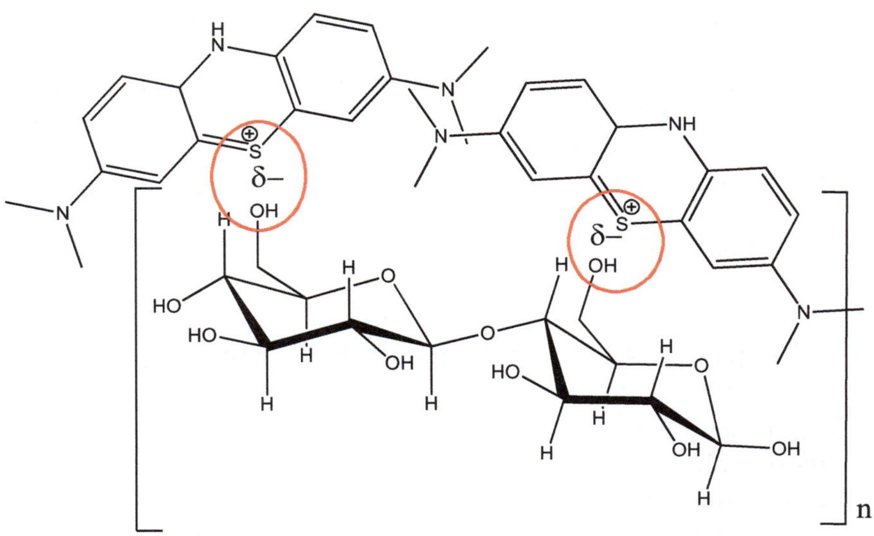

Figure 4.3 Representation of the adsorption of methylene blue by the cellulose-based stationary phase in paper chromatography. The S+ atoms of the methylene blue molecules are attracted to the O$\delta-$ atoms of the cellulose molecule (circled in red).

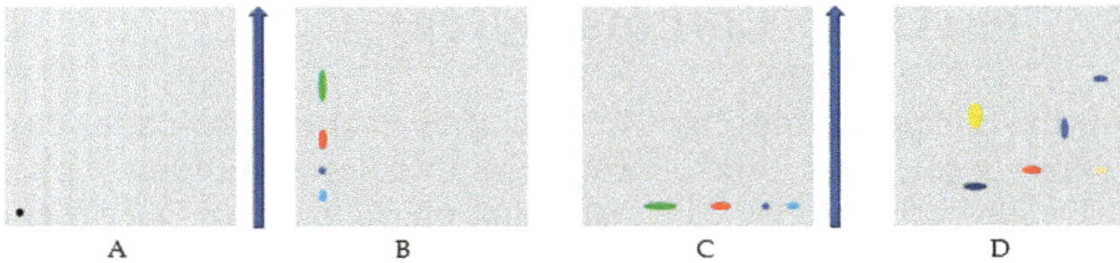

Figure 4.4 Two-dimensional TLC. The sample is applied to the bottom left of the TLC plate and the chromatogram is developed (the arrow shows the direction of solvent movement) in solvent 1 (A), the components of the test sample separate (B), the chromatogram is dried and then rotated 90° anticlockwise and developed in solvent 2 (C), resulting in two-dimensional separation of the components of the original test mixture (D). The pattern of spots on the chromatogram is characteristic of the sample (*e.g.*, a fibre dye).

4.4 Two-dimensional Chromatography

Both paper chromatography and TLC can be run in two dimensions (this was discussed in Chapter 1 *The Whittle Case – 1976*); this means that the sample is spotted at the bottom left corner of the paper or plate, run to the top of the chromatogram, dried and then the chromatogram is turned anti-clockwise 90° and re-run in a different solvent system. This means that the components separated in run 1 can be further separated in run 2 (*i.e.*, if any of the spots separated in dimension 1 comprise more than one component) (Figure 4.4).

Two-dimensional TLC uses small plates (*e.g.*, 50 mm × 50 mm) which develop quickly and so results can be obtained quickly. This is the method that was used in the Lesley Whittle case to compare the dyes from the Black Panther's balaclava helmet and the dye spot on the post office floor (see Shaw IC (1980) *Analyst* **103**, 729–730 for the actual method used).

4.5 High-performance Liquid Chromatography

HPLC is based on the same principles of chromatography as paper chromatography and TLC. In essence, it is liquid/liquid chromatography, involving partition between two liquids. One of the liquids is the HPLC solvent system (mobile phase) and the other is an immobilised liquid attached to a very high surface area support (stationary phase) packed into a metal tube (column). A typical column is 200 mm long with an internal diameter of 4.6 mm.

HPLC is an immensely complex topic, but since it is not the purpose of this book to present a treatise on HPLC (others have done that very well – see *Further Reading*), I will use a very commonly used type of HPLC, namely, reverse-phase HPLC to illustrate the process.

Reverse-phase HPLC uses a hydrophobic solvent bonded to microparticulate (*e.g.*, 5 µm particle diameter) porous (pore size approx. 100 Å) silica. The most common immobilised hydrophobic solvent is octadecyl (*i.e.*, $-C_{18}H_{37}$) silane (ODS). The silane group facilitates chemical bonding of the C_{18} liquid to the silica particles (Figure 4.5).

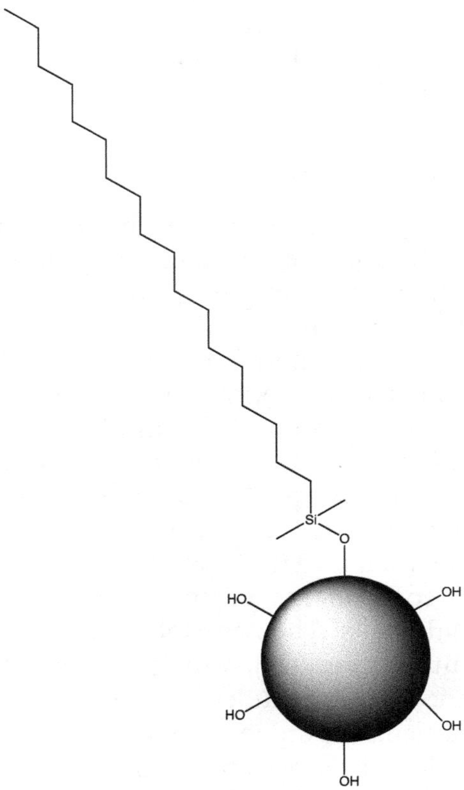

Figure 4.5 Octadecylsilane chemically bonded to a silica particle. A silica particle (grey sphere) is a complex silica crystal with many hydroxyl groups (OH) on its surface and lining its pores. Many millions of ODS molecules (just one is shown here) are chemically bonded to many of the hydroxyl groups on a single silica particle resulting in a huge ODS surface area. The long carbon chain (C_{18}) of the bonded ODS molecules represents an immobilised, very hydrophobic liquid. HPLC involves the partition of the molecules to be separated between the hydrophobic stationary phase liquid and the mobile phase (solvent system).

This type of HPLC is usually termed C18 or ODS. The ODS groups are bonded to both the outside of the silica particles and inside their pores. This results in a huge ODS surface area, which means that the chromatographic separation is incredibly efficient – even very similar molecules (including isomers) can be separated. The concept of plate number as a measure of chromatographic separation efficiency was introduced in Section 4.3 *Thin Layer Chromatography*: HPLC can have plate numbers (termed the theoretical plate number and given the symbol N) in the high thousands to tens of thousands, depending on the column type and the characteristics of the test chemical peak. The word 'performance' in HPLC refers to the column's high performance in terms of its resolution of closely related molecules (*i.e.*, its high theoretical plate number).

4.5.1 The HPLC System

The HPLC system comprises a column (more of this later), pumps (to pump the mobile phase through the column), an injection port (through which the sample is injected onto the column) and a detector (to detect the separated molecules at the bottom of the column) (Figure 4.6). There are many types of detectors used, but spectrophotometric detection in the ultraviolet (UV) range is perhaps the most common for simple HPLC analysis. This detection method relies on the molecules being separated absorbing light in the UV range (280–400 nm). The output from the detector can either be sent electronically to a computer, or directed to a printer from which an HPLC trace is printed (Figure 4.7).

The separated components of the sample are shown on the trace as individual peaks. The time taken for the 'peak' to elute is termed the retention time (t_R), which is characteristic of a particular compound. The area under the peak is proportional to the amount of that particular component of the sample. The void volume (V_0) is the unoccupied (the space around the stationary phase) volume of the column; nothing can elute during this elution time.

Mass spectrometry (MS) is also commonly used to detect molecules eluted from the HPLC column. This gives incredibly sensitive and precise detection based on the molecular masses (see Section 4.5.2 *Mass Spectrometry*) of the compounds being separated – this is termed HPLC–MS or just liquid chromatography (LC)–MS (see Section 4.5.3 *Liquid Chromatography–Mass Spectrometry (LC–MS)*).

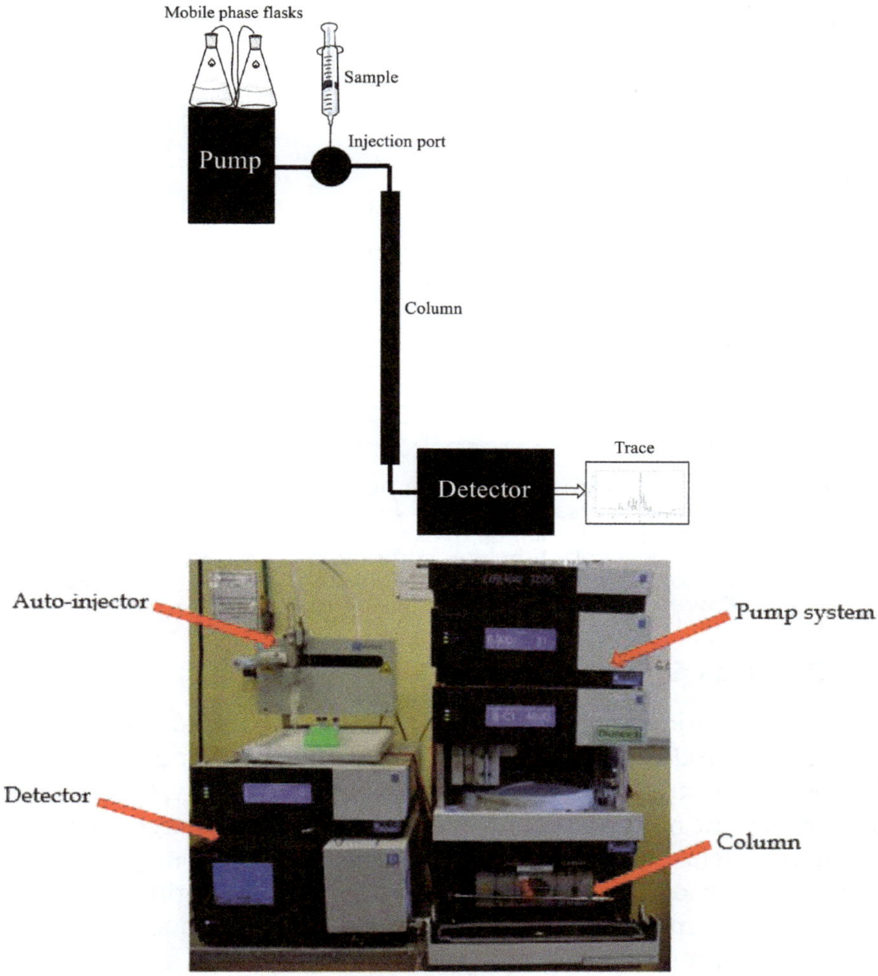

Figure 4.6 Top: the anatomy of an HPLC system: the mobile phase is made by mixing (automatically in the pump) the solvents from the mobile phase flasks (solvent gradients are often used), the mobile phase is pumped *via* the injection port (through which the sample is injected), through the column (containing the stationary phase; *e.g.*, ODS), and the components of the sample are separated on the column and passed through the detector which produces a trace which shows the components of the sample mixture as individual peaks. Bottom: Photograph of an HPLC system to show what it looks like in a lab setting – this system has an auto-injection device to facilitate the handling of a large number of samples, and the detector output is electronic, so there is no trace. HPLC trace from ref. 2, https://commons.wikimedia.org/wiki/File:Hplc-perfume-chromatogram.png, under the terms of the CC BY-SA 3.0 license https://creativecommons.org/licenses/by-sa/3.0/deed.en.

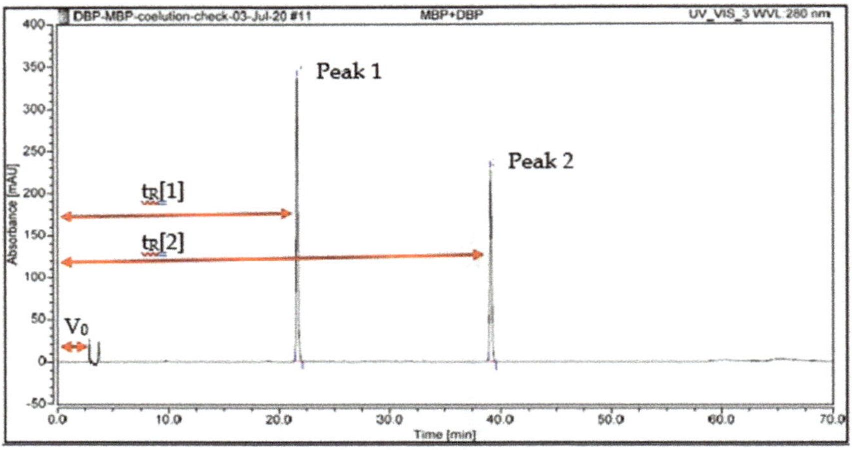

Figure 4.7 An HPLC trace showing two peaks corresponding to two components of a sample. Key elution parameters are shown on the trace: $V_O = 2.6$ min, $t_R[1] = 21.6$ min, $t_R[2] = 39.1$ min. Compound 1 = monobutyl phthalate, compound 2 = dibutyl phthalate. Detection was based on UV absorption at 280 nm. Courtesy of Pravesh Tyagi, University of Canterbury.

4.5.2 Mass Spectrometry

This is a technique used to identify compounds. In short, the compound is introduced into a machine in a vacuum and bombarded with either electrons (e^-) or positively charged ions (*e.g.*, CH_4^+) which smash the test molecule into fragments. The mass/charge (m/z) ratio of the fragments is characteristic of the original molecule, and so the fragmentation pattern can be used to identify the original molecule. The output from a mass spectrometer is a chart with a series of bars, with each bar representing a fragment ion. The pattern and m/z values of the ions are characteristic of a particular compound (Figure 4.8).

4.5.3 Liquid Chromatography–Mass Spectrometry (LC–MS)

HPLC–MS is usually termed LC–MS. The technique is the ultimate HPLC technique. It is very specific (determined by the high separating power of the HPLC column and molecular specificity of MS detection) and so identification of a compound by LC–MS is as close to certainty as it is currently possible to be.

To use MS as the detection method for HPLC (*i.e.*, LC–MS) presents a huge technical problem; HPLC works on the basis of a high-pressure solvent (approx. 200 bar; 2900 psi; 1.5×10^5 mmHg) forced through a

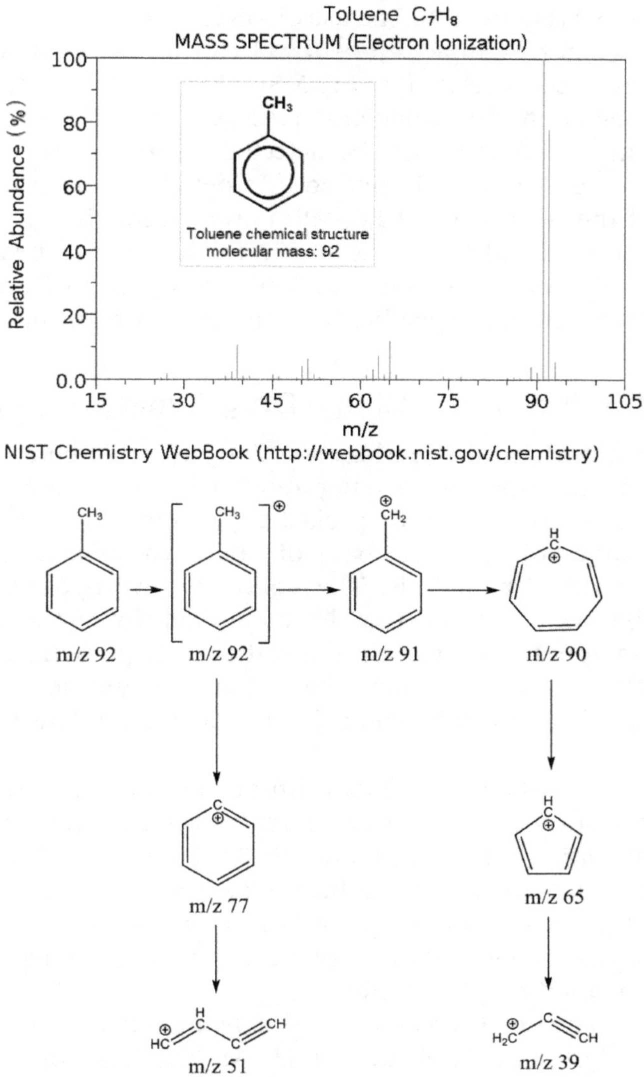

Figure 4.8 Electron impact mass spectrum for toluene (top) and a simplified electron impact fragmentation pattern for toluene showing the origins of the mass spectral peaks.

tightly packed microparticulate column feeding into a MS which requires a near absolute vacuum (10^{-7} mmHg) to function. I need to say no more than the engineering of the HPLC/MS interface is a triumph!

To use MS as an LC detector, the MS is usually tuned to the *m*/*z* of the parent ion of the molecule being sought; *i.e.*, the molar mass of the compound + 1 for positive ion fragmentation – for toluene this would be m/z 92 (Figure 4.8). The use of MS as an LC detector can be far more complex; for example, tandem MS can be used where two MS machines are used in series (LC–MS/MS). This enables the monitoring of the fragmentation of a particular ion (*e.g.*, the parent ion) as well. This gives even more certainty that the detected compound is what it was presumed to be. Again, I will not go into detail because this is not a book about the intricacies of LC–MS; however, for the toluene case, the parent ion's daughter ion at m/z 77 might be used because it is derived directly from the parent ion. Suffice to say, LC–MS can be both incredibly sensitive and specific. It carries great weight in court.

4.5.4 The Use of LC–MS in a Drug Smuggling Case

Consider a hypothetical situation: a country's border control officers arrest two people who are carrying about 1 kg of a white crystalline substance concealed in 100 g plastic packets beneath the false floor of a large suitcase. The bags of unknown substance are sent to the laboratory for analysis. The scientists investigating the case will probably be able to narrow down the identity of the substance based on its appearance. Continuing with our hypothetical case; the white crystalline substance might be methamphetamine (colloquially known as meth or P); this would be an ideal candidate for LC–MS analysis.

The specifics of the LC–MS/MS technique used will vary from lab to lab, but an example is to use the parent ion ($m/z = 150$) transitions (*i.e.*, fragmentation) to m/z 119 and m/z 91 (Figure 4.9). If the parent ion and the two transition (daughter) ions manifest and the appearance of the daughter ions is quantitatively linked to the disappearance of the parent ion, this is very strong evidence indeed that the compound is methamphetamine.

The importance of LC–MS/MS in our hypothetical methamphetamine smuggling case is clear, but LC–MS/MS has applications in many other forensic investigations, which will be discussed in greater detail in Chapter 9 *Forensic Toxicology*.

4.6 Gas Liquid Chromatography (GLC)

The principles of GLC are exactly the same as for all other types of chromatography: separation of molecules in a test sample based on their partition between a stationary and a mobile phase. For GLC, the stationary phase is a high boiling point, long chain hydrocarbon

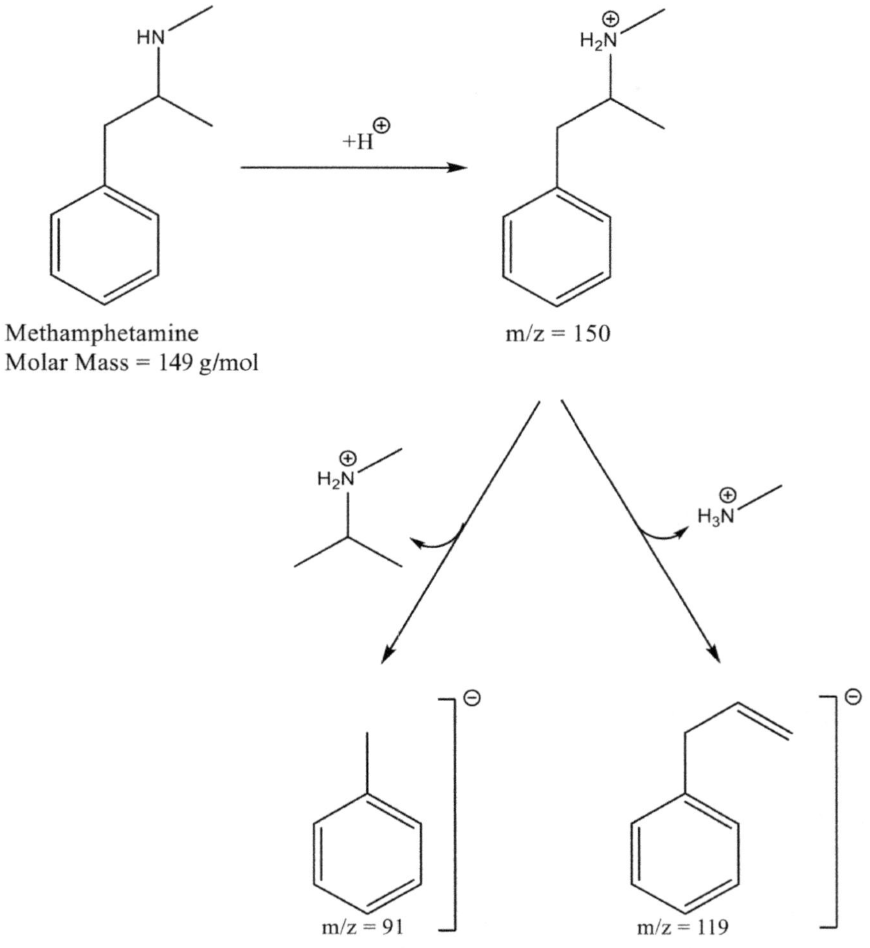

Figure 4.9 Mass spectral fragmentation of methamphetamine showing the ion transitions (*m/z* 150 → *m/z* 119 and *m/z* 150 → *m/z* 91) used for LC-MS/MS analysis.

liquid (*e.g.*, cyanopropylphenyl dimethyl polysiloxane; Figure 4.10) adsorbed onto an inert particulate solid (*e.g.*, diatomaceous earth) packed into a column, over which the mobile phase (a gas) is passed – commonly used mobile phase gases are helium, nitrogen, argon or hydrogen. The name *gas–liquid* chromatography is derived from the stationary liquid phase and the mobile gas phase.

A GLC setup (Figure 4.11) resembles an HPLC set up, but the column is usually very long with a tiny internal diameter (capillary column; gives a high surface area for partitioning), and because the mobile phase is a gas, the sample to be separated must also be a gas.

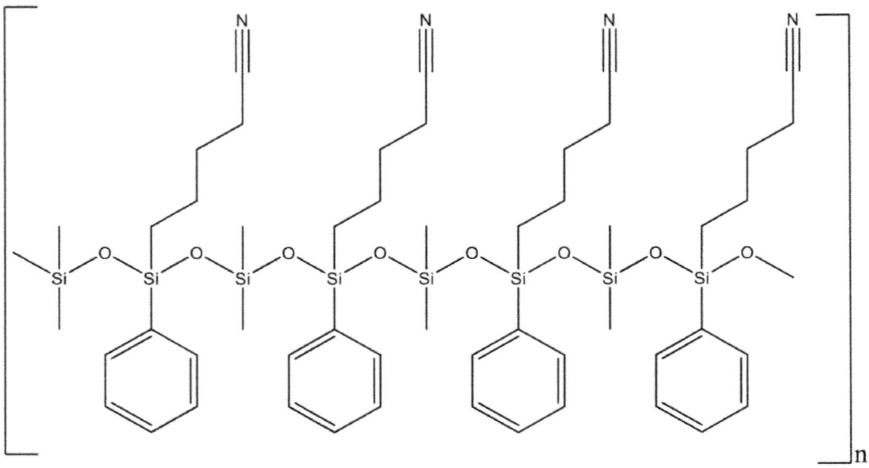

Figure 4.10 Structure of a typical GLC stationary phase – cyanopropylphenyl dimethyl polysiloxane (Si = silicon). The illustration shows a section of a repeating unit; *n* denotes many repeats to give a very long, low volatility, highly viscous liquid which is adsorbed onto diatomaceous earth to form the GLC stationary phase.

This is achieved by heating the injection port and by housing the column in an oven (Figure 4.11; oven temperature varies, but usually around 200 °C) to ensure that the sample remains in the gas state throughout its journey down the column. If a sample cannot be volatilised (*i.e.*, its boiling point is too high), it may be derivatised by adding a molecule to reduce its boiling point (Figure 4.12). If this is not possible, GLC cannot be used (in this case, HPLC might be the best choice).

Like HPLC, GLC uses several different types of detectors. The most common is flame ionisation detection (FID), where the effluent from the column is passed through a flame which ionises the molecules (*i.e.*, makes them charged) present in the sample (*i.e.*, the components of the sample); the ions are detected, producing peaks on either a paper trace or electronically captured and visualised on a computer screen. MS is commonly used as a detector for GLC – like HPLC, the method is then termed GC–MS. GC–MS/MS is also used for definitive identification. The principles of both GC–MS and GC–MS/MS are the same as described for LC–MS and LC–MS/MS (see Section 4.5.3 *Liquid Chromatography–Mass Spectrometry (LC–MS)*); even the technical problems associated with GLC MS detection are similar to LC–MS, but the GLC MS interface involves a gas flow running into a MS running at near absolute vacuum. Again, the solution to these challenges is a triumph of engineering!

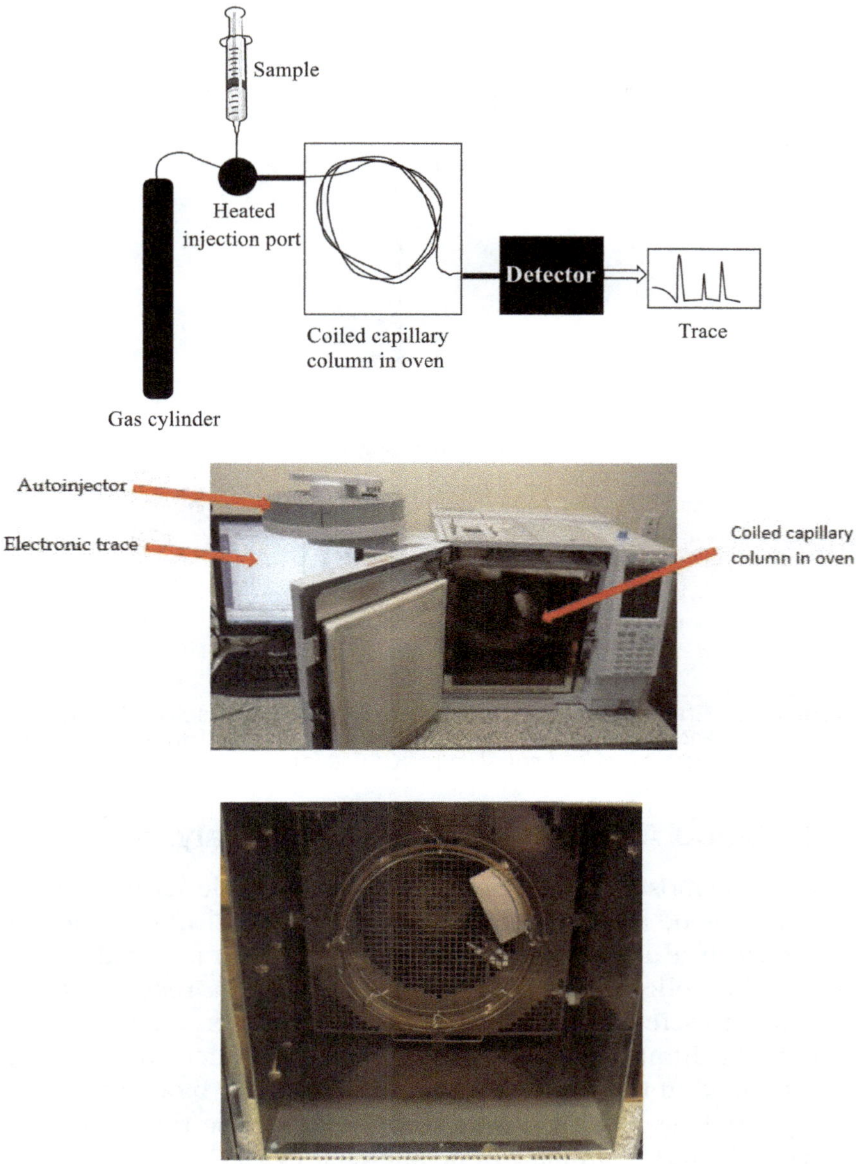

Figure 4.11 Top: The anatomy of a GLC system: the mobile phase is a gas (or mixture of gases), the mobile phase gas is pumped *via* the heated injection port (through which the sample is injected) through the column (containing the stationary phase), the components of the sample are separated on the column and passed through the detector, which produces a trace showing the components of the sample mixture as individual peaks. Middle: Photograph of a GLC set up. Bottom: Close up of a capillary column in the oven.

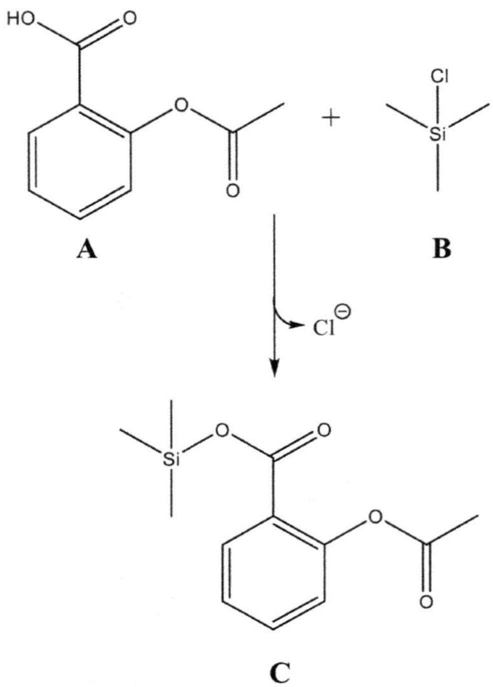

Figure 4.12 Chemical derivatisation of non-volatile aspirin (A) with trimethylsilyl chloride (B) to make a volatile trimethylsilylaspirin derivative (C) suitable for GLC analysis.

4.6.1 Blood Alcohol GLC Headspace Analysis

Arguably the most common use of GLC in a forensic setting is for the analysis of blood alcohol (specifically ethanol, the alcohol in alcoholic beverages). If a driver is stopped by police for erratic or dangerous driving, the police officer is likely to administer a roadside breathalyser test to screen for alcohol. If the breathalyser reading is above the national limit for alcohol (*e.g.*, breath equivalent of 80 mg per 100 mL blood in the UK, USA and New Zealand), a blood sample will be requested for analysis (to confirm or otherwise the breathalyser result). This blood sample will be analysed by GLC.

The blood sample is sealed in a vial, which is heated in an autoinjector device to evaporate the ethanol. The vapour above the blood sample is injected into the GLC machine *via* its injection port (this is termed headspace analysis). An internal standard (*e.g.*, propan-1-ol) at known concentration is added to the blood sample before analysis; this is used to calculate the concentration of ethanol present in the blood sample. Detection is carried out by FID – a typical trace is shown in Figure 4.13.

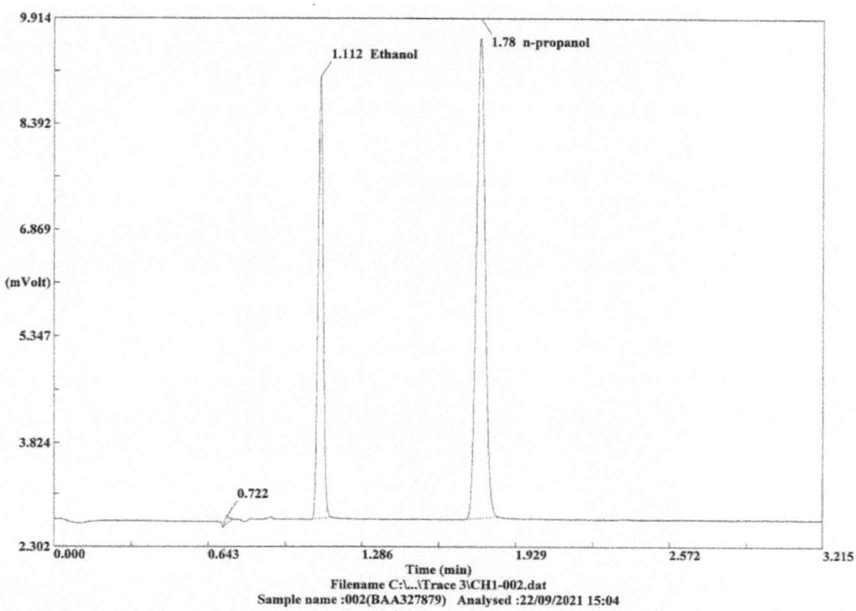

Figure 4.13 A blood alcohol headspace GLC analysis trace showing ethanol ($t_R = 1.112$ min), the internal standard, propan-1-ol (n-propanol) ($t_R = 1.75$ min) and $V_O = 0.722$ min. Courtesy of Dr Helen Poulson, ESR, Wellington, New Zealand.

It is, of course, very important that the ethanol concentration in the blood sample is correct because if it is above the statutory limit, a prosecution is likely. For this reason, each blood analysis is accompanied by blood samples with known added ethanol concentrations (standards), an internal standard (*e.g.*, propan-1-ol, described above) and blank (*i.e.*, containing no ethanol) blood samples to make sure there are no GLC peaks eluting at the same time as ethanol that might enhance the ethanol peak (and thus blood concentration).

Headspace analysis is also used in arson cases to identify and analyse the chemical make-up of accelerants that might have been used to start the fire.

4.6.2 GLC Headspace Analysis in Arson Cases

In cases of arson, identifying the accelerant used to start the fire might be useful in tracking the culprit. Consider an hypothetical arson case: a private house was found burning late one night. The fire brigade managed to extinguish the fire and on investigation of the scene, a wad of partly burned cloth was found on the badly burned carpet below the letterbox at the front door. The initial investigation suggested that the fire started in the hall in the vicinity of the front door. Clearly, the wad of cloth is of significance because it might have initiated the fire by being soaked in petrol (for example), ignited and pushed through the letter box. The identity of any accelerant still present on the cloth might be important evidence.

4.6.2.1 Headspace Analysis of the Accelerant-soaked Cloth

Regardless of the accelerant used (*e.g.*, petrol), it is likely to be volatile and so it can be evaporated off the sample of cloth in a sealed container (very similar to alcohol analysis; see *Blood alcohol GLC headspace analysis*) and the headspace is then injected onto the GLC column.

It is also possible to concentrate the evaporated accelerant by evaporating it off the sample (in this case a piece of cloth) in a sealed container, and then adsorbing the accelerant gas from the headspace onto a solid phase extraction (SPE) cartridge – this is a cartridge packed with a material that will adsorb compounds from the headspace and thus concentrate them on the adsorbent. The headspace atmosphere is passed through the SPE cartridge, allowing the volatile compounds to be adsorbed and concentrated. The adsorbed compounds can be released and injected onto the GLC column later. Interestingly, SPE packing materials are often very similar to GLC stationary phases (*e.g.*, cyanopropylphenyl dimethyl polysiloxane).

For arson investigations, it is likely that GC–MS would be used because it allows for the visualization of a GLC separation pattern and identification of its components. The pattern of the trace (*i.e.*, the mixture of compounds it contains) is characteristic of a particular accelerant (*e.g.*, petrol). It is, however, important to note that since our piece of cloth has been in a hot environment (*i.e.*, a fire in the hallway of the house) some of the more volatile components would have evaporated. Therefore, when comparing the pattern of separated compounds on the GLC trace with the pattern of compounds from a candidate accelerant, it is likely that the more volatile compounds will be missing from the cloth headspace analysis. The pattern of compounds on the GLC trace (Figure 4.14) is used as a fingerprint for a particular accelerant and can be compared, for example, with a candidate accelerant found in the home of a suspect. The identities of the peaks might be necessary to definitively show that two samples (*e.g.*, headspace analysis of the cloth in our example and the contents of a can found in the garage of a suspect) are the same or very similar, but often the fingerprint is sufficient.

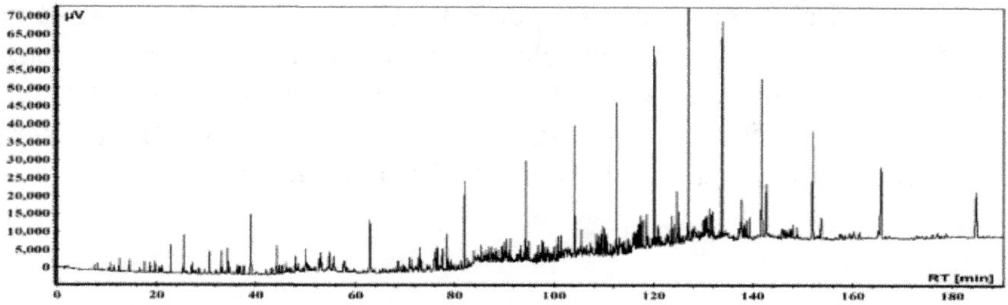

Figure 4.14 GC–MS analysis of a kerosene-diesel blend. Each peak represents a different chemical component of its complex hydrocarbon make up; the overall pattern of peaks (fingerprint) is characteristic of the accelerant. Reproduced from ref. 3 with permission from Springer Nature, Copyright 2019.

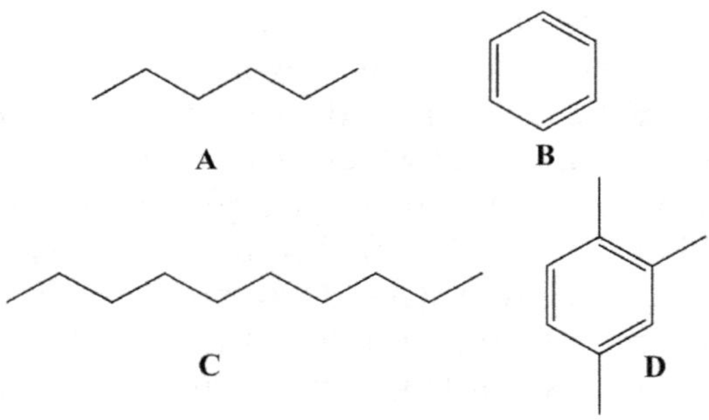

Figure 4.15 Some of the components of diesel, petrol and kerosene: hexane (A), benzene (B), decane (C) and 1,2,4-trimethylbenzene (D).

Nowadays, commercial petrol (gasoline) and diesel are likely to be very similar in composition, so key differences in their components might be important. There are very many individual chemical components in most potential accelerants (*e.g.*, motor vehicle fuels;) ranging from short chain alkanes (*e.g.*, hexane), through aromatic compounds (*e.g.*, benzene) to long chain alkanes (*e.g.*, decane) as well as more esoteric compounds (*e.g.*, 1,2,4-trimethylbenzene) that might serve as distinguishing features of a particular accelerant (Figure 4.15).

4.6.3 Accelerant Hydrocarbons in Blood

If a body found in a burned building (or other place) is sufficiently intact to allow the collection of a blood sample, the blood sample can be analysed by GC–MS to detect accelerant hydrocarbons. If the person died as a result of the fire, they would have inhaled hydrocarbon vapour, which will result in the presence of hydrocarbons in their blood. If, on the other hand, the person was dead before the building burned, there would be no accelerant hydrocarbons in their blood. This technique can also be used to investigate solvent abuse: GC–MS can be used to detect components of inhaled solvents in the abuser's blood (see *Further Reading* at the end of this chapter).

4.7 Infrared Spectroscopy

Infrared (IR) spectroscopy is a technique used to identify molecules; it involves passing a beam of infrared radiation ('light' at the far-red end of the spectrum; wavelength = 700–1000 nm) through the test sample. This process induces vibrations in the molecular bonds, resulting in absorption of energy. The wavelengths at which bonds absorb energy are characteristic of a particular bond and so a bond can be identified by its vibration energy at a particular wavelength, measured as wavenumber (proportional to wavelength and frequency) with the unit cm^{-1}. For example, the carbonyl bond (C=O) absorbs at 1700 cm^{-1} and the hydroxyl bond (O–H) absorbs at 3200–3400 cm^{-1}. So, if a molecule has both an hydroxyl and a keto group, it will exhibit IR spectral peaks at both 1700 cm^{-1} and 3200–3400 cm^{-1}. Similar to other analytical methods, forensic scientists often use the complex spectral fingerprint of a test substance for comparison with a control substance (Figure 4.16).

Figure 4.16 Left: IR spectrum of bromomethane (CH_3Br). Each of the peaks represents the energy absorbed to vibrate a specific bond in a specific way (*e.g.*, stretch and bend). The spectral fingerprint is characteristic of bromomethane. Reproduced from ref. 4 with permission from NIST, Copyright 2023 by the U.S. Secretary of Commerce on behalf of the United States of America. Right: Bromomethane molecule (CH_3Br: red = Br; grey = C, and white = H) showing the IR absorptions for the bonds.

4.8 Other Chemistry Techniques Used in Forensic Investigations

Chemical analysis and molecular identifications are becoming more and more sophisticated, and more certain. The methods I have outlined so far are those that might be regarded as the 'routine' techniques applied to forensic investigations. However, complex cases may necessitate complex methodologies that extend beyond the techniques usually found in a good forensic science laboratory. The utilization of non-routine techniques in a particular case depends on factors including, the importance of the case and the budget available. You could not justify the use of Raman spectroscopy, for example, in a case where a stolen TV set is incinerated in a burned garden shed. However, Raman spectroscopy might provide invaluable information in a major arson case. Raman spectroscopy is an excellent example of a non-routine forensic technique.

4.8.1 Raman Spectroscopy

Raman spectroscopy is named after the Indian physicist, Sir Chandrasekhara Venkata Raman (1888–1970), who was awarded the Nobel Prize in physics in 1930 for his discovery that when light passes through a transparent material its wavelength changes (Raman scattering) – Raman spectroscopy is based on this principle.

Modern Raman spectroscopy involves directing laser light of a known wavelength onto the sample for analysis. The light interacts with the molecules in the sample which changes its wavelength. The interaction, and thus the change in wavelength, of the light in the Raman spectrometer is determined by the chemical bonds in a molecule. A spectrum is produced based on the types and nature of the bonds in the molecule. This spectrum – the Raman spectrum – is characteristic of a particular molecule. In forensic investigations, Raman spectroscopy can be used to identify a molecule or serve as a fingerprint of the molecular components of a sample, allowing for comparison with a control sample. For example, Raman spectroscopy can be used to examine arson residues with a view to comparing the spectrum with a spectrum of a suspect accelerant (see *Further Reading*). An important aspect of Raman spectroscopy is that the sample is not destroyed during its analysis, which means that it can be analysed using several techniques to further ensure the results.

Raman spectroscopy can also be applied to the identification of textile fibres and plastics. It is even possible to identify burned fragments of plastics or textiles (Figure 4.17), and advances in

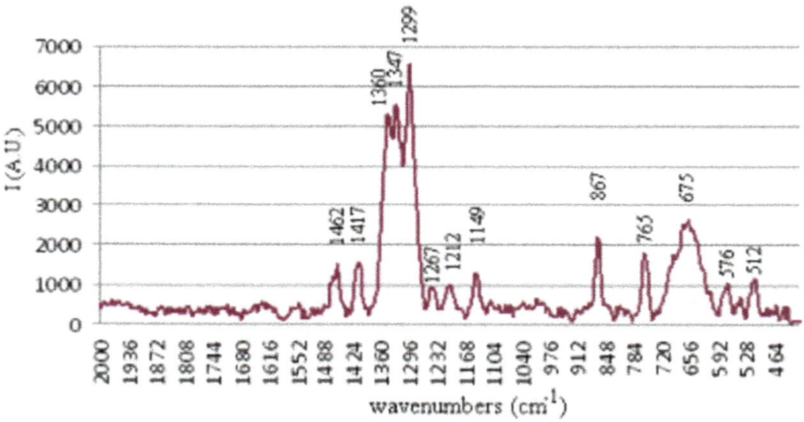

Figure 4.17 Raman spectrum of nylon from stockings showing its characteristic fingerprint. Reproduced from ref. 5 with permission from Elsevier, Copyright 2011.

technology have led to the development of hand-held Raman spectrometers that can be used at the crime scene.

4.9 Are the Results Correct and Will They Serve as Evidence in Court?

There are two aspects of forensic chemistry that might be presented as evidence in court:

1. Quantitative (numerical) data: *e.g.*, the concentration of alcohol in blood.
2. Identification of a molecule: *e.g.*, identification of methamphetamine in an imported white powder.

It is also, of course, possible that evidence might comprise a combination of **1** and **2**; for example, identification and concentration of methamphetamine in a blood sample.

The reliability of quantitative data is determined by validation of the analytical method, and of particular importance is the variability (*e.g.*, standard deviation) of a quantitative result. So, if a blood alcohol concentration of 50 mg/100 mL blood is found, what is the variability around that result? This is usually expressed at plus/minus (±) the standard deviation (*e.g.*, 50 ± 0.2 mg per 100 mL). This denotes the level of confidence we should place on the results; for this example, the true value is in the range 49.8–50.2 mg per 100 mL.

If the result is used by the court to determine whether someone has exceeded a statutory limit (*e.g.*, the blood alcohol limit of 80 mg per 100 mL in many parts of the world), what blood level would likely lead to a prosecution? And how would the results be presented to the court? In most jurisdictions, the analytical laboratory's accreditation requires that every result is reported with an analytical variation and a percentage certainty. In most countries, results are reported in favour of the defendant (to be as fair as possible). For example, if a blood alcohol concentration is 81 ± 4 mg per 100 mL, it will be considered as 77 mg per 100 mL (*i.e.*, $81 - 4 = 77$ mg per 100 mL) and therefore it would not be used to support a prosecution. However, if the result is 85 ± 4 mg per 100 mL, it would be considered as at least 81 mg per 100 mL (*i.e.*, $85 - 4 = 81$ mg per 100 mL) and would be used to support a prosecution. This is a typical example of how analytical variability is taken into account when reporting results to the courts, but of course, there are other methodologies used around the world.

Identifying a molecule is quite different. Standard substances are compared with the 'unknown' molecule. For example, if a definitive identification of a white powder seized at a port was sought, the analytical parameters of pure, authentic methamphetamine would be compared with the 'unknown' substance's parameters. If MS is used as the identification method, the fragmentation pattern of authentic methamphetamine would be compared with the 'unknown' substance's fragmentation pattern. If the mass ions and their relative abundance (*i.e.*, peak heights on the MS trace) were identical, this is very good evidence that the 'unknown' is methamphetamine. However, this could be extended further by using tandem MS in which the primary fragments are further fragmented and the transitions (*i.e.*, daughter ions and their origins) are observed. For methamphetamine (Figure 4.9), the transition m/z 150 → m/z 91 + m/z 119 is diagnostic. This can be taken even further, by measuring the accurate mass of a key ion (*e.g.*, m/z 150 for methamphetamine) on the mass spectrum. Accurate masses can be determined up to many decimal places (often 10) and if the ion derived from the standard compound has the same mass (to 10 decimal places) as the equivalent (*i.e.*, the same m/z value) ion from the 'unknown', there is a very high probability that the unknown has the same molecular structure as the authentic compound. The degree of proof necessary will differ from case to case, but this illustrates the level of certainty that forensic chemistry can achieve.

Further Reading

1. K. Kimura, T. Nagata, K. Hara and M. Kageura, Gasoline and kerosene components in blood – a forensic analysis, *Hum. Toxicol.*, 1988, 7, 299–305.
2. V. K. Yadav, K. Nigam and A. Srivastava, Forensic investigation of arson residues by infrared and Raman spectroscopy: from conventional to non-destructive techniques, 2020.
3. M. C. McMaster, *HPLC: A Practical User's Guide*, John Wiley & Sons, Inc., 2nd edn, 2007.

References

1. Paper chromatography [Wiki]. Available from: https://en.wikipedia.org/wiki/Paper_chromatography [accessed 4 September 2023].
2. File: Hplc-perfume-chromatogram.png [Wiki]. Available from: https://commons.wikimedia.org/wiki/File:Hplc-perfume-chromatogram.png [accessed 4 September 2023].
3. Vempatapu, *et al.*, *SN Appl. Sci.*, 2019, **1**, 164.
4. Methane, bromo- [Internet], webbook.nist.gov. Available from: https://webbook.nist.gov/cgi/cbook.cgi?ID=74-83-9&Type=IR-SPEC&Index=QUANT-IR.
5. J. González-Rodríguez, *et al.*, *J. Anal. Appl. Pyrolysis*, 2011, **91**, 210–218.

5 Forensic Biology

In this book, forensic biology is defined in its broadest context: the use of biology to investigate crimes.

Various aspects of biology have found applications in forensic investigations, including identification of pollen, traces of plant material, algae (including diatoms), and coccoliths (calcium carbonate plates formed by some algae) in putty. In combination with Locard's Exchange Principle, the identification of fragments of plant found on the soles of a suspect's shoes might suggest where he or she had walked, the identification of coccoliths in minute fragments of putty stuck to a burglar's trousers might link them to a particular window, diatoms in the lungs of a drowned body might pinpoint where they entered the water, or fragments of plant material in the stomach of a murder victim could help identify their last meal.

However, forensic biology is much broader, encompassing the use of biochemical methods to identify body fluid stains (*e.g.*, acid phosphatase to locate semen stains in sexual assault cases), the identification of tissues (*e.g.*, brain) and the species from which they originated, and the macabre use of maggots in a corpse to determine time since death.

We must also consider the techniques used by biologists, particularly microscopy – microscopy is fundamental to almost all forensic investigations. Improvements in microscopy over the years, culminating in highly sophisticated scanning electron microscopy have revolutionised its application to forensic biology. Microscopy is particularly important in hair and fibre identification and comparisons (synthetic fibres are also included in this chapter).

Forensic Science: The Science Behind the Truth
By Ian Shaw with Anna Sandiford

Published by the Royal Society of Chemistry, www.rsc.org

Forensic biology is now dominated by the use of DNA profiling to identify individuals: this is a huge and very important field and is covered separately in Chapter 6 *DNA*.

5.1 The Use of Biology in Forensic Investigations

Forensic biology is quite different from forensic chemistry. Forensic chemistry ostensibly involves the application of analytical techniques (*e.g.*, chromatography to identify drugs) to forensic casework, whereas the application of biology involves different experts (*e.g.*, palynologists – pollen experts; entomologists – insect experts), some of whom are not 'normal' members of a forensic team and are brought in from elsewhere (*e.g.*, universities). It is impossible to include every conceivable biological application to forensic investigations – there are far too many possibilities. In this chapter, I will focus on the more commonly encountered methods and areas of expertise.

Arguably, the most fundamental need for all forensic scientists is to magnify minute fragments of 'evidence' recovered from the crime scene; this makes microscopy a fundamental tool, not only for forensic biologists but also for all forensic scientists. The comparison microscope facilitates the direct comparison of two samples (*e.g.*, hairs) and adds significantly to the forensic scientist's magnification armoury.

5.2 Microscopy

Microscopy is fundamental to forensic biology. The identification of algae, pollen, plant debris, tissues, bacteria, and the identification and comparison of hairs and fibres would be impossible without a microscope. For this reason, I will discuss it upfront.

Instruments of magnification have been associated with forensic science – both in reality and fiction – from the beginning. Sir Arthur Conan Doyle's Sherlock Holmes (first published in 1887) is frequently depicted peering through his magnifying glass at the scene of a crime or examining specimens under a Victorian microscope back at his Baker Street lodgings. Samples collected from the crime scene are often small-scale and require magnification to fully appreciate the information they hold. Similarly, close scrutiny of handwriting on a slip of paper might reveal idiosyncrasies or additions indicative of fraud. A close examination of a human hair might show colour variations consistent with dying or, perhaps, that its 'owner' had spent time in the sun. This is the substance of a good Sherlock Holmes story but reflects the real-world forensic scientist's reliance on magnification to reveal details invisible to the naked eye.

A typical magnifying glass – still used by modern day forensic scientists – achieves a magnification of 3 and a good light microscope can achieve a magnification of 1000, while a high-spec electron microscope can achieve magnifications up to 100 000. Sometimes, all that is needed is a modest enlargement with a magnifying glass to see part of a typed letter on a fragment of paper (see Chapter 12 *Forensic Document Examination*) or, at the opposite end of the magnification spectrum, 10 000 times to observe the intricate biological design of a coccolith (see Section 5.3.1.3 *Coccoliths*), which is crucial for its identification.

5.2.1 The Magnifying Glass

A magnifying glass is a convex lens housed in a frame device often with a handle, although modern lab magnifying glasses are often mounted on an adjustable arm (Figure 5.1), allowing for hands-free use.

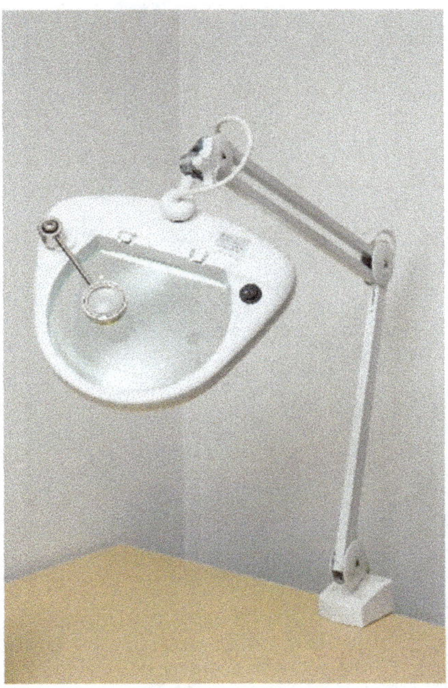

Figure 5.1 A typical benchtop, adjustable magnifying glass that might be found in a forensic science laboratory. Courtesy of Neil Macbeth.

The concept of magnifying lenses has been known for millennia – it was first mentioned by the Greek playwright Aristophanes in *The Clouds* in 424 BC and later in detail by the Arabic physicist Ibn al-Haytham in his *Book of Optics*, published in 1021.

Most magnifying glasses can achieve magnifications of 3 or 4. To put this into perspective, a difficult-to-locate 5 mm long fibre lifted from a pair of trousers would appear about 20 mm long under a magnifying glass – this would make it easily retrievable using sharp forceps.

5.2.2 Light Microscopy

The basic design of currently used light (or optical) microscopes was probably introduced by the Dutch engineer Cornelius Drebbel (1572–1633) in London in the early 17th century. It uses a series of lenses above the sample to magnify and a light source (originally directed daylight from a mirror, but now an electric light source is used) below the sample. Magnifications up to 1000 can be achieved with modern light microscopes (Figure 5.2). For high magnifications,

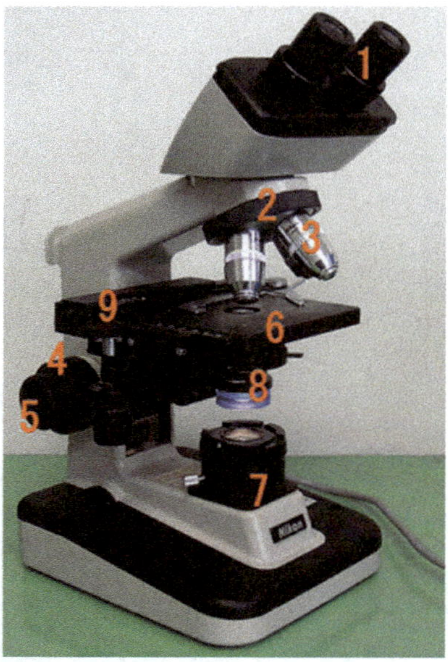

Figure 5.2 The anatomy of a modern light microscope: 1 eyepiece; 2 objective turret; 3 objective; 4 course focus adjustment; 5 fine focus adjustment; 6 stage – the slide housing the specimen is placed here and is held in place with two metal clips; 7 light source; 8 condenser to focus the light from the light source; and 9 the stage can be moved up and down mechanically.

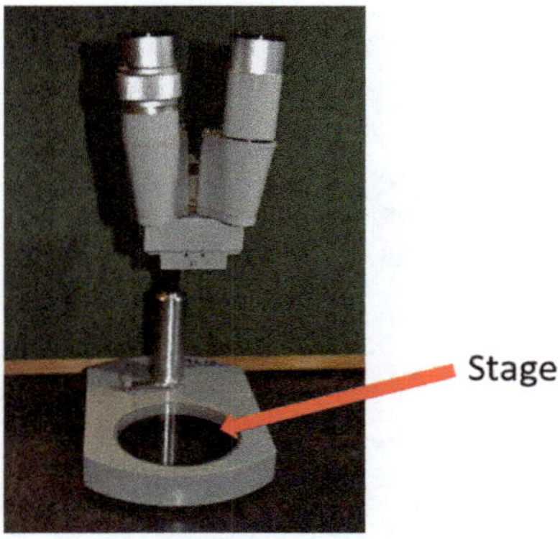

Figure 5.3 A simple, low magnification (×25) binocular microscope. The sample is placed on the stage and illuminated from above using daylight or an external light source (*e.g.*, an angle lamp).

oil immersion objectives (the lens immediately above the sample) are used because this improves the lens's resolution. The total magnification of the microscope is the magnification of the eyepiece lens multiplied by the magnification of the objective.

A simple low-magnification (up to ×25) binocular microscope (sometimes called a stereomicroscope; Figure 5.3) is also useful. These microscopes light the sample from above and allow a three-dimensional view. They might be used, for example, for identifying fragments of plants.

5.2.3 Comparison Microscopy

The comparison microscope is a clever light microscope adaptation that allows two samples to be viewed simultaneously and brought together for close comparison through sophisticated optical techniques (Figure 5.4). Basically, a comparison microscope comprises two separate microscopes joined together by a single eyepiece *via* an 'optical bridge'. Comparison microscopy is used in various facets of forensic science (*e.g.*, see Chapter 11 *Firearms*), but in a forensic biology setting, it plays a key role in fibre and hair comparisons (Figure 5.4).

5.2.4 Electron Microscopy

As discussed above, the maximum magnification achievable by light microscopy is of the order of 1000. This resolution is limited by the wavelength of visible light (310–1100 nm) because a structure cannot

Figure 5.4 A comparison microscope basically comprises two microscopes joined by clever optics and a common eyepiece (left), with two hairs being compared (right). Image of the microscope: courtesy of Radical Scientific Equipments Pvt. Ltd, India. Hair comparison: reproduced from ref. 1 with permission from Forensic Testing Services.

be seen if its dimensions are less than the wavelength of visible light. Therefore, to increase a microscope's resolution (*i.e.*, magnifying power), a shorter wavelength is required. Accelerated electrons have wavelengths about 100 000 times shorter than the shortest visible light wavelength. This means that the resolving power of a microscope that uses accelerated electrons instead of visible light is hundreds of thousands of times greater – this, of course, is the electron microscope.

The first electron microscope was built by a physicist, Professor Ernst Ruska (1906–1988), and an electrical engineer, Professor Max Knoll (1897–1969), at Berlin Technische Hochschule (Berlin Technical University) in 1931. Ruska won the Nobel Prize in Physics in 1986 for his work on electron optics.

Ruska's and Knoll's electron microscope achieved a magnification of only 400, which compared with a magnification of 2 000 000 that can be achieved using today's electron microscopes is trivial; however, the concept of electron microscopy was conceived. Following its patent in May 1931 by Reinhold Rüdenberg (1883–1961), the then Scientific Director of Siemens-Schuckertwerke, the first commercial electron microscope that surpassed the magnifying power of light microscopes became available in 1938.

Thus far, electron microscopy has involved passing electrons through a very thin sample, similar to how a light microscope uses light passed through the sample. This is termed transmission electron microscopy. However, in 1937, Manfred von Ardenne (1907–1997; a controversial

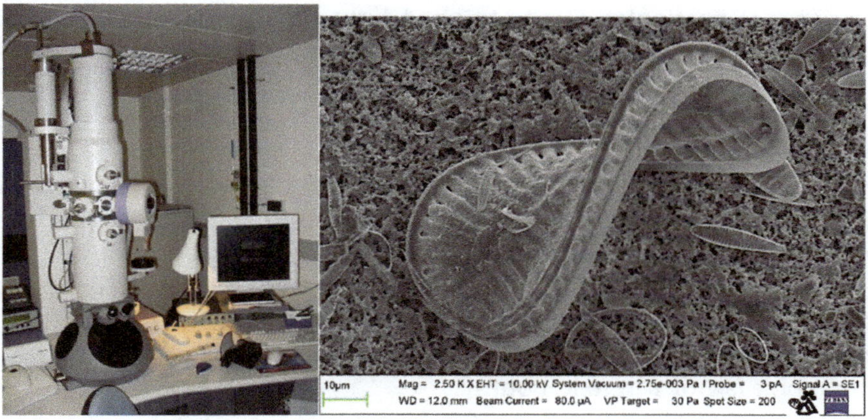

Figure 5.5 A SEM (left) and a SEM image of a diatom (*Surirella spiralis*) (2500×). SEM image: reproduced from ref. 2, https://en.wikipedia.org/wiki/Electron_microscope#/media/File:Electron_Microscope.jpg, under the terms of the CC BY-SA 2.0 license, https://creativecommons.org/licenses/by-sa/2.0/deed.en; image of diatom: reproduced from ref. 3, https://en.wikipedia.org/wiki/Diatom#/media/File:Surirella_spiralis_-_SEM_MUSE_-_sharpened.jpeg, under the terms of the CC BY-SA 3.0 license https://creativecommons.org/licenses/by-sa/3.0/deed.en.

scientist who was also part of the Soviet atomic bomb project) pioneered scanning electron microscopy (SEM; Figure 5.5). This was a significant advancement involving electrons 'illuminating' the sample from above. The image obtained shows the three-dimensional surface topography of the sample in astounding detail. Modern SEMs can achieve magnifications of 100 000.

Transmission electron microscopes (TEMs) are not frequently used by forensic scientists, but SEMs are commonly used; for example, to identify diatoms (see Section 5.3.1.1 *Diatoms*; Figure 5.5) from the lungs of drowning cases and coccoliths (see Section 5.3.1.3 *Coccoliths*) from putty in robbery investigations.

5.3 Forensic Botany

Botany is the science of plants; forensic botany is the application of plant science to criminal investigations. Botany is a very broad science encompassing plant structure, function, evolution and biochemistry; forensic botanists tend to focus on plant structure and its use for identifying plants, often tiny fragments of plants (*e.g.*, from the sole of a shoe). Microscopes are the tools of their trade; they use distinctive features of particular plants to shed light on the identity of a plant fragment under their microscope.

Forensic botanists use every conceivable part of a plant to aid in its identification. They might only have the tiniest fragment of a leaf recovered from a shoe, a grain or two of pollen from the hair of a murder victim, some diatoms recovered from the lungs of a drowned child, the spore of a fungus or the beautiful calcium carbonate-based structural architecture of a coccolith from an accused burglar's trousers...

In this section, I will only scratch the surface of forensic botany to place it within the broader context of forensic science. I could fill an entire book with the work of forensic botanists alone, but the focus of this book is not forensic botany!

5.3.1 Algae

The term 'algae' encompasses a vast array of eukaryotic (*i.e.*, their cells have nuclei) photosynthetic organisms (*i.e.*, they have chlorophyll and manufacture sugars using the energy from sunlight), these can be single-celled (*e.g.*, diatoms) or multiple-celled (*e.g.*, *Chlorella* sp.) or huge megastructures comprising millions of cells (*e.g.*, kelp). Algae *per se* are not often investigated by forensic botanists, however, the structures created by some algae are used in specialized forensic investigations; for example, the strange calcium carbonate plates termed coccoliths formed by some algae and the incredibly resilient silicon cytoskeletons of diatoms.

5.3.1.1 Diatoms

Diatoms comprise a large group of algae and a significant proportion of the earth's biomass generating up to 50% of the earth's oxygen annually. They are usually single-celled, but some species form clumps – they range in size from 2–200 μm. They are unusual because their cell walls (termed frustules) are silicon-based – most plant cell walls are carbon-based (*i.e.*, made up of cellulose, a glucose polymer). Diatom frustules are composed of silicon dioxide (SiO_2) polymers, which are incredibly resilient; in fact, they are not even dissolved by concentrated nitric acid (HNO_3), which is used as the basis of their isolation from tissues in forensic casework.

5.3.1.2 Diatoms in the Lungs of Drowning Victims

If a person drowns in sea or freshwater, they may inhale water containing diatoms into their lungs. Due to the characteristic spectrum of diatom species inhabiting a particular body of water, the diatoms isolated from the victim's lungs could provide clues about the location of drowning. For example, if a body is found washed up on a seashore and the diatoms found in its lungs are of freshwater origin, it indicates that the person drowned in a river and that their body was transported *via* the river to the sea.

On the other hand, for example, if a body is found in a river, but there are no diatoms in the lungs, this might indicate that the person was not drowned, but was dumped into the river after death. This conclusion might be drawn because most waterways contain diatoms and upon drowning, water may be inhaled, but a dead body dumped into a waterway would not inhale diatom-containing water.

To investigate diatoms in the lungs, *post-mortem* samples of lung tissue are taken and either heated with concentrated nitric acid or treated with a protease (an enzyme that breaks down proteins; *e.g.*, subtilisin Carlsberg, see Chapter 9 *Forensic Toxicology*) to dissolve the tissue. The acid or protease digest is centrifuged to sediment the diatom frustules (the rest of the diatoms' cells are dissolved by the acid or protease) and the frustules prepared for SEM. The characteristic structures of the frustules are used to identify the diatoms (Figure 5.6) and compare the diatoms in the lungs with those from the water in which the body was found. Identification of the species of diatoms found is often not necessary because comparisons of the signature frustule structures might be sufficient.

Studies have clearly shown that the number of diatoms present in the lungs of drowning victims is highly variable and is related to the number of diatoms in the water, but the diatom count in lung tissue is usually significantly greater than the number found in the lungs of bodies immersed after death. This means that a large number of diatoms in the lungs (*i.e.*, multiples of thousands per g tissue) is indicative of drowning, but a number around or below 1000 diatoms per g tissue does not necessarily indicate drowning (Table 5.1).

5.3.1.3 Coccoliths

Coccoliths are both fascinating and beautiful microstructures (2–25 μm across; Figure 5.7) made by some algae – they were first noted in 1858 by the evolutionary biologist, Thomas Huxley (1825–1895).

Coccoliths are made from calcium carbonate ($CaCO_3$) in the algal cell's Golgi apparatus (the place where cell excretions are packaged for externalisation) and when they are pushed out of the cell, they either adhere to the algal cell wall forming surface plates or fall from

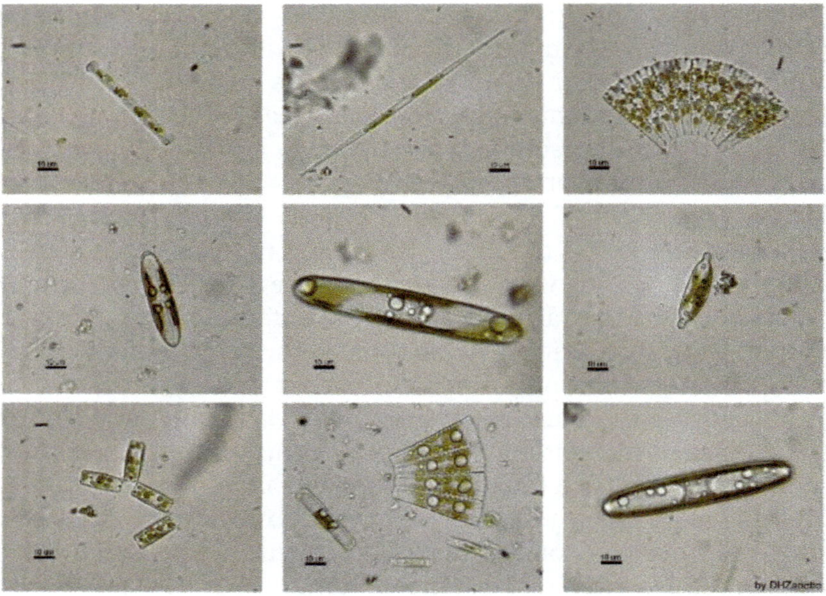

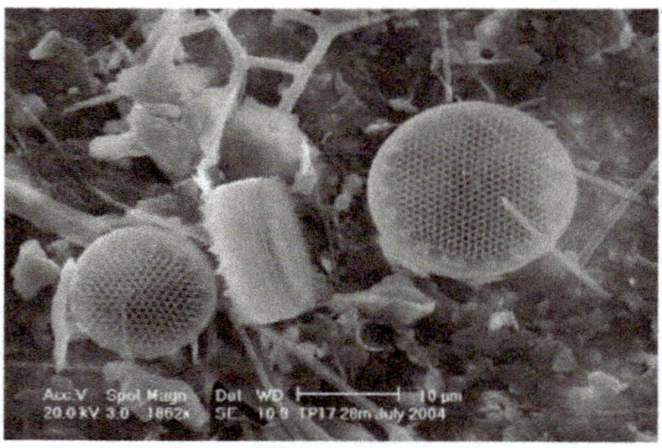
Acc.V Spot Magn Det WD 10 µm
20.0 kV 3.0 1862x SE 10.9 TP17 28m July 2004

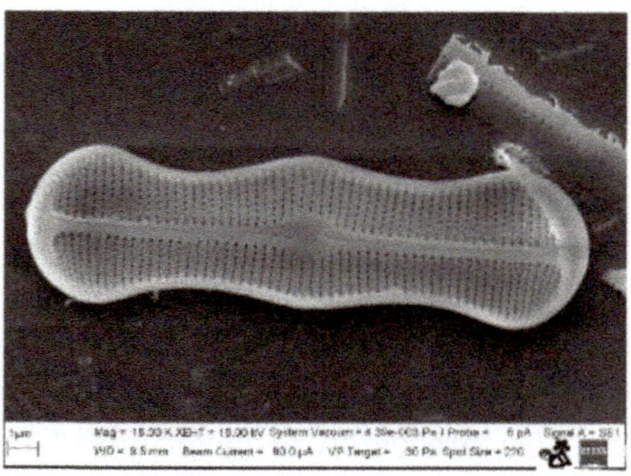

the cell. The coccoliths that fall from the algal cells comprise a major component of sediments; those that adhere to the cell wall end up in the sediment when the cell dies. These calcium carbonate-rich sediments are the main constituent of chalk (the British White Cliffs of Dover are composed of many millions upon millions of coccoliths).

There are many different coccolith structures that are easily visualised and distinguished using SEM (see Section 5.2.4 *Electron Microscopy*) (Figure 5.8) – as you will see later, this is useful in forensic investigations.

Chalk is a key component of glazing putty which is traditionally made from chalk and linseed oil (modern putties often use polybutene in place of linseed oil). Therefore, if a glass pane from a traditionally glazed window is broken or forced out, its putty might distribute and adhere to the clothing of the person breaking the window (Locard's Exchange Principle – see Chapter 2 *The History of Forensic Science*). If the window-breaker was a burglar, coccolith evidence might help identify the putty because putties made from different chalks likely have a different range of coccolith types.

In addition to their use in glazing putty for investigating burglaries, the presence of coccoliths in chalk might also be useful more generally in forensic investigations. For example, in 2002, two 10-year-old girls were murdered in the small town of Soham near Cambridge in the UK; a path composed partially of chalk and its coccoliths became important aspects of the forensic investigation of the case, later known as the Soham Murders.

5.3.1.4 Coccoliths and the Soham Murders

On 4th August 2002, Holly Wells and Jessica Chapman, two 10-year-old girls from Soham went missing. They had been at the Wells' home for a barbeque and at about 6.15 pm, they decided – without telling Holly Wells' parents – to go to the local sports centre to buy sweets.

Figure 5.6 The characteristic microscopic morphology of diatoms is used for their identification. Top: freshwater diatoms under a light microscope (approx. ×1000). Middle: SEM – *Thalassiosira* sp. (×1860). Reproduced from ref. 4, https://en.wikipedia.org/wiki/Diatom#/media/File:Diatoms-HCMR.jpg, under the terms of the CC BY-SA 4.0 license https://creativecommons.org/licenses/by-sa/4.0/. Bottom: SEM – *Achanthes trinodis* (×15 000). Reproduced from ref. 5, https://en.wikipedia.org/wiki/Diatom#/media/File:Achanthes_trinodis_-_SEM_MUSE.tiff, under the terms of the CC BY-SA 3.0 license https://creativecommons.org/licenses/by-sa/3.0/deed.en.

Table 5.1 Diatom count in lung tissue from drowning and submerged after death bodies. Reproduced from ref. 6 with permission from Elsevier, Copyright 2016.

Body type	Diatom count range in lungs per g tissue	L/D ± SD[a]
Drowned	29–757,742	13.50 ± 17.47
Immersed after death	15–945	0.51 ± 0.43

[a] Ratio of the number of diatoms in lung tissue/number of diatoms in drowning or immersion water ± standard deviation.

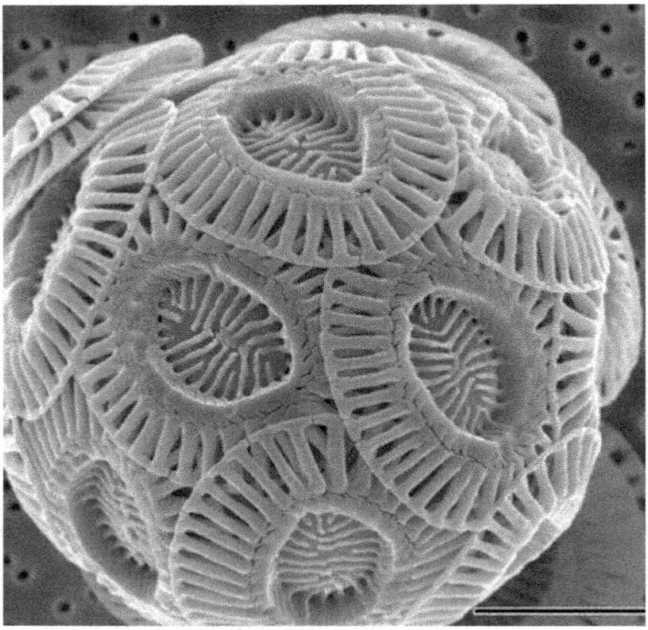

Figure 5.7 The appropriately named alga, *Emiliania huxleyi*, shows coccoliths attached to its cell wall. Reproduced from ref. 7, https://doi.org/10.1371/image.pbio.v09.i06, under the terms of the CC BY 4.0 license https://creativecommons.org/licenses/by/4.0/.

On their way home from the sports centre, the girls were lured into the home of Ian Huntley, the caretaker at the local secondary school, under the pretext that his girlfriend, Maxine Carr, who was a teaching assistant at the girls' school, was there. It was not until 8.00 pm that Holly Wells' mum went to Holly's bedroom, where Holly and Jessica had been playing, and found them both missing. Mrs Wells telephoned Jessica Chapman's parents to check if the girls had gone there. They had not. The girls were reported missing at 9.55 pm. They never returned home.

Figure 5.8 Coccolith from glazing putty visualised using SEM (approx. ×6000), showing their wide range of structures. Courtesy of Norman Downes, Birmingham Forensic Science Laboratory.

About two weeks later on 17th August, their asphyxiated bodies were found about 12 miles (19 km) from Soham in a 5 feet (1.5 m) deep irrigation ditch near a track called Common Drove, which runs along the perimeter of RAF Lakenham in Suffolk (a county bordering Cambridgeshire).

During the time between the girls' disappearance and the discovery of their bodies, Huntley became the unofficial spokesman for the Soham community. He was far from being a suspect. Indeed, Carr corroborated Huntley's story that they had talked to the girls as they passed Huntley's house on their way home from the sports centre. They were both lying – Carr had been at her mother's home in Grimsby, Lincolnshire, about a 2.5 h drive away from Soham at the time of the girls' disappearance while Huntley had been murdering the two girls. In an interview with the press, Carr referred to the girls in the past tense which prompted the investigators to think since at that time the girls were still thought to be alive.

The girls' clothes were found at Soham Village College (the school where Huntley worked as caretaker). Huntley and Carr were arrested on suspicion of abduction and murder of the girls on 17th August. Carr later admitted lying to protect Huntley.

5.3.1.5 Coccoliths Were Found on the Suspect's Car...

The track along Common Drove was made of quartz, builders' rubble and chalk. Dust was recovered from the top of Huntley's car's suspension arm. This was believed to have occurred because the rubble had been bulldozed into ridges when the track was made, and when the car was driven over the ridges, the material had been deposited onto the suspension arm underneath the vehicle. Lab investigation of chalk fragments from the debris recovered from the suspension arm and traces of debris from the carpet in the footwell on the driver's side of the car revealed micropalaeontological (the study of microfossils, including coccoliths) evidence that connected Huntley's car with the Common Drove track, next to where the girls' bodies were dumped. In addition, there was other evidence linking Huntley to the case, including his fingerprints on a bin containing the girls' burned clothing at Soham Village College, traces of blood and saliva on the clothing, and blood patterns in the hallway of Huntley's home. A BBC news report on 17th December concluded, '*As a result of the evidence gathered, including the micropaleontological submissions, Ian Huntley was found guilty of the two murders...*'

On 17th December 2003, Huntley was convicted of the murder of Holly Wells and Jessica Chapman; he was sentenced to double life imprisonment. Carr was sentenced to $3\frac{1}{2}$ years in prison for perverting the course of justice.

5.3.2 Higher Plants

Our environment, whether it is rural or urban, is filled with plants. Wherever you walk, you will see plants; even in the most built-up urban settings. Plants are survivors and will find a suitable niche to allow them to survive. This means that when someone walks around, they may collect traces of plant material on their shoes, or pollen grains in their hair or stuck to their clothes. For the most part, these plant fragments are of little forensic use because they are from very common plant species that have little or no evidential value. Sometimes, there might be a rare plant growing at a place of interest in the investigation of a crime, or perhaps a particular combination of plants that makes a fragment of the rare plant or a combination of plant fragments matching the plant species growing at the site of interest of significant interest to the forensic botanist.

5.3.2.1 Plant Debris

Plant debris attached, for example, to the soles of a shoe or lodged in the treads of a tyre can be recovered and identified microscopically. The forensic botanist will likely begin by examining the debris under a low power binocular microscope (see Section 5.2.2 *Light Microscopy*) to see if there are any features that might help in identifying the plant (*e.g.*, parallel veins in a monocotyledonous plant, *e.g.*, grass). Then, a thin section of the plant fragment might be cut, perhaps stained with a dye to accentuate its structural features, and mounted on a microscope slide for higher power light microscopy (see Section 5.2.2 *Light Microscopy*). This is likely to guide an experienced forensic botanist towards the identification of the plant or at least identification of its family or genus. Comparison microscopy might also be used to compare sections of plants from the crime scene with debris found, for example, on a suspect's shoes.

5.3.2.2 Plant Debris in Stomach Contents – Identifying the Last Meal

Identifying stomach contents *post-mortem* might be important in determining the components of the deceased's last meal. Knowing the last meal can help identify where the deceased might have been before their death. Publishing the last meal might stimulate someone's memory; a landlady, perhaps...

Since plants are an important component of our diet, plant fragments are often present in stomach contents. These fragments can be recovered and investigated microscopically in much the same way as described in Section 5.3.2.1 *Plant Debris*.

Some plants have highly characteristic structures (*e.g.*, pineapple; Figure 5.9), which would be very useful in identifying food in stomach contents.

In the early 1980s, a young woman was murdered in Denver, USA. She had eaten her last known meal (lunch) at a fast-food restaurant

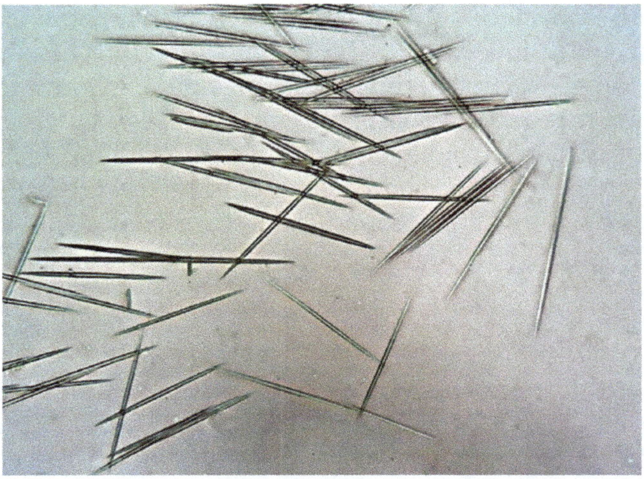

Figure 5.9 A characteristic feature of some plants, including pineapples, is the presence of microscopic crystals of calcium oxalate or calcium carbonate called raphides. Figure is in the Public Domain.

with her boyfriend – the restaurant had a very limited menu. Investigation of the woman's stomach contents revealed fragments and cells consistent with red cabbage, red kidney beans, green peppers (capsicum), onions, tomatoes and lettuce being part of her last meal. Red cabbage, beans and green peppers were not served by the fast-food restaurant; therefore, it was highly unlikely that her last meal was from there.

The woman's parents were expecting her for dinner at their home and they called the police when she did not turn up. Her character was such that she would not have let them down and she would not have dined with a stranger. This suggested that she knew her killer and that she ate with him. Her boyfriend was ruled out as her killer because he had an alibi for the evening and thus could not have been with her.

The suggestion that she had dined with someone she knew was very important in finding and convicting the young woman's murderer. It turned out that she thought she had met her murderer at her brother's house some months before, but she was mistaken. Her murderer was, in fact, a serial killer whom she had never met, but he looked like one of her brother's friends. The man took her to a salad bar for dinner, where the menu included salads containing cabbage, red kidney beans, green peppers, onions, tomatoes and lettuce.

5.3.2.3 Using DNA to Identify Plant Debris

The traditional microscopical methods used by forensic botanists might soon be replaced by DNA analysis (see Chapter 6 *DNA* for background information about DNA). Each plant species has specific genes that are unique to the species and therefore it is possible to sequence the genes from miniscule plant fragments and identify the plant species (or even sub-species or cultivar). DNA plant identification evidence of this type would be unequivocal and hold great value in court. Recently, techniques have been developed that use a selection of key genes as a 'sort of' bar code to identify a plant. This holds great promise for forensic botany investigations (see *Further Reading*).

5.3.2.4 Tree Rings

Other, more esoteric, plant investigations sometimes provide important evidence in the investigation of a case. A good example is tree rings (Figure 5.10), which might be used to match a felled tree to its stump – especially if the felled tree has been stolen!

Many years ago (as a forensic science student), I was involved in the forensic investigation (with Dr Nicholas Prance, Birmingham

Figure 5.10 The cut end of *Pinus taeda* showing tree rings. Each ring represents a growth year, and the thickness of the rings is determined by the conditions in that year. Reproduced from ref. 8, https://commons.wikimedia.org/wiki/File:Pinus_taeda_crossx7358.JPG, under the terms of the CC BY-SA 3.0 license https://creativecommons.org/licenses/by-sa/3.0/deed.en.

Forensic Science Laboratory, UK) of a crime involving the stealing of felled trees from a forestry block. Trees were found on the property of the accused and slices were cut from the cut end of the tree trunk and compared in the laboratory with slices cut from the stumps remaining at the forest block. After a great deal of painstaking work, matches were found between the stumps and the tree trunks from the suspect's property. At this point the suspect became the accused!

The significance of the tree ring match evidence depends on the 'degree of uniqueness' of the ring patterns. Trees usually form one growth ring per year and the width of the ring is related to the climatic conditions for that year (*i.e.*, a thicker ring means more growth which equates to good growing weather). This means that the ring pattern (Figure 5.10) is influenced by the local growth conditions. Therefore, a match is very good evidence and would carry significant weight in court. Tool marks from, for example, a saw are also good evidence that might be used to connect a suspect to the crime.

5.3.2.5 Pollen

Pollen produces the male gamete (sperm cells) of seed-producing plants. Pollen grains range in size from 3 μm (alpine forget-me-not (*Myosotis alpestris*)) to 200 μm (cucumber (*Cucumis sativus*)). A pollen

Figure 5.11 Electron micrographs of pollen grains from *Lilium auratum* (×5000; left) and Arabis sp. (×1500; right), showing their structural differences. Image (left): Figure is in the Public Domain. Image (right): Reproduced from ref. 9, https://en.wikipedia.org/wiki/File:Arabis_voch1-4.jpg, under the terms of CC BY-SA 3.0 license https://creativecommons.org/licenses/by-sa/3.0/deed.en.

grain comprises an outer wall (exosperm) and an inner wall (endosperm) to protect the sperm cells. The distinctive morphology of the outer wall (Figure 5.11) can be used by forensic palynologists (from the Greek for 'study of dust') to identify the plant species from which the pollen originated. It is even possible to identify sub-species and cultivars using pollen grain morphology.

The rough exosperms of pollen grains allow them to adhere to many surfaces, particularly natural fabrics, including wool and human hair – wool fibres and human hair (see Section 5.8.2 *Hairs and Fibres*) have rough outer scales to which pollen grains can 'stick'. This can be very useful in a forensic setting because pollen grains attached to clothing or trapped in a person's hair might indicate where the person had been.

5.3.2.6 A Drug Trafficking Case Solved Using Pollen Grains

In 1992, 124 g of cocaine hydrochloride was seized by the New York City Police. Microscopical examination of the cocaine revealed pollen grains, which on further investigation were identified as originating from both spring flowering South American plants and species that grow in north central and north east USA, but flower later in the year. Putting these palynological data together suggested that the cocaine had been made or trafficked through South America in the spring, and during the following year, it was taken through the eastern USA and perhaps packaged for sale. The identification of the plants from which the pollen originated allowed the police to develop a good working hypothesis that they could further investigate; this case clearly illustrates the power of forensic palynology.

5.4 Entomology

Forensic entomology is the use of insects (and sometimes other arthropods, *e.g.*, spiders) in criminal investigations. In its broadest sense, it might include insects present in foods as evidence of poor storage or spoilage (stored-product forensic entomology), insect pest infestations in buildings, gardens, *etc.* (urban forensic entomology) that could be the subject of litigation between, for example, a tenant and a landlord, or insects found on corpses, which might be used to estimate time since death or the place of death or where a body was dumped (medico-legal forensic entomology).

Forensic entomologists identify insects (both adults and larvae) and using their knowledge of the habitats of identified insects might, for example, be able to pinpoint where a body has been after death. Perhaps the most important role of the forensic entomologist is to estimate time since death based on the life cycle of the blow flies (*Calliphora* sp.) which can infest corpses very quickly.

5.4.1 The First Use of Entomology to Estimate Time Since Death

Dr Louis François Étienne Bergeret (1814–1893), a French physician, was the first to use insect infestation of a corpse to estimate time since death. He carried out a *post-mortem* examination on a mummified infant body found in a Paris house in 1855. He observed different stages of the lifecycles of insects remaining on and in the corpse. From his knowledge of insect lifecycles and successions, he concluded that the child had been dead for 7 years. The suspected killers were a couple who had lived in the house 7 years before the mummified body was found. Berget's entomological evidence was important in their conviction for the child's murder.

Berget's evidence was far below the level of robustness we demand in modern forensic entomology, but his ingenuity initiated an important means of estimating time since death that is still used today.

5.4.1.1 Blowfly Maggots and Time Since Death

Blowflies have an incredibly sensitive sense of smell. They can smell and locate a corpse almost immediately after death. As soon as they find the corpse, they lay their eggs on its softer parts (*e.g.*, the eyes) from where the newly hatched larvae (maggots) can gain access to the corpse's highly nutritious flesh. The maggots pass through a series of very well-defined stages before they pupate and eventually hatch to release more blowflies. The released blowflies might fly off in search of new cadaveric pickings or, if there is still meat on 'their' corpse, they will lay eggs there. The time between each larval stage of the blowfly is temperature-dependent but is fixed (Figure 5.12). So, for example, at 70°F (21°C), it takes 23 h for the eggs to hatch into the first larval stage. If a corpse is found with only one blowfly developmental stage present, it is likely that the time taken for the blowfly eggs to reach that stage of development is approximately the time since death. However, if more than one larval stage is found, this likely means that several lifecycles have passed and the time since death is at least one lifecycle period plus the time taken to reach the next larval stage found.

If a body is found outdoors at an ambient temperature of about 20°C and the temperature has not changed very much over the preceding few weeks (this can be checked *via* the country's meteorological office – if it has changed considerably, corrections will be made to the life cycle time periods) and only stage 3 maggots are

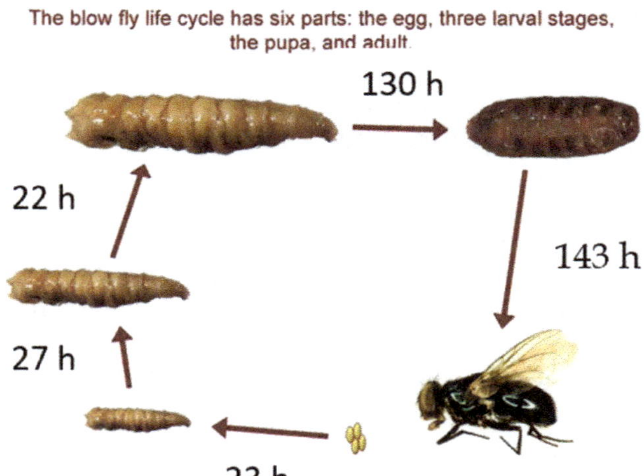

Figure 5.12 Lifecycle of the blowfly at 70°F (21°C). Reproduced from ref. 15 with permission from The Cleveland Museum of Natural History.

Figure 5.13 This brushtail possum (*Trichosurus vulpecula*), common in New Zealand, had been dead at the roadside near to the author's home for about 23 h as evidenced by the infestation with larval stage 1 maggots, which take about 23 h to develop from eggs at 20°C (see Figure 5.12).

found on the corpse; this means that the body has been dead for approximately 23 h + 27 h + 22 h = 72 h, but if only eggs were found, the time since death would be less than 23 h. As an example, Figure 5.13 shows a possum infested with first larval stage maggots, which means that it had been dead for about 23 h.

5.4.1.2 Other Insects in Forensic Casework

Flies are the insects most commonly studied by forensic entomologists; however, beetles, wasps, ants, bees and mites (mites are not insects, but are studied by forensic entomologists for convenience) have all featured in forensic cases over the years. For example, rove beetles (Staphylinidae) are commonly found burrowing into corpses. They have a long lifecycle, particularly the egg hatching time of about 4 days; therefore, the presence of larvae signifies at least 4 days since death.

5.5 Biochemical Methods

5.5.1 Acid Phosphatase and Semen Detection

Acid phosphatase is secreted by the prostate gland and is present in semen at very high activity. Its function in semen is not known, but from the forensic perspective, this does not matter; the fact that it is a biochemical marker for semen is what is important. However, it is important to recognise that acid phosphatase is also present in other bodily secretions (including vaginal fluid), but at very much lower activity. Therefore, tests for semen that rely on the presence of acid phosphatase are not definitive and their interpretation depends on the training and experience of the scientist. Confirmation is often achieved by finding spermatozoa in the acid phosphatase positive stain – this confirms that it was semen. Since semen stains contain spermatozoa, unless the man from whom the semen originated suffers from azoospermia (no sperm in the semen) or was vasectomised, they are an excellent source of DNA (see Chapter 6 *DNA*) and very important in determining the identity of sexual offenders.

5.5.1.1 The Acid Phosphatase Test

One method to check for semen on a garment (*e.g.*, underwear in a rape case) involves detection of acid phosphatase, which is present at high activity in semen. The garment is laid down on a flat surface and dampened (with water) blotting paper is laid on top of the garment and the paper gently pressed down. If semen stains are present, acid phosphatase will migrate onto the wet blotting paper (acid phosphatase is water soluble). The outline of the garment is marked in pencil on the blotting paper, which is then removed and sprayed with a reagent containing 1-naphthyl phosphate and Brentamine Fast Blue B in a buffer (at the correct pH for acid phosphatase optimum activity) – this reagent is usually purchased ready-made. The acid phosphatase from the semen hydrolyses the naphthyl phosphate, releasing naphthol which reacts with red Brentamine Fast Blue B to form a purple-blue azo-dye (Figure 5.14). Purple-blue regions on the blotting paper indicate the presence of semen in the corresponding position on the test garment.

The test can also be carried out in a test tube. The stain is extracted with saline and the extract treated with 1-naphthol/Brentamine Fast Blue B reagent; a blue-purple colour indicates acid phosphatase, which in turn suggests that the stain was semen.

If the acid phosphatase test is positive, the stain can be gently extracted with saline (to reduce spermatozoa damage) and a small amount of the extract is placed onto a microscope slide, dried and stained before looking at it under a light microscope (Figure 5.15). Spermatozoa will be seen if the stain is semen. Spermatozoa are definitive confirmation of the presence of semen, but the lack of spermatozoa does not mean that

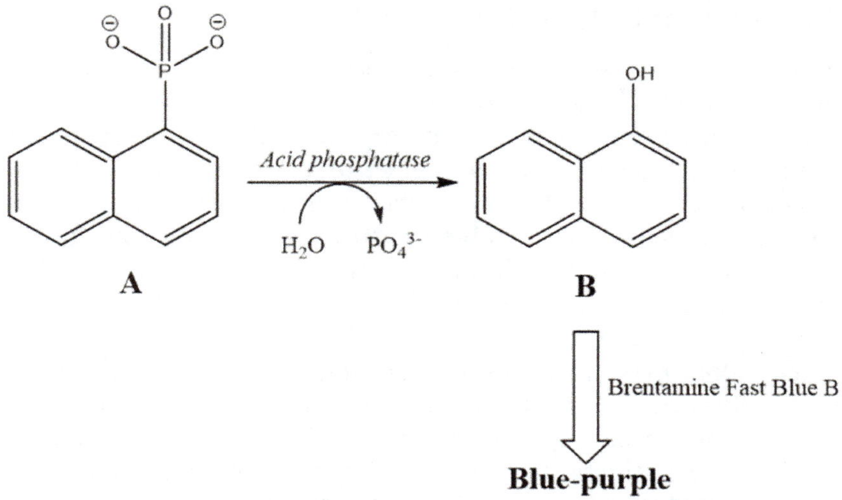

Figure 5.14 1-Naphthylphosphate (A) is hydrolysed by acid phosphatase in semen to form 1-naphthol (B). The 1-naphthol reacts with Brentamine Fast Blue B to form a blue-purple azo dye.

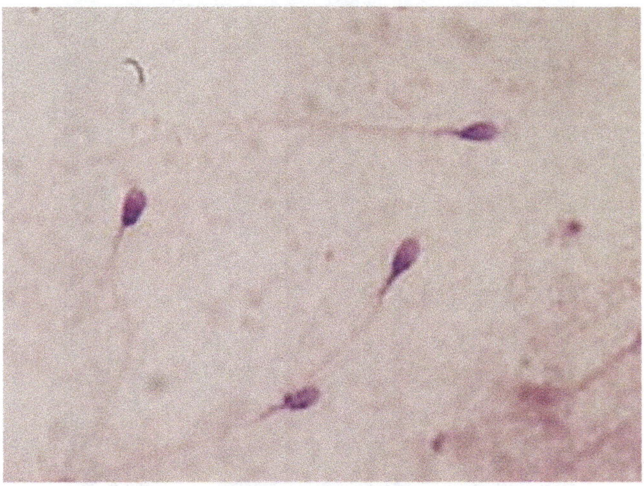

Figure 5.15 Human spermatozoa stained with eosin and haematoxylin viewed under a light microscope (×400). Reproduced from ref. 10, https://commons.wikimedia.org/wiki/File:Sperm_stained.JPG, under the terms of CC BY-SA 4.0 license https://creativecommons.org/licenses/by-sa/4.0/.

the stain is not semen because they might have been lost if the clothes were washed (acid phosphatase can remain after washing) or the man might have azoospermia (approx. 2% of cases investigated in sexual assaults are azoospermatic) or have had a vasectomy.

5.5.2 Luminol Blood Detection

Luminol exhibits chemiluminescence in the presence of an oxidising agent (Figure 5.17). The chemistry of luminol's chemiluminescence is complex and I am not going to attempt to explain it here; suffice to say, an oxidising agent donates an electron to the luminol molecule to produce a highly unstable excited-state molecule; this molecule then relaxes to a closely related stable form of the molecule and loses energy in the form of blue light (425 nm) – this is termed chemiluminescence. The blue chemiluminescence lasts for about 30 s and can only be seen in the dark (Figure 5.16).

Haemoglobin (Hb), the oxygen carrying red pigment in red blood cells (erythrocytes) contains an iron ion (Fe^{2+}) at its centre, which facilitates the formation of oxyhaemoglobin (HbO_2) to carry oxygen from the lungs to the tissues. Fe^{2+} can be reversibly oxidised to Fe^{3+}; it is this oxidoreduction reaction that, in the presence of hydrogen peroxide (H_2O_2), facilitates luminol's chemiluminescence (Figure 5.17).

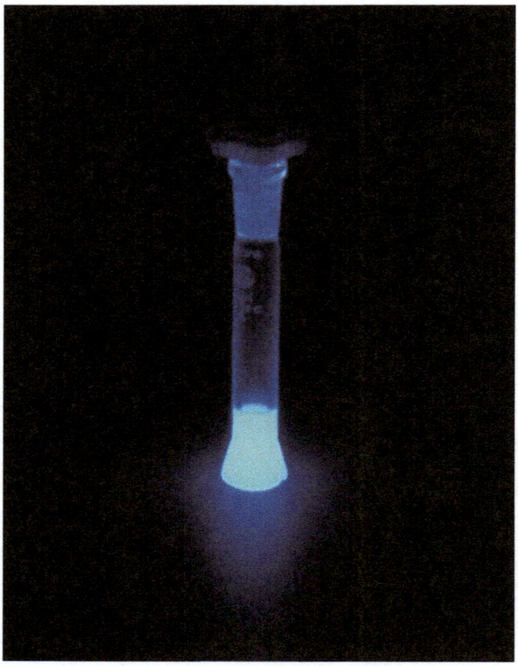

Figure 5.16 Luminol in solution in the presence of an oxidising agent, showing blue chemiluminescence. Reproduced from ref. 11, https://commons.wikimedia.org/w/index.php?curid=1254510, under the terms of the CC BY-SA 2.5 license https://creativecommons.org/licenses/by-sa/2.5/deed.en.

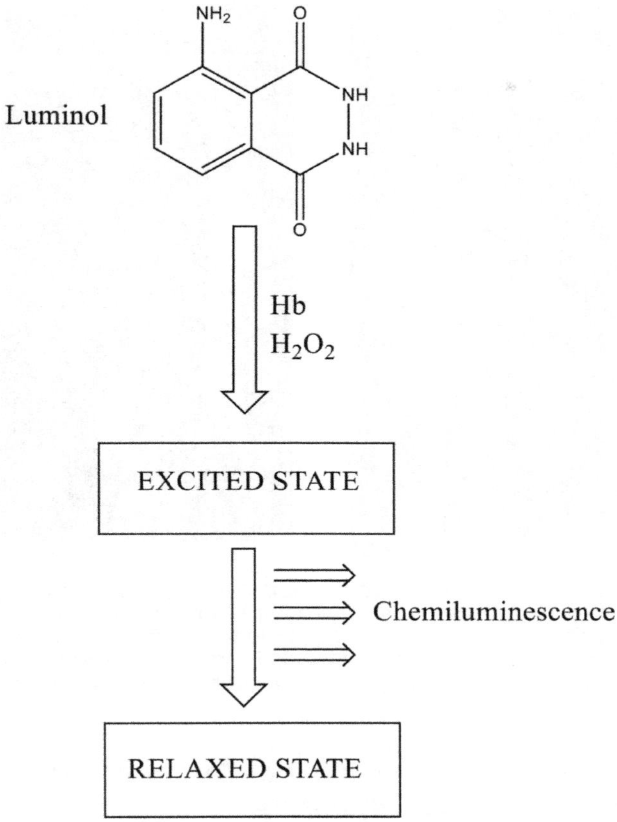

Figure 5.17 Luminol reacts with haemoglobin (Hb) in blood in the presence of hydrogen peroxide (H_2O_2), forming an excited state molecule which rapidly relaxes releasing energy in the form of blue (425 nm) chemiluminescence.

Therefore, blood gives a positive luminol reaction. Since the luminol reaction is very sensitive, it can pick up the tiniest traces of blood (1 : 5 000 000 diluted blood gives a positive test), and it is even possible to detect blood spots on fabrics or other surfaces that have been washed.

As luminol responds to oxidising agents generally, luminescence in the luminol test does not definitively identify blood, it simply shows the presence of an oxidising agent. Problematically, luminol gives a positive test for iron, copper, bleach (contains sodium hypochlorite (NaOCl) a powerful oxidising agent), faecal stains, urine and horseradish (which contains the enzyme peroxidase). Despite these drawbacks, the luminol test is an excellent screening test for the presence of bloodstains. It is quick, simple, can be carried out on large areas of fabrics and other surfaces, and it is incredibly sensitive.

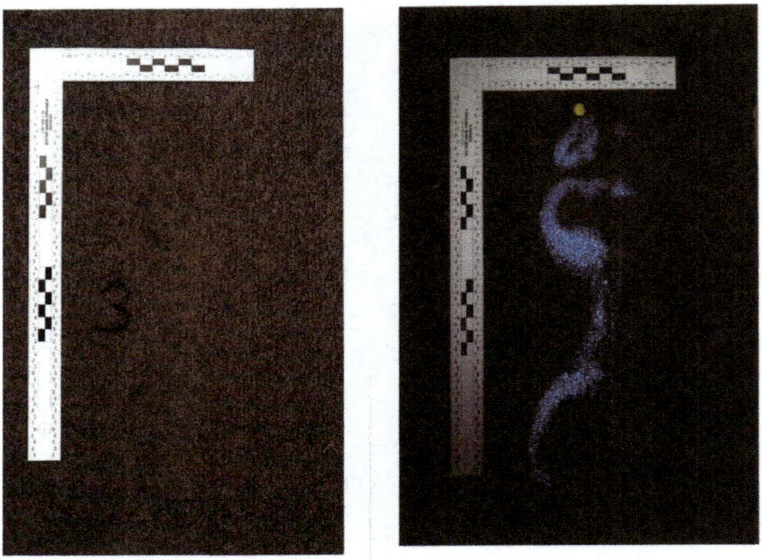

Figure 5.18 Left: Carpet taken from a house in which a murder was committed. Right: Carpet sprayed with luminol showing a bloody footprint. From this, it was possible to determine the foot size of the person who left the print and to identify the source of the blood using DNA profiling (see Chapter 6 *DNA*). Courtesy of Dr Nick Powell.

5.5.2.1 Application of the Luminol Test at a Crime Scene or in the Forensic Laboratory

The area to be tested is sprayed with the luminol reagent, which contains luminol and hydrogen peroxide in an aqueous solution at alkaline pH. To observe chemiluminescence, the testing should be completed in the dark or a special light sensitive camera might be used to detect the specific wavelength (425 nm) light emitted. The chemiluminescence lasts for only 30 s, but respraying the area may reproduce the chemiluminescence reaction – this can be carried out several times. Nowadays, it is usual to record the chemiluminescence as a digital image so that it can be analysed fully later (Figures 5.18 and 5.19).

5.5.3 Other Tests for Blood

Although the luminol blood test is most commonly applied in forensic settings, various other tests for blood have been used over the years. Most are presumptive tests, which means that a positive result *suggests*, but does not *prove*, the presence of blood. These tests rely on

Figure 5.19 A luminol test on the floor of a kitchen in a forensic investigation. It is clear that an attempt has been made to clean up the blood - look at the area around the floor by the mop and bucket and what are likely to be footprints or footwear marks between the large blood-stained area and the bucket! Image: courtesy of the Minnesota Bureau of Criminal Apprehension Forensic Science Services.

a reaction between a blood component (*e.g.*, Hb) and a chemical reagent. A good example is the Kastle–Meyer test.

5.5.3.1 The Kastle–Meyer Test for Blood

An American chemist, Joseph Hoeing Kastle (1864–1916), noted in 1901 that phenolphthalein reacts with blood to form a bright pink colour in the presence of hydrogen peroxide (Figure 5.20); in 1903, Erich Meyer (1874–1927), a German physician and chemist, perfected the chemistry and created the Kastle–Meyer test for blood.

The Kastle–Meyer test involves using a small swab to lift some of the 'blood' from a suspect stain; a drop of phenolphthalin (note phenolphthalin not phenolphthalein; see Figure 5.20) reagent is then applied to the swab, followed by a drop of hydrogen peroxide. If the swab turns pink, this indicates the presence of blood. However, there are many other 'things' (*e.g.*, copper and nickel salts) that produce a positive reaction and so the Kastle–Meyer test is prone to false positives. Therefore, a positive test is not proof of blood, but a negative test is proof of the absence of blood (*i.e.*, the test does not give false negative results). The Kastle–Meyer test is incredibly sensitive, it can detect a 1 : 10 000 000 (10^{-7}) dilution of blood.

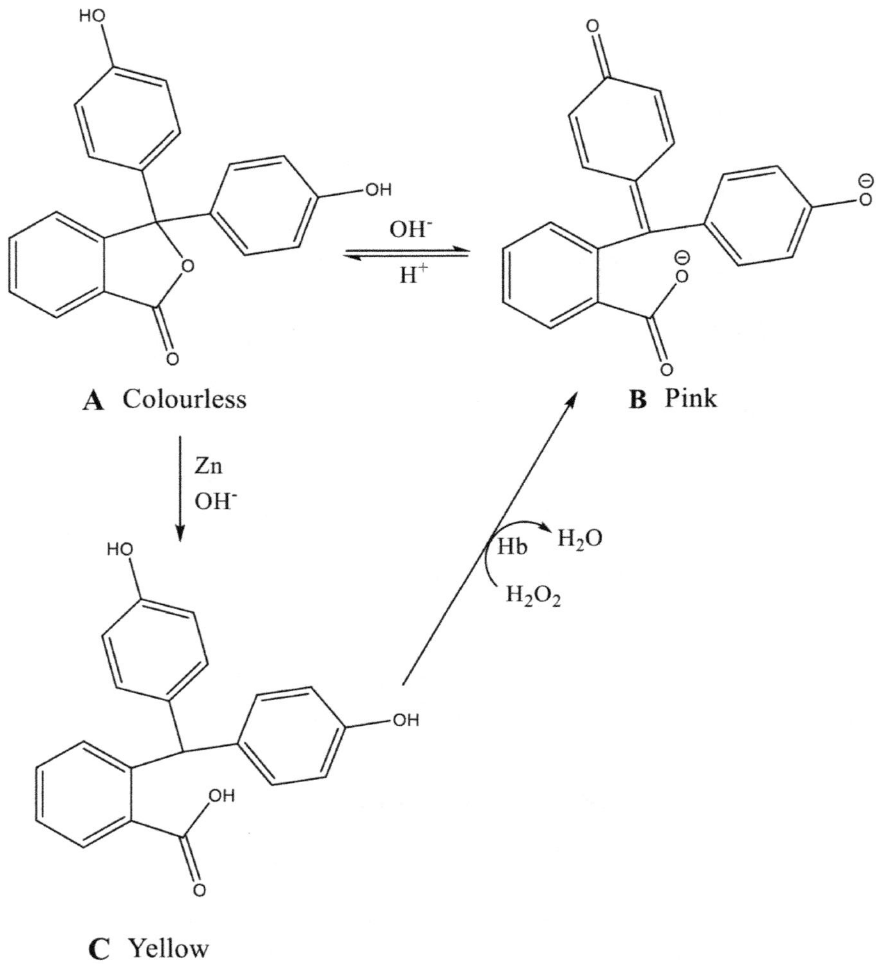

Figure 5.20 Chemistry of the Kastle–Meyer test for blood: phenolphthalein (A) is an indicator, which is colourless at acid pH (A) and pink at alkaline pH (B). The Kastle–Meyer test involves reducing phenolphthalein (A) with zinc (Zn) at alkaline pH (OH^-) to form yellow phenolphthalin (C). The conversion of hydrogen peroxide (H_2O_2) to water (H_2O) catalysed by haemoglobin (Hb) in blood converts phenolphthalin to pink ionised phenolphthalein.

5.6 Tissues

Fragments or smears of tissues might be found in, for example, a murder case. They could be crucial evidence in the case. Knowing the tissue is human is important if it is suspected as having originated from a murder victim. DNA testing (see Chapter 6 *DNA*), of course, will be of paramount importance in identifying the origins of the tissue, but knowing whether the tissue is of human origin might be of great importance. Similarly, if the case involved an animal - perhaps the killing of a prize animal - identifying the species of tissues found on a suspect could be important evidence.

5.6.1 Identifying Tissue Type

If a large enough sample of tissue is available, a thin section can be cut, mounted on a microscope slide, stained and observed under a light microscope (termed histology). The structural organisation of the tissue is often sufficient for a pathologist to identify its origin (Figure 5.21).

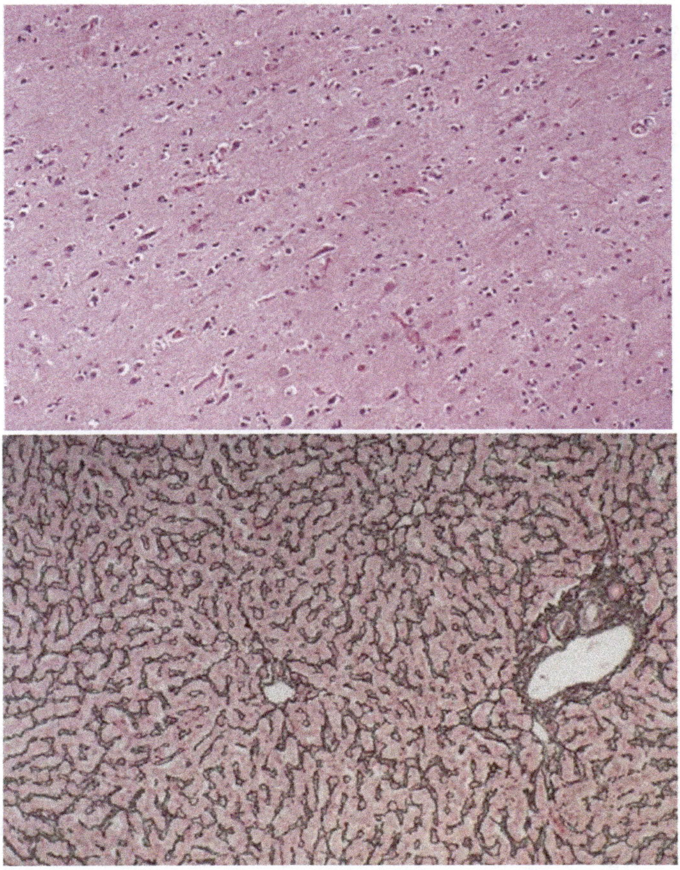

Figure 5.21 Histology of the brain (haematoxylin and eosin stain - top) and liver (reticulin stain - bottom), showing the great differences in structural organisation that is diagnostic of these two tissues. Both sections are approx. ×100. Courtesy of Edward C. Klatt MD.

However, if the sample is too small to allow a histological investigation, it is possible to use tissue specific protein markers present on cell surfaces using fluorescent-tagged antibodies or an antibody that catalyses a colour-producing reaction that recognise these tissue-specific proteins (termed immunohistochemistry – IHC). Very small fragments of tissue can be investigated by immunohistochemistry; the sample is placed onto a microscope slide and, if the fluorescent method is being used, treated with a fluorescent-tagged antibody directed against a tissue specific protein. If the specific protein is present on the cells of the test tissue the antibody will bind to them, and the antibody not bound to the sample is washed off and the sample is observed under a fluorescence microscope (Figure 5.22). A fluorescence microscope uses UV light instead of visible light, which causes the fluorescent probe attached to the antibody to fluoresce. If the tissue fluoresces this means it has the tissue specific protein and thus is the corresponding tissue type. Non-fluorescent (termed chromogenic) tags can also be used. In this case, instead of using a

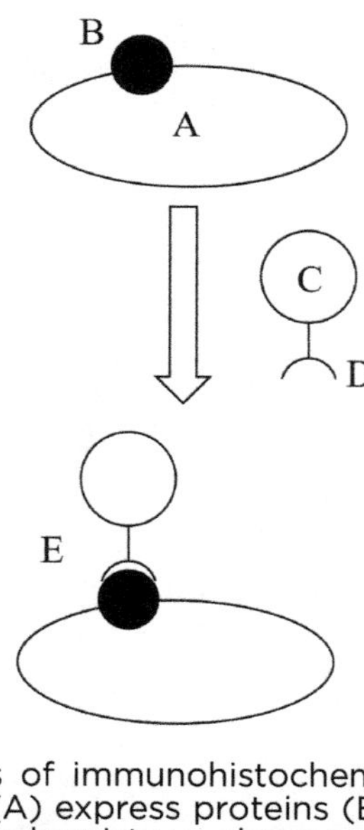

Figure 5.22 The process of immunohistochemical tissue type determination. Cells (A) express proteins (B) specific to their origin. In immunohistochemistry a chromophore or fluorescent tag (C) is attached to an antibody (D) directed against the specific cell protein. The test tissue is treated with the tagged antibody (An antibody is a protein made by the immune system that specifically recognises and binds to a large molecule (*e.g.*, another protein)) which binds to the protein (E), thus indirectly attaching the chromophore or fluorescent tag to the cell and causing it to fluoresce or change colour – both are visible under a microscope.

fluorescence microscope to pinpoint the tag's fluorescence, a 'normal' light microscope is used, and the tissue appears stained if it expresses the tissue specific protein diagnostic of a particular tissue (Figure 5.23).

IHC in a clinical setting is widely used for identifying abnormal cells (such as cancer cells and tumours) in amongst otherwise healthy tissue. This is also one of the techniques utilised by forensic neuropathologists who specialise in diseases of, and injuries to the brain, spinal cord and nerves. Examination of the brains of deceased children to visualise injury from shaking or impact is a common issue addressed by forensic neuropathologists, although forensic neuroradiology (examination of the brain by non-invasive imaging such as computerised tomography, CT) is becoming more common.

Forensic neuropathologists can use macroscopic and microscopic examination of tissues removed during *post-mortem* examination or biopsy to inform their interpretations. Using IHC, they can accentuate various biological structures within brain and central nervous system (CNS) tissue to visualise damage and changes within these tissues.

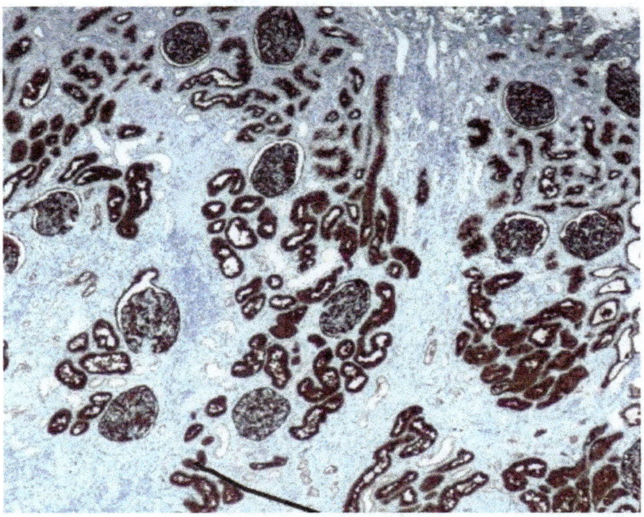

Figure 5.23 Immunohistochemically stained section of the human kidney using a brown chromophore-tagged antibody against a specific protein (nephrilysin) found in specific parts (glomeruli and proximal tubules) of the kidney structure. There is sufficient tissue here for a pathologist to identify the tissue from its structure alone, but if only a very small section was available the brown staining would be sufficient to identify the tissue as kidney. Reproduced from ref. 12, https://commons.wikimedia.org/wiki/File:Kidney_cd10_ihc.jpg, under the terms of the CC BY-SA 3.0 license https://creativecommons.org/licenses/by-sa/3.0/deed.en.

This allows them to assess causes or contributions to death, natural (such as causes or effects of aneurisms, which are localised swelling of the wall of a blood vessel) or otherwise. In addition to the non-accidental injury questions raised in child head injury cases, other common questions that IHC testing is used to address in criminal cases are whether neck pressure (strangulation *via* hands or ligatures) has been applied that interrupted blood flow into or out of the brain or rupturing of brain matter and vasculature caused by blunt force impact (such as an impact from a weapon).

For a brain sample for example, IHC for a specific brain protein (glial fibrillary acidic protein – GFAP) can be carried out. A sample of brain is cut, preserved (10% neutral buffered formalin) and mounted in paraffin wax. This is termed formalin-fixed paraffin-embedded (FFPE). Once the paraffin has solidified, the sample is cut into thin sections (4 mm thick), stained (haematoxylin and eosin), mounted onto a microscope slide and then examined under a high power ($\times$1000) light microscope.

The sections are then treated with antibodies (Table 5.2) and the antibody bound to the 'brain' proteins (Table 5.2) is tagged using a commercial kit that specifically binds to antibodies and facilitates a chemical reaction with a coloured molecule so that the presence of the original protein can be seen under the microscope.

5.6.2 The Lundy Case

In August 2000, Christine Lundy and her 7-year-old daughter, Amber, were murdered in their home in Palmerston North, New Zealand. Mark Lundy, Christine's husband and Amber's dad, was accused of the murders. Two small areas of questioned material were found on Mark Lundy's polo shirt. These were examined using standard techniques used in forensic cases to enhance the cellular structures. As these tests suggested the material from the shirt was CNS tissue (either brain or spinal cord), the question at the first trial was whether the tissue was from Christine or Amber. However, a crucial step was missed: was the tissue human?

The IHC methodology in question was developed using known human tissue samples. Clinical laboratories tasked with analysing human tissue samples were therefore using processes designed to enhance and identify tissues from known human sources. However, the samples from the Lundy shirt did not have a confirmed origin; it was *assumed* they were human. Once the IHC testing was explored further,

Table 5.2 Brain protein antibodies and functions of proteins that the antibodies are directed against. Reproduced from ref. 13 with permission from Wolters Kluwer Health, Inc., Copyright 2002 Lippincott Williams.

Antibody against	Function of protein	Brain[a] specific?
Glial fibrillary acidic protein (GFAP)	Important in cytoskeleton of astrocytes (central nervous system – CNS cells).	No. Found in brain, but also in kidney, bone, testis, pancreas and liver cells.
Synaptophysin	Function unknown.	Yes. Present in all nerve cells (neurones) in the brain and spinal cord.
Neurofilament	Structural – they 'seem' to support the shape of axons. Structural changes might lead to diseases like Alzheimer's.	Yes. Found only in neurons.
S-100 protein	Uncertain, but linked to biochemical control mechanisms in cells (*e.g.*, calcium homeostasis).	No. Derived from embryological neural crest. Present brain and other cells (*e.g.*, macrophages).
Cytokeratin-LMW[b] Cytokeratin 5/6[c]	Important in the cytoskeleton.	No. Expressed in epithelial cells, including cells in the CNS.
CD45	Protein tyrosine phosphatase, receptor type C. Regulates cell growth and differentiation (*e.g.*, blood cells)	No.

[a] Brain and/or other places in the CNS.
[b] A mixture of low molecular weight (LMW) cytokeratins.
[c] Two high molecular weight (HMW) cytokeratins.

it became apparent that the IHC testing was not human-specific but, rather, had been developed for use on known human tissues.

The appeal process determined that IHC as a method is appropriate for use on human tissues and that CNS tissue was present on the shirt. However, it could not be used to assess that the CNS material was human. Specialist mRNA testing was then developed by a laboratory instructed by the prosecution for the retrial to try to answer this question. Unfortunately, the testing was not undertaken in a sufficiently rigorous manner and although the test results were presented by the prosecution during the retrial, a subsequent appeal determined that that testing should not have been put before the jury because it was not scientifically valid, there was no independent verification of the testing regimen and the results were therefore unreliable.

We are left in the position where the IHC results indicate CNS (either brain or spinal cord) but it is not possible to say that the material was human. The topic is still being investigated.

5.7 Blood

Blood has 'always' been important in crime investigations. However, some television depictions of forensic science definitely overplay the value of blood – convicting a person of murder based on one minute blood spot is typically not possible! In the past, blood groups were often important facets of the portfolio of forensic science. If a murder victim had a rare blood group (*e.g.*, AB: approx. 3% of the UK population is group AB) and an AB blood stain was found on the shirt of the suspect, this was not definitive, but added to the information linking the suspect to the victim. If, on the other hand, the victim had a common blood group (*e.g.*, O: approx. 37% of the UK population is group O) and a group O stain was found on the suspect, this would be of little evidential value because it is very common.

Nowadays, blood groups are rarely used in forensic casework because DNA is present in, and can easily be extracted from, blood and DNA profiling can be highly discriminatory (see Chapter 6 *DNA*). Blood group analysis is, however, still evident in historical cases. For this reason, I will take a brief foray into blood grouping for completeness.

5.7.1 Blood Groups

There are several blood group systems (*e.g.*, ABO and MN). I will use the ABO system as an example here; however, the methods used to determine other groups (*e.g.*, MN) are similar.

There are two antigens expressed by red blood cells (RBCs), A and B. These antigens are proteins that stick out of the surface of the RBC cell membrane. Humans can express, A (group A) or B (group B), both A and B (group AB) or neither (group O). Injecting group A blood into a group B person would prompt the production of antibodies (proteins tailored to bind only to another specific protein) against the A antigen, which would lead to a severe life-threatening immune response (this is why before blood transfusion the donor's and recipient's blood groups are matched). When the antibody against the A antigen (termed anti-A) binds to RBCs expressing the A antigen, the antibody joins the RBCs together *via* their A antigens causing the cells to clump (agglutination). The same applies to group B blood injected into a group A person or group AB blood injected into either a group A or group B person. Group O blood has no antigens and so can be injected into a group A, B or AB person with no ill effects (*i.e.*, group O is the universal donor). This antibody response followed by agglutination is the basis of blood grouping.

In a forensic context, a blood stain may be lifted from fabric or any other surface using saline (0.9% w/v NaCl (aq.) – this prevents the breaking of blood cells), an aliquot of the extract is mixed with anti-A antibodies and another aliquot is mixed with anti-B antibodies (these antibodies are prepared by injecting rabbits with human RBCs of a known group – the serum from the rabbits is termed rabbit anti-human blood group A, for example). The reaction between the anti-serum and the saline blood stain extract is observed. If agglutination (Figure 5.24) occurs, it indicates that the blood stain extract contains the antigen corresponding to the antiserum used; for example, if agglutination occurs with the anti-B serum, the RBCs have the B antigen. The blood group of the sample is determined from the coagulation pattern with the different antisera (Table 5.3).

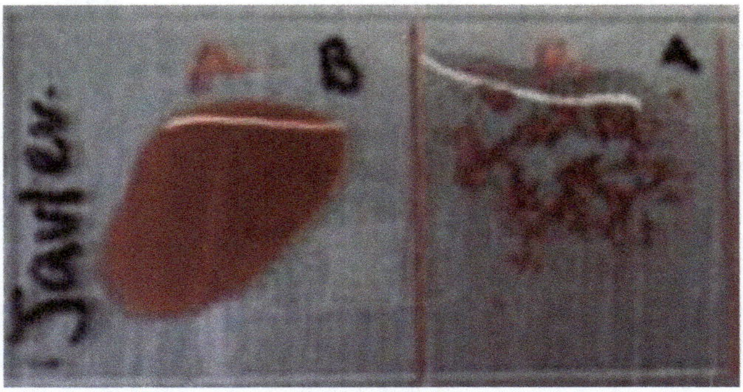

Figure 5.24 An example of agglutination. Left: blood sample mixed with anti-B – no agglutination. Right: blood sample mixed with anti A – agglutination. The RBCs in the sample therefore express A antigen, the blood sample is therefore group A. Figure is in the Public Domain.

Table 5.3 Agglutination (X) patterns for different blood groups.

Sample group ↓	Anti-serum →	A	B
A		X	
B			X
AB		X	X
O			

5.7.2 Determining the Species of Origin of a Blood Stain – The Precipitin Test

Determining the species of origin of a blood stain might also be important. Consider a situation where blood stains are found on the clothes of a murder suspect and the suspect works in an abattoir. In this case, determining the species of origin of the bloodstains becomes of paramount importance.

Often whether the blood stain is human or not is the important question. Anti-human serum is one method that can be used (this is produced in the same way as described in Section 5.7.1 *Blood Groups* human blood is injected into an animal – often a rabbit – to generate anti-human antibodies). When the anti-human serum is mixed with a saline extract of the blood stain, a precipitate appears (hence the test's name – precipitin test) if the blood is human. There are other precipitin tests for different species; for example, a rabbit anti-sheep serum would be used to check whether the blood stain originated from a sheep.

The precipitin test (sometimes called the ring test because the interface precipitate appears as a ring) is carried out by adding a small volume of an appropriate antiserum (*e.g.*, rabbit anti-human for determining whether the blood is human or not) to a test tube, and then carefully layering the blood extract on top (to give two distinct layers). The tube is incubated for about 4 h; a white precipitate appears at the interface of the two liquids if the test is positive (Figure 5.25).

5.8 Forensic Microbiology

Forensic microbiology covers a broad array of investigations from the application of the sequence of bacteria colonising a corpse as a means of estimating time since death to the identification of specific bacteria or their spores used by criminals; for example, anthrax spores as agents of terrorism.

5.8.1 Anthrax and Terrorism

Bacillus anthracis is a very dangerous bacterium. If inhaled, it can cause severe lung infection; if it occurs on the skin, it can cause severe dermal infections, and if it is ingested, it can cause severe gastrointestinal infections. All of these infections can be fatal. The clinical manifestation of infection with *B. anthracis* is called anthrax.

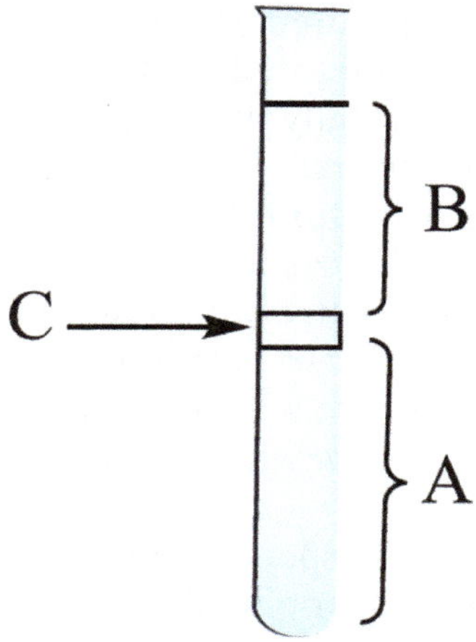

Figure 5.25 The precipitin or ring test: the appropriate antiserum (A) is put into a test tube and the test blood sample extract (B) carefully layered on top. The tube is incubated for about 4 h, the test is positive if a ring of off-white precipitate (C) appears at the interface between the two liquids. To determine whether a blood sample is human, rabbit anti-human serum (A) would be incubated with the test blood sample extract (B); an interface precipitate (C) shows that the blood sample is human.

To make things worse, under adverse conditions (*e.g.*, extreme dryness), *B. anthracis* can form spores which are able to survive environmental adversity and germinate later, when the conditions are better, liberating viable bacteria. Anthrax spores are incredibly resilient.

Anthrax spores have been used by criminals (including terrorists) as threats to individuals or institutions. Clearly, if such a threat occurs it is very important that the spores or the bacteria are identified quickly so that the response to the threat can be managed safely and effectively.

I will use an example – the 2001 USA anthrax letter attack – to illustrate how an anthrax threat is dealt with by forensic microbiologists.

5.8.1.1 2001 USA Anthrax Letter Attack

Letters containing *B. anthracis* spores were mailed from New Jersey to people in the news media and to politicians beginning on 11th September 2011 (the anniversary of the terrorist attack on the New York Twin Towers – 9/11). Twenty-two of the recipients developed anthrax and 5 of them died – this was a serious event; it later came to be known as Amerithrax. The incident triggered what the Federal Bureau of Investigation (FBI) described as '*one of the largest and most complex in the history of law enforcement*'.

The finger of blame was pointed at a bioweapon expert, Dr Steven Hatfill, who was later exonerated; a government microbiologist, Dr Bruce Ivins (who was initially involved in investigating the case), and even Al-Qaeda and Iraq were blamed for the attacks. I will leave the ins and outs of this whodunnit to the experts and focus on the forensic microbiology of the case.

Initially, the spores were likely to have been identified microscopically (this, however, has not been reported) to distinguish them from other materials (*e.g.*, milk powder and yeast extract) or less harmful *Bacillus* sp. (*e.g.*, *B. thuringiensis*) spores that might have been used as a hoax.

Having determined that the spores were from *B. anthracis*, it became necessary to determine their origin. There was a suggestion that the spores might have originated from an American research institute (hence the two scientist suspects). To investigate this, DNA was isolated from the spores and its genome was sequenced (see Chapter 6 *DNA*). There were four *B. anthracis* mutants (*i.e.*, four different genome sequences) found in the samples from the letters and they were indistinguishable from the mutants in the spore stock (RMR1029) kept by the US Army Medical Research Institute for Infectious Diseases – this is where Dr Ivins worked. It was suggested that the 'letter spores' had originated from Ivins' lab which implicated him as the perpetrator. Dr Ivins committed suicide (paracetamol overdose) when he learned that the FBI would issue charges against him.

The FBI concluded that '*evidence developed from [the] investigation established that Dr. Ivins, alone, mailed the anthrax letters.*'

5.8.2 Hairs and Fibres

Hairs and fibres may be included in the remit of the forensic biologist. Hairs are, of course biological, but many fibres are synthetic (*e.g.*, nylon) and so are most certainly not 'biological'. However, fibre investigations utilise microscopy as a key part of the fibre identification and comparison process. Microscopy serves as the fundamental technique for forensic biologists and so fibres fit naturally within their remit (this varies around the world; for example, sometimes scientists specializing in physical evidence also examine fibres, but the methods used are the same).

A hair found on the clothing of a suspect might have come from the victim. Comparison microscopy is used to compare the hair with an authentic sample taken from the victim (see Section 5.2.3 *Comparison Microscopy* and Figure 5.26).

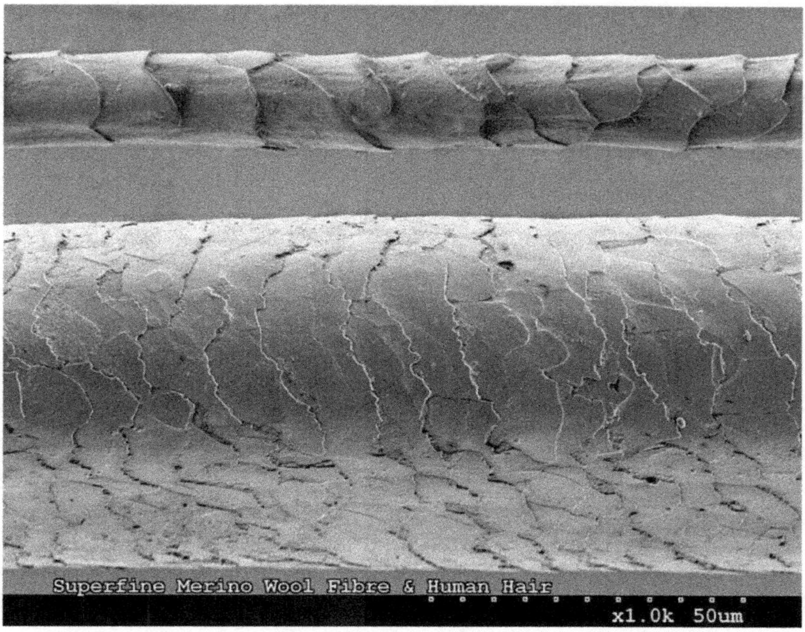

Figure 5.26 Scanning electron microscopy (×1000) of sheep wool (top) and human hair (bottom) showing the morphological differences that can be used to distinguish them. Reproduced from ref. 14, https://commons.wikimedia.org/wiki/File:CSIRO_ScienceImage_8115_Human_hair_and_Merino_wool_fibre.jpg, under the terms of the CC BY 3.0 license https://creativecommons.org/licenses/by/3.0/deed.en.

Of course, not all hairs are human. Dog hairs might link a suspect with a crime scene or wool fibres (hairs from sheep) might be useful circumstantial information to link a suspect with a victim who was wearing a woollen jumper. Hairs from different species have microscopical characteristics that help in their identification (Figure 5.26).

Fibre types can also be distinguished microscopically. Some fibres are highly characteristic and easily distinguished from other fibres (Figure 5.27).

Fibres, including hairs, might be dyed. Dyes can be extracted from the fibres and compared using chromatography. Such dye comparisons are a crucial aspect of investigating the origin of and comparing fibres (see Chapter 4 *Forensic Chemistry*).

Fibres are important in contact transfer (Locard's Exchange Principle; see Chapter 2 *The History of Forensic Science*) and are often part of the detailed investigation of clothing in forensic casework. Recovering (lifting) fibres from, for example, clothing in a murder case

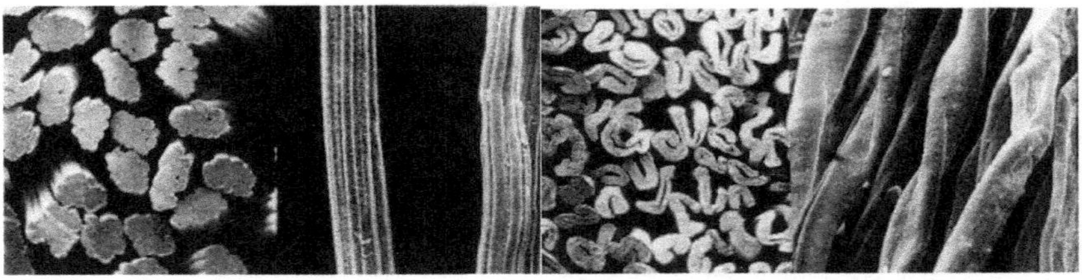

Figure 5.27 Electron micrographs (approx. ×1000) of cross-sectional and longitudinal views of viscose (left) and cotton (right). These two fibres are easily distinguishable because of the clear differences in their surface morphology. Reproduced from ref. 16, https://doi.org/10.1155/2016/4614168, under the terms of the CC BY 4.0 license https://creativecommons.org/licenses/by/4.0/.

might provide important information about fibre transfer from the victim to the suspect. The examination and recovery of fibres from clothing in the laboratory is both methodical and very thorough.

5.8.2.1 Recovering Fibres from Clothing

Transparent sticky tape housed in a specially designed device is used for fibre recovery (this is termed ‘taping’). The tape is pressed onto the article of clothing methodically to cover the entire garment area of interest – the tape is changed several times during the process to make sure that it does not get ‘saturated’ with fibres. The tape is then adhered to an acetate transparency (to prevent any atmospheric fibres settling on the tape) and viewed under a low power microscope (see Section 5.2.2 *Light Microscopy*) so that fibres, hairs and anything else attached can be seen (Figure 5.28). Using fine forceps, fibres, *etc.* are removed and mounted on a microscope slide for more detailed observation under a higher-powered light microscope or, perhaps, an electron microscope. There might be particular fibres of interest; for example, a murder victim might be wearing a bright red woollen jumper which would mean that bright red fibres recovered from the suspect's clothing would warrant further scrutiny.

It is important to remember that just because the victim was wearing a bright red woollen jumper, a bright red fibre recovered from the suspect's clothing may hold little value in an evidential context; however, if the fibre recovered turns out to be wool, it becomes more ‘interesting’, but it is still far from constituting strong evidence of contact between the victim and accused. Furthermore, if the dye from the victim's jumper and the dye from the recovered fibre are identical chromatographically, this is ‘more interesting’. And if hundreds of red

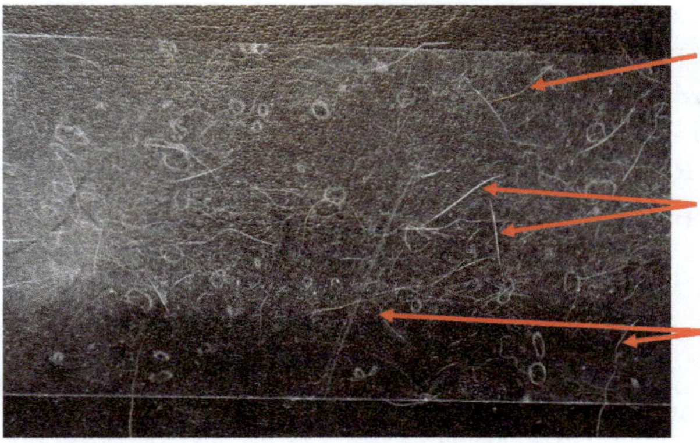

Figure 5.28 A tape from the author's jumper showing a multitude of fibres and hairs. The author has three dogs: a brown Staffordshire crossbreed (Sybil), a black, white and tan fox terrier (Elsie) and a long wavy-haired brown chihuahua (Pico). The evidence of contact with these canine friends abounds!

woollen fibres with chromatographically identical dye are recovered, the likelihood of contact or transfer between the victim and the suspect increases considerably. However, if the victim's bright red woollen jumper was available from a major chain store and the store had sold tens of thousands of the same jumper over the past year or so, the evidential value of the red wool fibres would drop considerably and if the victim had had legitimate contact with the suspect's clothing, then the evidential value would potentially be zero.

Further Reading

1. H. W. Bailey, L. T. Gallagher, A. Moncreiff and C. J. Wood, Calcareous micropalaeontology in forensic investigations, with particular reference to the so-called 'Soham murder case', in *The Archaeological and Forensic Application of Microfossils*, ed. M. Williams, T. Hill, I. Boomer and I. P. Wilkinson, The Geological Society, Bath, UK, 2017, pp. 278–289, ISBN 978-1-78620-305-2.
2. G. Ferri, M. Alù, B. Corradini and G. Beduschi, Forensic botany: species identification of botanical trace evidence using a multigene barcoding approach, *International Journal of Legal Medicine*, 2009, **123**, 395–401.
3. J. H. Bock and D. O. Norris, Cases Using Evidence from Plant Anatomy, *Forensic Plant Science*, Elsevier, Amsterdam, 2016, ch. 5, pp. 85–94.

4. R. T. Miller, R. Grantham, B. Lockett, C. Temple-Camp and J. Pang, Immunohistochemical analysis of forensic evidence from a double homicide, 2002, *American Journal of Forensic Medicine and Pathology*, **23**, 321–325.
5. J. Robertson, C. Roux, K. C. Wiggins, ed., *Forensic Examination of Fibres*, CRC Press, 3rd edn, 2018, ISBN 978-1-4398-2864-9.

References

1. Hair comparison – Forensic Testing Services, 2023. Available from: https://forensic-testing.net/proficiency-tests/hair-comparison/.
2. Electron microscope [Wiki], 2023. Available from: https://en.wikipedia.org/wiki/Electron_microscope#/media/File:Electron_Microscope.jpg.
3. Diatom Surirella spiralis [Wiki], 2023. Available from: https://en.wikipedia.org/wiki/Diatom#/media/File:Surirella_spiralis_-_SEM_MUSE_-_sharpened.jpeg.
4. Diatoms Thalassiosira sp on a membrane filter, pore size 04 μm [Wiki], 2023. Available from: https://en.wikipedia.org/wiki/Diatom#/media/File:Diatoms-HCMR.jpg.
5. Diatom Achanthes trinodis [Wiki], 2023. Available from: https://en.wikipedia.org/wiki/Diatom#/media/File:Achanthes_trinodis_-_SEM_MUSE.tiff.
6. Zhao, *et al.*, *J. Forensic Leg. Med.*, 2016, **42**, 75–78.
7. PLoS Biology Issue Image, *PLoS Biol.*, **9**(6), ev09.i06.
8. File: Pinus taeda crossx7358.JPG [Wiki], 2023. Available from: https://commons.wikimedia.org/wiki/File:Pinus_taeda_crossx7358.JPG.
9. File: Arabis voch1-4jpg [Wiki], 2023. Available from: https://en.wikipedia.org/wiki/File:Arabis_voch1-4.jpg.
10. File: Sperm stainedJPG [Wiki], 2023. Available from: https://commons.wikimedia.org/wiki/File:Sperm_stained.JPG.
11. File: Luminol2006jpg [Wiki], 2023. Available from: https://commons.wikimedia.org/w/index.php?curid=1254510.
12. File: Kidney cd10 ihcjpg [Wiki], 2023. Available from: https://commons.wikimedia.org/wiki/File:Kidney_cd10_ihc.jpg.
13. Miller, *et al.*, *Am. J. Forensic Med. Pathol.*, 2002, **23**, 321–325.
14. File: CSIRO ScienceImage 8115 Human hair and Merino wool fibrejpg [Wiki]. 2023, Available from: https://commons.wikimedia.org/wiki/File:CSIRO_ScienceImage_8115_Human_hair_and_Merino_wool_fibre.jpg.
15. Visible Proofs: Forensic Views of the Body: Galleries: Technologies: Life cycle of the black blow fly. Available from: https://www.nlm.nih.gov/exhibition/visibleproofs/galleries/technologies/blowfly.html.
16. A. Kumari and K. Khurana, Regenerated cellulose-based denim fabric for tropical regions: An analytical study on making denim comfortable, *J. Text.*, 2016, **2016**, 4614168.

6 DNA

The application of DNA technology has revolutionised forensic science. DNA profiling can now provide significant scientific support that an individual has been in a particular place or been in contact with a particular object. The use of DNA technology in forensic investigations has made some previously important methodologies nearly redundant. Thanks to DNA, gone are the days of the forensic scientist peering down his or her microscope looking only for hairs or fibres that might link the suspect to the victim in a murder case. Nowadays, the 'root' of a single hair can yield enough DNA to generate a profile. A drop of blood on the floor of a murder scene is no longer painstakingly lifted and blood grouped to see if it might be the same group as the injured suspect. Instead, a few blood cells from the dried drop of blood can provide enough DNA for a full DNA profile. This revolution has inordinately and permanently transformed forensic science.

6.1 The Deoxyribonucleic Acid (DNA) Revolution

When Dr Francis Crick (1916–2004) and his PhD student, James Watson (b. 1928), rather modestly began their 1953 *Nature* paper (see *Further Reading 2*)

> *We wish to suggest a structure for the salt of deoxyribose nucleic acid (D.N.A). This structure has novel features, which are of considerable biological interest.*

Forensic Science: The Science Behind the Truth
By Ian Shaw with Anna Sandiford

Published by the Royal Society of Chemistry, www.rsc.org

they realised the potential importance of their findings to our understanding of biological sciences, particularly genetics and inheritance, hinting at this in the paper's final sentence:

> *It has not escaped our notice that the specific pairing [of DNA bases] we have postulated immediately suggests a possible copying mechanism for the genetic material.*

Elucidating the structure of DNA revolutionised our understanding of cells, genetics and cancer. It paved the way for molecular biology, sparked debates on genetic manipulations, and addressed the treatment of hitherto untreatable diseases. Its impact cannot be exaggerated. James Watson, Francis Crick and Maurice Wilkins (1916–2004) were awarded the Nobel Prize in Physiology or Medicine in 1962 for '*their discoveries concerning the molecular structure of nucleic acids and its significance for information transfer in living material.*'

Thirty-one years later (1984), Dr Alec Jeffreys (b. 1950; now Professor Sir Alec Jeffreys) working at Leicester University, UK, had a brilliant idea that the person-uniqueness of DNA could be applied to criminal investigations: genetic fingerprinting (DNA profiling) was born and the biggest revolution in criminalistics since Locard's Exchange Principle began. Jeffreys and his co-workers, Dr Peter Gill and Dr David Werrett both from the UK Forensic Science Service wrote in their *Nature* paper (see *Further Reading 3*):

> *It is envisaged that DNA fingerprinting will revolutionise forensic biology particularly with regard to the identification of rape suspects.*

They were rather less modest than Watson and Crick had been three decades earlier, but nevertheless, they were right!

6.2 Structure and Function of DNA

In order to understand the application of DNA in forensic investigations, a basic understanding of DNA biochemistry is necessary.

DNA is a nucleic acid (the other is ribonucleic acid, but more of this later...) present in the nucleus of all nucleated cells from all living things. It carries the genetic code, which determines the morphology and biochemistry of cells, as well as the physiology, characteristics and traits of the organism.

DNA comprises two strands made up of a phosphorylated sugar (deoxyribose) backbone, to which four bases (Figures 6.1 and 6.2) (guanine, cytosine, adenine and thymine) are attached in a specific sequence, all wound up into a double helix (Figure 6.3). The DNA polymer is basically a sugar-phosphate-sugar-phosphate backbone with a base sticking off every sugar residue (Figure 6.1). This is termed single-stranded DNA (ssDNA). Double-stranded DNA (dsDNA) comprises two ssDNA polymers joined together by hydrogen bonds (weak bonds that can be broken relatively easily) between their bases. The base–base hydrogen bonds are specific: only guanine can bond to cytosine (3 hydrogen bonds) and adenine can bond to thymine (2 hydrogen bonds; Figure 6.2).

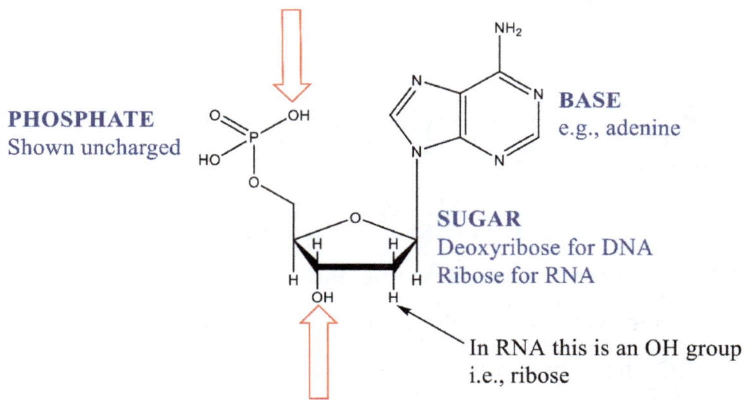

Figure 6.1 The structure of a single unit (nucleotide) that makes up DNA. Many thousands of these units are joined together, forming a DNA molecule. The red arrows show the positions of the links between nucleotides to form the DNA polymer.

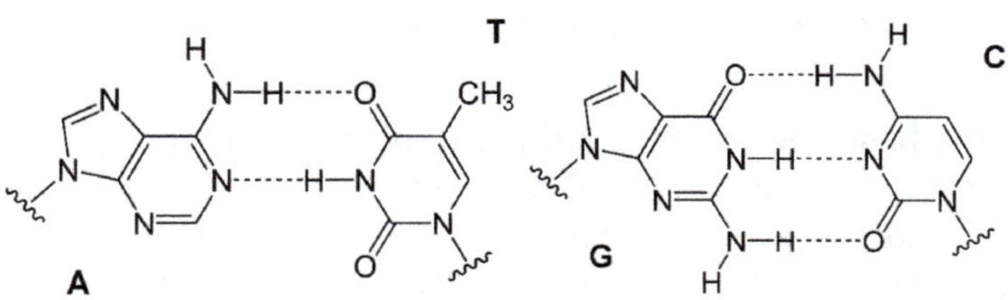

Figure 6.2 The specific base to base hydrogen bonding (-----) between DNA bases: adenine (A) and thymine (T) (left) and guanine (G) to cytosine (C) (right). This is the bonding that joins two ssDNA polymers to form dsDNA. The base sequence is the genetic code. Figure is in the Public Domain.

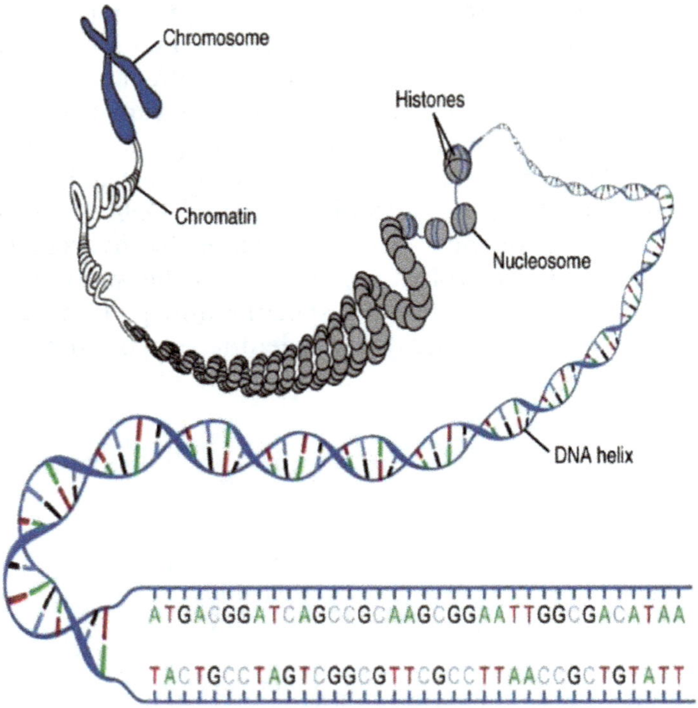

Figure 6.3 Structure of a chromosome and its DNA: DNA is a double helix comprising a phosphorylated sugar (deoxyribose) backbone (blue) with one of the four bases attached to the deoxyribose backbone-thymine (T, red), adenine (A, green), cytosine (C, grey) and guanine (G, black). The bases bind specifically, A to T and G to C, joining the two halves of the double helix. Three bases comprise a codon (3 bases), which codes for a specific amino acid, thus determining the amino acid sequence of proteins. The DNA double helix folds around histones (a type of protein), forming nucleosomes (grey) like a string of beads. The string of nucleosomes coils up to form chromatin, which further coils and folds to form a chromosome. There is an approx. 2.5 cm long strand of DNA per chromosome comprising about 140 million base pairs. This is a huge amount of genetic information! Created with Bing Image.

This DNA base sequence is unique to an individual organism (this is the genome). The sequence of bases codes for amino acids (3 bases make up a codon which codes for an amino acid), which are the building blocks of proteins. The DNA base sequence therefore specifies the amino acid sequence in a particular protein. Proteins play a crucial role in determining the cellular function (*e.g.*, enzymes) and morphology (*e.g.*, structural proteins). The base sequence is the genetic code.

The supercoiled protein-associated DNA double helix forms chromosomes, which pass the genetic material onto the offspring *via* eggs and sperm. The uniqueness of the DNA sequence determines the uniqueness of individuals. Even identical twins do not have identical DNA – this will be discussed in more detail below.

While this is not a treatise on nucleic acids (many have been written; see *Further Reading*), it is important at least to be aware of the fundamentals of nucleic acid biochemistry to understand the rationale for, and application of, DNA technology to forensic casework. Perhaps the most important aspect of DNA biochemistry to appreciate is how the genetic code – carried on DNA – is translated into the protein structure. This involves another nucleic acid, ribonucleic acid (RNA), which is also important in forensic science.

6.2.1 Protein Biosynthesis – Translating the Genetic Code into the Protein Structure

The translation of the genetic code into a protein sequence involves the production (transcription) of an RNA template from the DNA code – this is messenger (m)RNA. The mRNA serves as the vector for the DNA code; the mRNA is 'read' by the cell's protein biosynthesis apparatus (ribosomes), faithfully converting each codon into an amino acid and joining the amino acids together (with peptide bonds) to form a protein (Figure 6.4).

RNA (including mRNA) differs from DNA in two ways: DNA's sugar backbone is made up of deoxyribose units, whereas RNA's sugar backbone is made up ribose; and the DNA base, thymine, is replaced with uracil in RNA.

Understanding the role of nucleic acids in forensic science needs a little more background biochemistry to fully appreciate its powerful applications; particularly the role, origin and cellular sequestration of mitochondrial DNA (mtDNA), which can be used to determine maternal relationships.

6.2.2 Mitochondria and Mitochondrial DNA (mtDNA)

Mitochondria are cell organelles (*i.e.*, sub-cellular bodies; Figure 6.5) that have a specific and very important biochemical role. They are often termed 'the powerhouse of the cell' because they contain many of the enzymes necessary for energy metabolism – in simple terms, they liberate energy from glucose 'fuel' to drive cellular functions. Mitochondria have a complex invaginated membrane structure

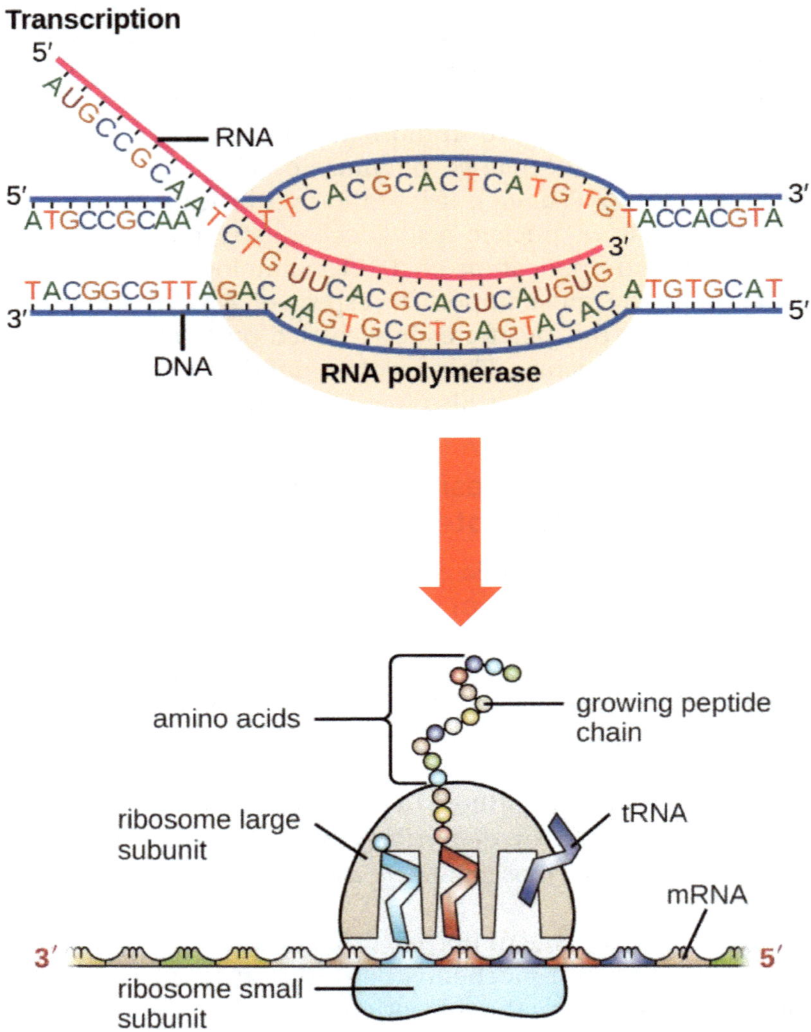

Figure 6.4 Protein biosynthesis: Top: DNA in the nucleus unwinds and RNA polymerase translates a section (containing genes) of the unwound DNA into mRNA. The base sequence of the mRNA reflects the DNA base sequence, except thymine (T) is replaced by uracil (U). Bottom: The mRNA leaves the nucleus *via* nuclear pores and binds to ribosomes (tiny organelles – consisting of large and small protein subunits – in the cytoplasm). The ribosome reads the mRNA base sequence and translates each codon into a specific amino acid delivered on a short-chain specialised RNA – transfer (t)RNA. The amino acids are joined together by peptide bonds, forming a peptide chain (protein). Reproduced from ref. 1, https://openstax.org/books/microbiology/pages/14-3-mechanisms-of-antibacterial-drugs, under the terms of the CC BY 4.0 license https://creativecommons.org/licenses/by/4.0/.

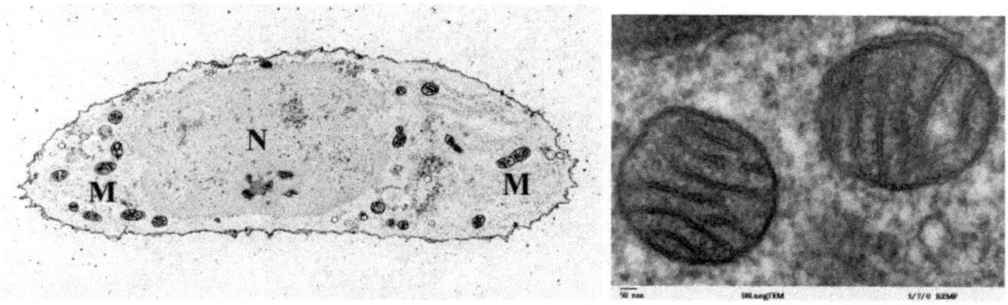

Figure 6.5 Left: Low-power (approx. ×75 000) electron micrograph (TEM) of a cell (chondrocyte) showing mitochondria (M) and the nucleus (N). Reproduced from ref. 2, https://commons.wikipedia.org/wiki/File:Chondrocyte-_calcium_stain.jpg, under the terms of the CC BY-SA 3.0 license https://creativecommons.org/licenses/by-sa/3.0/deed.en. Right: High-power (approx. 1 000 000×) electron micrograph (TEM) of two mitochondria (in a human lung cell) showing their detailed structure. Figure is in the Public Domain.

(cristae; you can see these in Figure 6.5 right), providing a huge surface area to accommodate the enzymes necessary for the mitochondria's complex biochemistry.

So, why is there a focus on mitochondria? Why are they relevant to forensic science? To clarify their role in forensic science, it is essential to provide a brief explanation of the evolutionary origins of mitochondria.

Mitochondria originated as bacteria engulfed by cells many millions of years ago. The engulfed bacteria provided a 'service' to the cell and the cell protected the bacteria (endosymbiosis). Over the millennia, the engulfed bacteria evolved into mitochondria. The bacteria brought their own circular strands of DNA to the cells, and to this day mtDNA remains circular and separate from the cell's nuclear DNA (Figure 6.6). Furthermore, mitochondria are inherited along the female line (*i.e.*, your mitochondria originated from your mother) – the reason for this is fascinating: while mitochondria are present in both ova and sperms, the sperm mitochondria are only in the sperm flagellum (tail) and when the sperm fertilises the ovum, it sheds its tail along with the male-derived mitochondria (Figure 6.7). The mitochondria in the ovum (from the mother) remain as part of the zygote (fertilized ovum) which develops into the embryo, foetus and offspring; therefore, the mitochondria in adults are derived only from the few mother-derived mitochondria present in the ovum at the very outset of their existence. While this mother-derived mtDNA dogma is broadly accepted, it has been shown that mtDNA can be derived from the father in some circumstances (see *Further Reading* 6). The scientists who discovered paternal transmission of mtDNA accept that it is

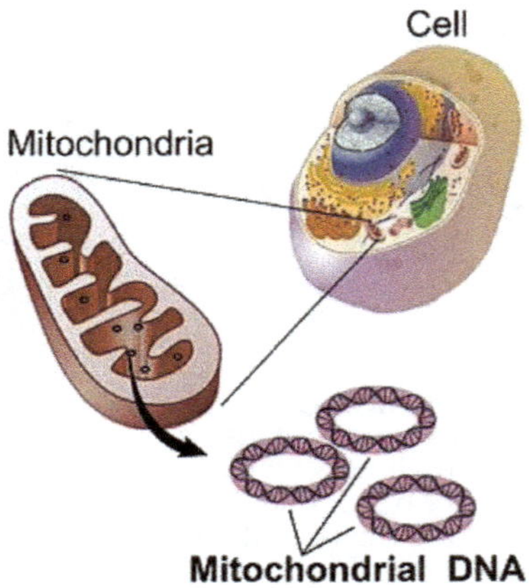

Figure 6.6 Mitochondrial DNA (mtDNA) is a circular strand and is separate from nuclear DNA because of the evolutionary origins of mitochondria from bacteria. Figure is in the Public Domain.

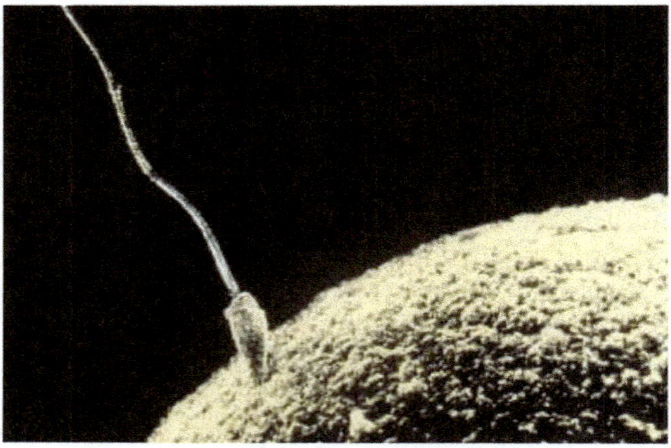

Figure 6.7 The point of fertilisation: the sperm head (contains no mitochondria) penetrates the ovum (contains mitochondria from the mum) then the sperm tail (containing the mitochondria from the dad) is lost; therefore, the zygote and thus the offspring only have mum-derived mitochondria. Figure is in the Public Domain.

an 'unusual mode of transmission'. But this possibility clearly raises some doubt regarding the use of an mtDNA sequence to identify someone's mother – more on this will be discussed later...

There is just one more aspect of DNA that we need to understand before we explore how it is used in a forensic science laboratory: there are two types of DNA in the nucleus, exons and introns. They are all part of the same double helix strand, but form different, varying length regions with markedly different functions. Introns are used in forensic science because they are believed to be unique to an individual.

6.2.3 Exons and Introns

There is far more DNA in the nucleus than is needed to produce all of the proteins required for cell function. This means that a proportion of the DNA must be redundant (*i.e.*, noncoding). This is sometimes called junk DNA or more correctly intronic DNA (the idea that this DNA is 'junk' is very controversial; it might be important in gene regulation, but that is controversial too – I will not cloud this simple discussion with the complexities of this scientific debate). The coding region of DNA is the exon.

When mRNA is translated from the DNA template, the intron sequences are removed before the mature exon-derived mRNA migrates from the nucleus to the ribosomes for translation into proteins.

The function of introns is not fully understood and, as mentioned above, is the subject of significant debate, but they likely play a key role in the regulation of gene expression and, importantly, in DNA forensic applications, they might protect exons from damage (*e.g.*, mutations) by being positioned in such a way that they bear the brunt of any mutational pressure. This is important in an evolutionary context because it means that genes are not altered with consequent harmful effects on the cell (mutations can be positive; indeed, they are the basis of Darwinian evolution, but this a subject for another, very large, book!). Since introns are non-coding, there is no selective pressure to fix mutations. In a forensic context, the myriad mutations (*i.e.*, changes in the base sequence) that occur in introns make the intron sequence unique to an individual. Mutations can occur simply due to errors in translation or errors in DNA repair (DNA is the only biological molecule that is constantly surveilled and any defects repaired) or they can be caused by environmental exposures (to radiation or chemicals). Changes in the base sequence, therefore, reflect an individual's environment. This is why identical twins have different intron sequences – it is extremely unlikely for them to have lived in identical environments and to have been exposed to exactly the same mutagens.

It is for these reasons that intronic DNA is important in forensic investigations – it is believed to be unique to an individual. Furthermore, intronic DNA has short sequence repeats (*e.g.*, guanine-adenosine-thymine-adenosine; GATA) on both DNA strands of the double helix – they are termed short tandem repeats (STRs) which form microsatellites (*i.e.*, collections of STRs). These repeats could be

Person 1
CGACGATAGATAGATAGATAGATAGATAGATACGACCGACGGACCGAC

Person 2
GGACGATAGATAGATAGATAGATAGATAGATAGATAGATACGACCGAC

Person 3
GGACGATAGATAGATAGATAGATACGACCGACCGACCGACCGACCGAC

Figure 6.8 STRs in single strand sections of DNA from 3 people. The length of the GATA STR locus (red; microsatellite) is different for each person: Person 1 has 7 GATA repeats in their microsatellite, person 2 has 9 repeats, and person 3 has 5 repeats. G = guanine; A = adenosine; T = thymine; and C = cytosine.

replicated in response to, for example, mutations and when replication occurs the wrong number of STRs might be produced in the replicated DNA, resulting in microsatellites with different sequence lengths. Therefore, the number of repeats and size of microsatellites vary considerably from person to person (Figure 6.8).

DNA can be isolated from forensic samples (*e.g.*, a semen stain) and amplified (to make enough DNA to analyse) using the polymerase chain reaction (PCR); the DNA can then be cut (using enzymes) into fragments of varying lengths determined by the base sequence and the fragments are separated by gel electrophoresis – the pattern produced is the DNA fingerprint.

6.2.4 DNA Fingerprinting

6.2.4.1 Precautions

Contamination of samples or DNA extracts with extraneous DNA must be fiercely guarded against. Forensic DNA labs operate under the strictest conditions to prevent contamination from any source (including the analytical scientist). This is incredibly important because the DNA fingerprinting process involves amplification (*i.e.*, producing millions of copies of a tiny amount of DNA extracted from the sample), so the slightest contamination would lead to a significant amount of contaminant DNA. A cough can spread DNA-containing cells from the mouth and throat, and a few cells might fall from the scientist's skin or, perhaps, transfer from their finger when they pick up a vial. For these reasons, DNA scientists wear gloves, masks, all-body suits with hoods and overshoes to minimise any transfer during the extraction and analysis process. Of course, it is of paramount importance that the 'case' sample(s) (*e.g.*, a hair found at the crime scene) and the reference sample(s) (*e.g.*, a buccal sample from the suspect) are not extracted at the same time so as to prevent cross-contamination.

6.2.4.2 Extracting the DNA

The tiniest amount (at least 25 ng) of DNA is needed. DNA extracted from just a few nuclei (*e.g.*, from cells from the mouth left on the rim of a glass) will suffice. For it to be useful, the extraction method must be efficient (you cannot afford to lose any DNA from the sample) and allow rapid throughput (many labs process hundreds of samples a day). Nowadays, kits are available to make the job easier, and in high throughput labs, fully automated technology is used. I will describe only the basics of DNA extraction here.

First, if the sample is on a substrate (*e.g.*, a semen stain on fabric), it is 'lifted' using saline; the saline then contains cells from the sample. Next, the cells in the sample must be shattered (lysed) to release their DNA. This is achieved through physical impact (*e.g.*, shaking the sample with tiny metal beads), osmotic shock (adding water to the cells, the cells take up the water, expand and explode) or the use of a protease (an enzyme that breaks down proteins) to cleave the proteins in the cell membrane thus disrupting and rupturing it to release the cell contents.

The cell lysate is centrifuged to form a pellet which includes the cell nuclei containing the DNA. The pellet is resuspended in water and treated with a mixture of phenol and chloroform. The phenol breaks down cell debris (including proteins) and the chloroform mixes with the phenol, ensuring that when the sample is centrifuged there is a clear separation of the phenol/chloroform and aqueous phases (phenol is more soluble in chloroform than water and chloroform is practically insoluble in water). The DNA is in the aqueous phase (top layer). The DNA is precipitated from the aqueous phase with ethanol and centrifuged to form a pellet (the DNA extraction process is summarised in Figure 6.9).

The DNA extracted from a few cells in this way will be an invisible smear at the bottom of the tube; but it *is* there!

A blank sample (*i.e.*, containing no biological material; *e.g.*, a piece of uncontaminated fabric) is extracted at the same time as the forensic sample to make sure that no contaminating DNA entered at any point during the extraction process.

6.2.4.3 Amplifying the DNA

The amount of DNA extracted is likely to be vanishingly small (tens of nanograms), far too little for further analysis. For this reason, the

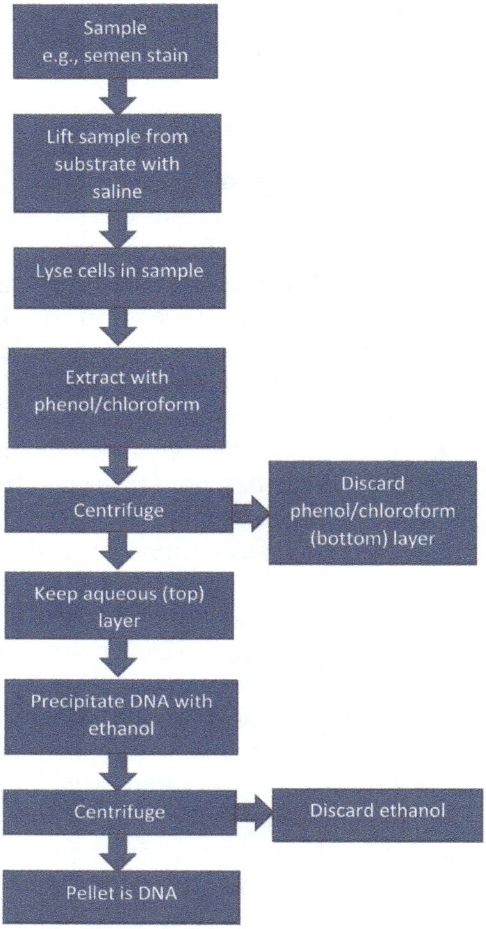

Figure 6.9 The process of DNA extraction from a forensic sample (*e.g.*, a semen stain).

DNA is amplified using an enzyme (DNA polymerase) in the presence of nucleotides (a DNA base attached to deoxyribose phosphate; *i.e.*, the basic DNA building block) – this is termed the polymerase chain reaction (PCR). PCR machines are used to amplify the extracted DNA automatically. Many millions (if not billions) of copies are made in a short time (minutes). At the end of the PCR process, there will be sufficient DNA for analysis.

The use of PCR is the reason why even the slightest DNA contamination is unacceptable. The tiniest amount of contaminating DNA amplified by PCR would become a major, misleading, component of the amplification process. If there is DNA in the amplified blank (see Section 6.2.4.2 *Extracting the DNA*), the results are void.

6.2.4.4 Cutting the DNA

The amplified DNA is chopped into fragments using specific enzymes (restriction endonucleases) which cut the DNA between specific bases. For example, the restriction endonuclease BamH1 cuts between two guanine residues (G–G) (Figure 6.10) and AluI cuts between a guanine and a cytosine residue (G–C). There are many restriction endonucleases available to cleave between specific base residues.

The restriction endonuclease fragments have a particular number of base pairs (bp) and thus a specific molecular weight (Figure 6.11).

Using a range of restriction endonucleases leads to base-sequence specific cuts at many places on the DNA molecule; this, in turn means that the range of fragment molecular weights reflects the original DNA sequence, and since the sequence is absolutely person-specific, the range of DNA fragments is also unique to an individual.

Gel electrophoresis is used to separate (based on molecular weight) and visualise (using a DNA stain) the DNA fragments.

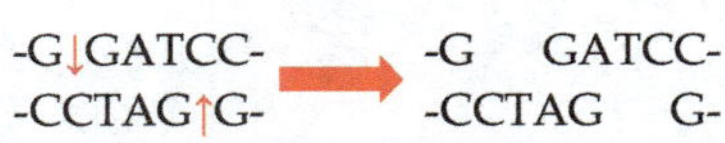

Figure 6.10 Restriction fragmentation of a section of double stranded DNA using the restriction endonuclease BamH1 which cleaves between two adjacent guanine residues (G–G).

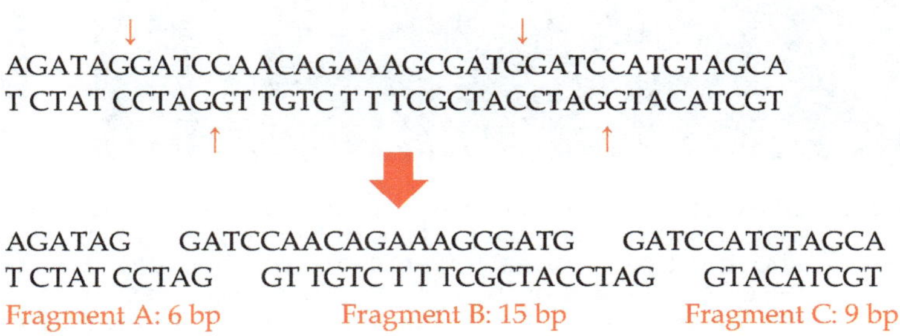

Figure 6.11 BamH1 cleaves (↓) a piece of double stranded DNA, leading to the production of three part double-stranded DNA fragments comprising different numbers of base pairs (bp) and therefore different molecular weights.

6.2.4.5 Gel Electrophoresis

Gel electrophoresis is akin to chromatography. The sample is applied to a gel (*e.g.*, agarose – a sugar polymer from seaweed), but instead of using a mobile phase to separate the sample components, an electric current (about 35 V) is applied. Positively-charged molecules move to the negative pole and negatively-charged molecules move to the positive pole. The speed of movement of the molecular components of a mixture through the gel depends on both their size (molecular weight), shape (smooth spheres would move fastest and rods would move more slowly) and their overall charge. Therefore, a highly charged small molecule will migrate faster than a large molecule with a small charge. DNA is negatively charged and so it always migrates to the positive pole – DNA's speed of migration is largely influenced by the size of the DNA molecule, and so DNA fragments in a mixture appear as bands separated according to their molecular weights (Figure 6.12).

The DNA fragment separation shown in Figure 6.12 is very simple and intended only to illustrate the principle. The equivalent separations for a forensic sample would exhibit many more DNA bands (*i.e.*, many more DNA fragments; Figure 6.13) which increases the person specificity of the results). The original DNA fingerprinting

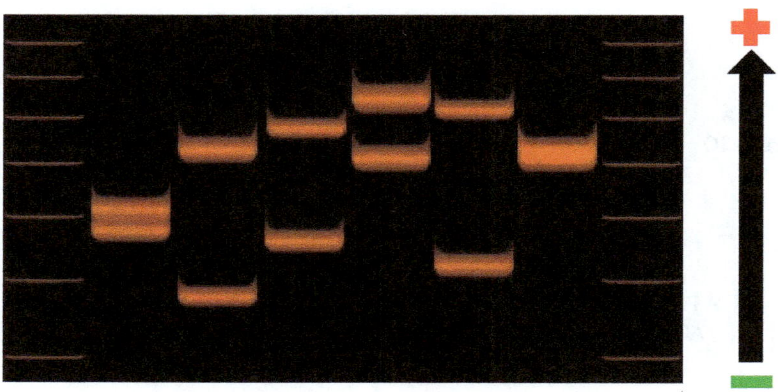

Figure 6.12 DNA separation by gel electrophoresis. The samples are applied in narrow bands to the bottom of the gel, the current is applied, and the DNA moves towards the positive (+). Separation is based on the size of the DNA fragment – smallest at the top and the largest at the bottom. Far left and right are molecular weight markers; the 6 lanes in the middle are STRs from different individuals (stained orange to show the DNA). Reproduced from ref. 3, https://commons.wikipedia.org/wiki/File:D1S80Demo.png, under the terms of the CC BY-SA 3.0 license https://creativecommons.org/licenses/by-sa/3.0/deed.en.

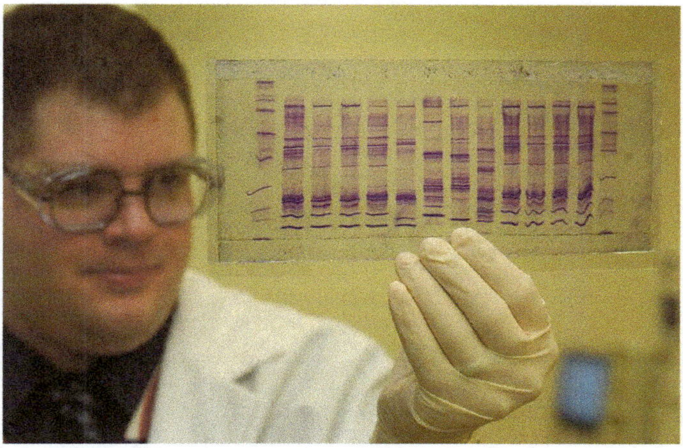

Figure 6.13 A scientist holding up an agar gel with 12 DNA separations (the far right and left lanes are DNA molecular weight markers) showing the complex pattern of DNA fragments (stained purple). The pattern is unique to an individual. Figure is in the Public Domain.

studies (see *Further Reading 3*) suggested that the probability of chance association (*i.e.*, DNA fingerprints from two different people appearing the same) is $\ll 5 \times 10^{-19}$ (*i.e.*, much less than 1 in 50 000 000 000 000 000 000); since the population of Earth is currently (2021) about 7 577 130 400 (*i.e.*, approx. 7.6×10^{9}), this means that finding two people with the same DNA fingerprint on earth would be as near impossible as it is possible to be! This will be discussed further in Chapter 13 *Probability in Forensic Science*.

DNA profiles are now often scanned and represented as peaks on a graph. The position of the peaks represents the position of the DNA fragment on the gel and the area under the peak reflects the intensity of the stained DNA fragment on the gel (*i.e.*, the amount of DNA present) (Figure 6.14).

6.3 DNA Stability

DNA is rapidly broken down by nucleases (DNases), but these DNA degrading enzymes can only function in aqueous environments and optimally in a narrow pH range. Nucleases are produced by most cells, including bacteria; this means that DNA under wet conditions, which might attract microbial growth, is likely not to last very long. However, under dry conditions, DNA can remain stable for decades or very much longer…

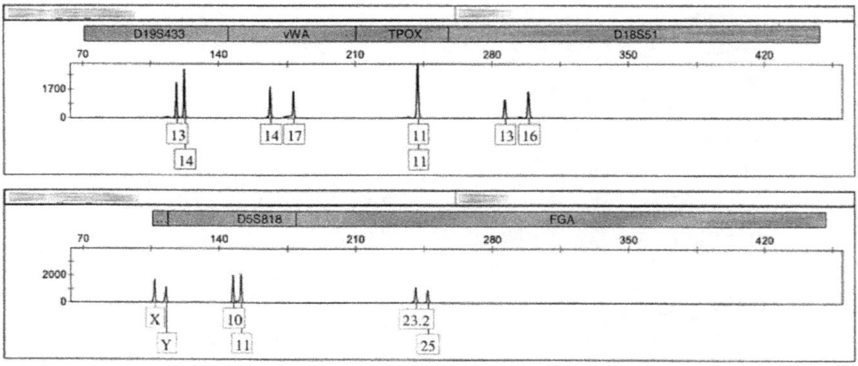

Figure 6.14 DNA profiles represented graphically. The position of the peaks represents the position of the DNA fragment on the gel (*i.e.*, its molecular weight) and the area under the peak reflects the intensity of the stained DNA fragment (*i.e.*, the amount of DNA). The codes at the top of each graph (*e.g.*, TPOX) are different STRs and the numbers below the peaks are the number of STR repeats in the fragment (*e.g.*, 11 repeats for TPOX). Reproduced from ref. 4, https://commons.wikimedia.org/wiki/File:Str_profile.jpg, under the terms of the CC BY-SA 3.0 license https://creativecommons.org/licenses/by-sa/3.0/deed.en.

6.3.1 DNA Identification of Tutankhamun's Father

In 1907, a mummy was excavated from an inauspicious tomb (KV55) in the Valley of the Kings in Egypt (Figure 6.15). Speculation about the identity of the mummy reigned until 2010 when scientists examined the remains of mummy KV55 and successfully isolated DNA from the bones. A DNA profile was obtained and when it was compared with profiles from other mummies, it looked like the unknown mummy was Tutankhamun's father (reigned approx. 1332–1323 BC), Akhenaten (reigned 1353–1336 BC). Egyptologists are still arguing about whether or not mummy KV55 is Tutankhamun's father; nevertheless, the investigation shows DNA's incredible stability (in this case 3357 years) under the right storage conditions.

6.4 DNA Databases

A DNA database (or databank) is a repository of DNA profiles stored electronically. Such databases can be kept for many reasons including, the analysis of genetic diseases, genetic fingerprinting for criminology, or genetic genealogy. Their use in criminology will be

Figure 6.15 The coffin recovered from tomb KV55 from the Valley of the Kings, Egypt. DNA extracted from the mummified remains inside identified the body as Tutankhamun's father Akhenaten. This showed that DNA is stable under the right conditions for at least 3357 years. Reproduced from ref. 5, https://www.flickr.com/photos/menesje/709237794/in/set-72157603861561726/, under the terms of the CC BY 2.0 license https://creativecommons.org/licenses/by/2.0/.

discussed here, but DNA databases maintained by other organisations (*e.g.*, genealogy companies) might be accessed by the police as part of their investigations, but this is ethically very sensitive and the subject of considerable heated debate – people who submitted their DNA for genealogical purposes did not usually give consent for its use by the police. This is not the place to delve into this debate, but it must be kept in mind when considering the very effective use of DNA databases to locate potential DNA matches and thus identify suspects in a crime. In short, the DNA profile (in digital format) of a sample recovered from a scene of crime is compared with a database of DNA from known people, the database searches for a profile match and if there is one, identifies the person with the potential matching DNA profile and calculates the statistics (*e.g.*, the percentage base pair matches) to give an idea of the degree of probability of the match.

The UK was the first country to set up a DNA database, which in 2020 held about 6.6 million DNA profiles; this gives an idea of the magnitude of national DNA databases.

The DNA profiles held in many DNA databases around the world are from people who have provided samples for elimination purposes, have been investigated by police or convicted of a crime. The middle group engenders significant ethical concern because, as Sir Alec Jefferies (see Section 6.1 *The Deoxyribonucleic Acid (DNA) Revolution*) pointed out, it presupposes the guilt of individuals, which is not acceptable in a democratic society. Similarly, storing DNA profiles from convicted criminals presupposes that they are more likely to commit crimes in the future; while that might be the case for some, it's certainly not for all of them.

The ethical debate will rumble on for a long time to come, but the usefulness of DNA databases to the police will very likely ensure their continuation.

6.5 The Bobby Dunbar Case

This is the ultimate 'cold case' (defined as an unsolved criminal investigation which remains open pending the discovery of new evidence) and serves as an excellent example of DNA profiling solving a case, albeit a hundred or so years after the event.

Lessie and Percy Dunbar took their 4-year-old son, Bobby, on a fishing trip to Swayze Lake near to their home in Louisiana, USA. Bobby disappeared a few days later on 23rd August, 1912. A huge search was undertaken by police – the search was extensively reported in local and national newspapers. Eight months after the search for Bobby began the police found an itinerant handyman, William Cantwell Walters, in Mississippi with a boy about Bobby's age who matched Bobby's description. Walters insisted that the boy was Bruce Anderson, the son of one of his family's employees, Julia Anderson.

Lessie Dunbar was brought in to identify (or otherwise) her son, Bobby (Figure 6.16). A newspaper of that time reported that on seeing Lessie, the young lad shouted 'Mother!' This proved that Bruce Anderson was in fact Bobby Dunbar! Similarly, a newspaper reported that when 'Bobby' was introduced to the Dunbars' other son, Alonzo, he immediately recognised him as his brother and shouted his name – further proof that Bruce was Bobby!

The police brought Julia Anderson from North Carolina to identify her son, Bruce (or was he the Dunbar's son, Bobby?). They presented Julia

Figure 6.16 Images of Bobby Dunbar before he disappeared (left) and the boy found with William Cantwell Walters and later identified as Bobby Dunbar by Lessie Dunbar (right) from a newspaper published in 1914. It is difficult to see why there was an issue of identity, it is clear that the two images are not of the same boy. The misidentification was proved in 2004 using DNA from descendants of the Dunbar family. Figure is in the Public Domain.

with five boys (including Bruce or was he Bobby?) of about the same age and asked her to identify which boy was her son. She was uncertain (Figure 6.16). This combined with the fact that Julia was deemed to have had low moral standards because she had had three children out of wedlock prompted the police to believe Lessie Dunbar and reunited Bobby (or was he Bruce?) with his rightful mum and dad (but were they?). Julia Anderson did not have the financial means to contest the police's decision and so Bobby went back to Louisiana with his 'mum and dad', and William Cantwell Walters was convicted of kidnapping Bobby. That was the end of it as far as the police were concerned.

The little lad grew up with the Dunbars, married and he and his wife had four children. He died in 1966 (age 58) as Bobby Dunbar.

6.5.1 Fast Forward to 1967...

A year after Bobby Dunbar's death his granddaughter, Margaret Dunbar Cutright, took an interest in her grandfather's complex story. She pored over newspaper articles from the time of his disappearance and took a particular interest in the reunion with his parents (her great-grandparents). She became suspicious about her grandfather's identity and decided to investigate further.

Four decades passed, during which DNA technology was introduced to forensic investigations, and in 2004 DNA samples were taken from

Bobby's son, also named Bobby (the case is confusing enough already; this didn't help! I'll refer to him as Bob Dunbar Jr) and compared with Bobby's brother Alonzo's son's DNA. Bob Dunbar Jr's DNA was unrelated to Alonzo's son's DNA; therefore, 'Bobby Dunbar' was definitely not Alonzo Dunbar's brother since there is no familial DNA commonality.

So, using DNA technology, it was proven 91 years after the little lad identified as Bobby Dunbar and reunited with his 'mother' that he was, in fact, not Bobby Dunbar. Two questions remain unanswered: was 'Bobby' really Bruce Anderson and if so, what happened to the real Bobby Dunbar? Suffice to say, the fishing spot to which Mr and Mrs Dunbar took little Bobby was the perfect habitat for alligators...

Further Reading

1. *Forensic DNA Analysis – A Primer for the Courts*, The Royal Society and The Royal Society of Edinburgh, 2017, ISBN: 978-1-78252-301-7.
2. J. D. Watson and F. H. C. Crick, A structure for deoxyribose nucleic acid, *Nature*, 1953, **171**, 737–738.
3. P. Gill, A. J. Jeffreys and D. J. Werrett, Forensic application of DNA 'fingerprints', *Nature*, 1985, **318**, 577–579.
4. P. Gill, O. Bleka, O. Hansson, C. Benschop and H. Haned, *Forensic Practitioner's Guide to the Interpretation of Complex DNA Profiles*, Academic Press, 2020, ISBN: 9780128205624.
5. The Ghost of Bobby Dunbar, A USA radio broadcast, part of This American Life, 2008, https://www.thisamericanlife.org/352/the-ghost-of-bobby-dunbar
6. S. Luo, C. A. Valencia, J. Zhang, N.-C. Lee, J. Slone, B. Gui, X. Wang, Z. Li, S. Dell, J. Brown, S. M. Chen, Y.-H. Chien, W.-L. Hwu, P.-C. Fan, L.-J. Wong, P. Atwal and T. Huang, Biparental inheritance of mitochondrial DNA in humans, *Proc. Natl. Acad. Sci. U. S. A.*, 2018, **115**, 13039–13044.

References

1. 14.3 Mechanisms of Antibacterial Drugs – Microbiology | OpenStax [Internet], openstax.org. Available from: https://openstax.org/books/microbiology/pages/14-3-mechanisms-of-antibacterial-drugs
2. File:Chondrocyte- calcium stainjpg [Wiki], 2023. Available from: https://commons.wikimedia.org/wiki/File:Chondrocyte-_calcium_stain.jpg [accessed 4 September 2023].

3. File:D1S80Demopng [Wiki], 2023. Available from: https://commons.wikimedia.org/wiki/File:D1S80Demo.png [accessed 4 September 2023].
4. File:Str profile.jpg [Wiki], 2023. Available from: https://commons.wikimedia.org/wiki/File:Str_profile.jpg [accessed 4 September 2023].
5. H. Ollermann, 2004_0315_133353AA [Internet]. Flickr. 2007. Available from: https://www.flickr.com/photos/menesje/709237794/in/set-72157603861561726/

7 Fingerprints

Fingerprints provide a means of assessing identification, because the ridge arrangement on every finger of every human being is considered unique and does not alter with growth or age.[1] This is why, until the introduction of DNA profiling, fingerprints were the mainstay of many forensic investigations. Most people would think of fingerprints in connection with criminal investigations. Since their introduction to criminal cases in the mid-19th century, fingerprints have remained a key facet of many forensic investigations, even though DNA 'fingerprints' have now become at least as important.

7.1 The History of Fingerprints

It is difficult to pinpoint exactly when fingerprints were introduced into criminal investigations. However, the idea of using fingerprints to identify criminals was put forward by a provincial surgeon, Dr Robert Blake Overton, during the investigation of the murder of the British politician, Lord William Russell in 1840, but the police ignored Dr Overton's advice (see Section 7.2 *The Lord William Russell Murder Investigation (1840): A Missed Fingerprint Opportunity*).

Fifty years later, Sir William Hershel opined that fingerprints are unique to an individual and do not change throughout life. He used these two important properties of fingerprints to identify individuals in an Indian civil service setting (see Section 7.3 *William Hershel's Use of Fingerprints to Identify Indian Villagers*). In 1880, the missionary, doctor and scientist Henry Faulds realised that the suggested uniqueness of fingerprints would be invaluable in forensic investigations.

Forensic Science: The Science Behind the Truth
By Ian Shaw with Anna Sandiford

Published by the Royal Society of Chemistry, www.rsc.org

Charles Darwin's second cousin and polymath, Sir Francis Galton took Hershel's and Fauld's ideas (without crediting them!) and devised a classification system for fingerprints (this is still used today: see Section 7.9 *Analysing Fingerprints (Dactyloscopy)*) and in 1882, he published his book *Fingerprints*. Then, Ivan Vučetić (1858–1925; he changed his name to Juan Vucetich Kovacevich), a Croatian anthropologist, who emigrated (1882) to Argentina where he became a police official, took Sir Francis Galton's ideas and devised a robust fingerprint identification system. He became the director of the Centre for Dactyloscopy in Buenos Aires and Argentina became the first country to apply fingerprint identification to criminal investigations (see Section 7.11.1 *Rojas Infanticide, Argentina 1892 – The First Use of Fingerprints in a Criminal Case Worldwide*). Vucetich published *Dactiloscopía Comparada* (Comparative Dactyloscopy) in 1904.

The stage was set for fingerprints to revolutionise criminal detection and identification worldwide.

There are four phases in the history of applying fingerprints to forensic investigations:

1. the realisation that fingerprints occur,
2. that they are unique to an individual and do not change throughout a person's life,
3. that they can be applied to criminal investigations, and
4. the development of technology to detect and compare fingerprints.

7.1.1 The First Observations and Use of Fingerprints

The ancient Chinese (8000–2205 BC) used fingerprints to authenticate official documents. Whether they thought that fingerprints were person-unique is impossible to know, but they must at least have ascribed them with some discriminatory properties; otherwise, they would not have used them for authentication purposes. Similarly, in Babylon around 200 BC, fingerprints were used to 'sign' written contracts. And in 851 AD, an Arab, Abu Zayd Hasan, noted that Chinese merchants used fingerprints to authenticate loans. So, in ancient times, fingerprints were definitely known and were used to discriminate between the people from whom they originated.

7.1.2 The Beginning of Fingerprints as Tools for Criminal Justice

During the reign of the Babylonian king Hammurabi (1792–1750 BC), fingerprints were taken from suspects when they were arrested.

Chinese officials between 221 and 206 BC expanded the idea of using 'prints' for criminal investigation purposes. They used hand and foot prints rather than just fingerprints. But it was not until the Iranian physician Rashid-al-Din-Hamadani (1247–1318 AD) wrote about the Chinese practice of using fingerprints to identify individuals that the concept of person uniqueness of fingerprints was realised; he wrote, '*Experience shows that no two individuals have fingers exactly alike.*'

7.2 The Lord William Russell Murder Investigation (1840): A Missed Fingerprint Opportunity

Lord William Russell (1767–1840) was a British member of parliament (1789–1807) and a member of the aristocratic Russell family (his father was the Marquess of Tavistock and his grandfather was the 4th Duke of Bedford). After he lost his parliamentary seat in 1807, his life was not at all newsworthy until he was murdered in his sleep by his valet, François Courvoisier.

On the morning of 6th May, 1840, Lord Russell's housemaid found rooms on the lower floors of the Russell London residence in disarray. She assumed that a robbery had taken place overnight and asked Russell's valet, François Courvoisier, to take a look. They went into Russell's bedchamber to check that all was well. Courvoisier opened the window shutters as he always did. As the room was lit by the morning sun, the housemaid found Lord Russell dead in his bed with his throat cut. The police were called and after their initial search of Lord Russell's bedchamber, Courvoisier pointed out marks on the door to his room (the butler's pantry), which suggested that the robbers might have entered the house this way.

The police found a small parcel of silver and gold items from the house plus a £10 note (worth about £700 (US$940) in 2021) in the butler's pantry in the house. They found this strange because surely a burglar would have taken his spoils with him. They concluded that the robbery had been staged by someone within the Russell residence. Their suspicion turned to Courvoisier. This suspicion was heightened when they found that the marks on the butler's pantry door matched a screwdriver found in Courvoisier's possession. The police charged Courvoisier with Lord Russell's murder.

A highly controversial trial followed, wherein Courvoisier's counsel, Charles Phillips, discovered partway through the trial that his client was guilty; despite this revelation, he continued with his defence.

Part of Courvoisier's defence was that there were no bloodstains on his clothing despite Lord Russell's throat having being cut. It turned out that Courvoisier had committed the murder naked!

Courvoisier was convicted of Lord Russell's murder and on 6th July, 1840, he was publicly hanged outside London's Newgate Prison. The novelist William Makepeace Thackeray attended the hanging and wrote in his essay *On Going to See a Man Hanged*, '*I came away that morning with a disgust for murder, but it was for the murder I saw done.*'

There is an important postscript to this case involving a letter from Dr Robert Blake Overton, a Norfolk surgeon, to the London police headquarters (Scotland Yard). In his letter, Dr Overton extolled the virtues of fingerprints as a means of identifying an individual, particularly the individual who had murdered Lord Russell:

> *It is not generally known that every individual has a peculiar arrangement [on] the grain of the skin...*
>
> *I would strongly recommend the propriety of obtaining impressions from the fingers of the suspected individual and a comparison made with the marks on the sheets and pillows.*

He illustrated his point by showing fingerprints from two different people in his letter and highlighting their differences. The police ignored Dr Overton.

Dr Overton's suggested use of fingerprints in the Russell murder case investigations was 50 years before Sir William Hershel formally proposed fingerprints as a means of identification, which led to their use in forensic cases.

7.3 William Hershel's Use of Fingerprints to Identify Indian Villagers

Sir William Hershel (1833–1917) was a British Indian civil servant who carried out significant and long-term studies on fingerprints (Figure 7.1). He showed that fingerprints are unique to an individual and that they remain unchanged throughout a lifetime (he recorded his own fingerprints at regular intervals until a few years before his death). He used the person uniqueness of fingerprints to authenticate contracts instead of relying on signatures.

Despite the use of fingerprints as person identifiers by the ancient Chinese some 10 000 years prior to Hershel, Hershel is credited with

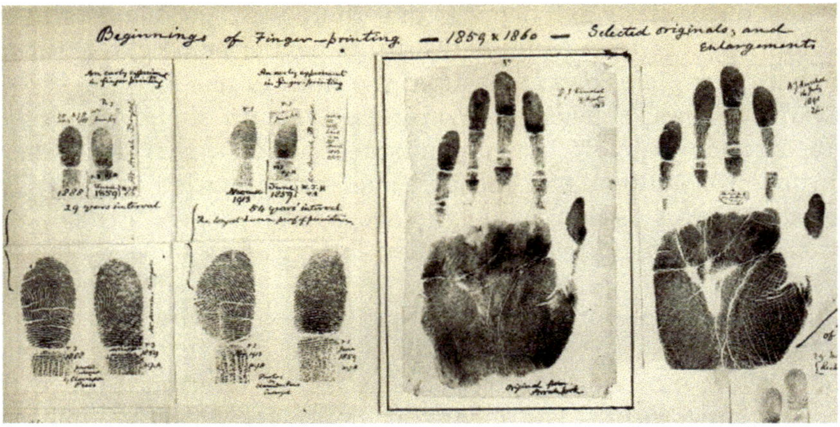

Figure 7.1 Pages from Sir William Hershel's notebook showing his finger and palm print studies. Hershel is credited with showing the individual uniqueness of prints, which underpinned their introduction to forensic investigations at the end of the 19th century. Figure is in the Public Domain.

realising their value in identification and underpinned their consequent use by the police.

7.4 The Transition from Anthropometry to Fingerprints

Until the police were convinced of the importance of fingerprints in identifying an individual, they used photographs and a series of bodily measurements (anthropometry) as their means of identification. An anthropometric system was introduced by the French police officer Alphonse Bertillon (1853–1914; termed the Bertillon system) (Figure 7.2), which included ear, foot and arm span measurements, as well as eye, hair and beard colour (Figure 7.2) – in the USA, 11 anthropometric measurements were included in a profile. If two anthropometric profiles were compared and all of the parameters matched, it would have been concluded that the two profiles were of the same person.

Anthropometry as a means of identification was fairly short-lived because of a mix-up of identities of two criminals with the same name in the USA in 1903 (see Section 7.4.1 *The Will West Case – USA 1903*).

Once the authorities realised that an individual could be identified by their fingerprints (even if they changed their name or appearance),

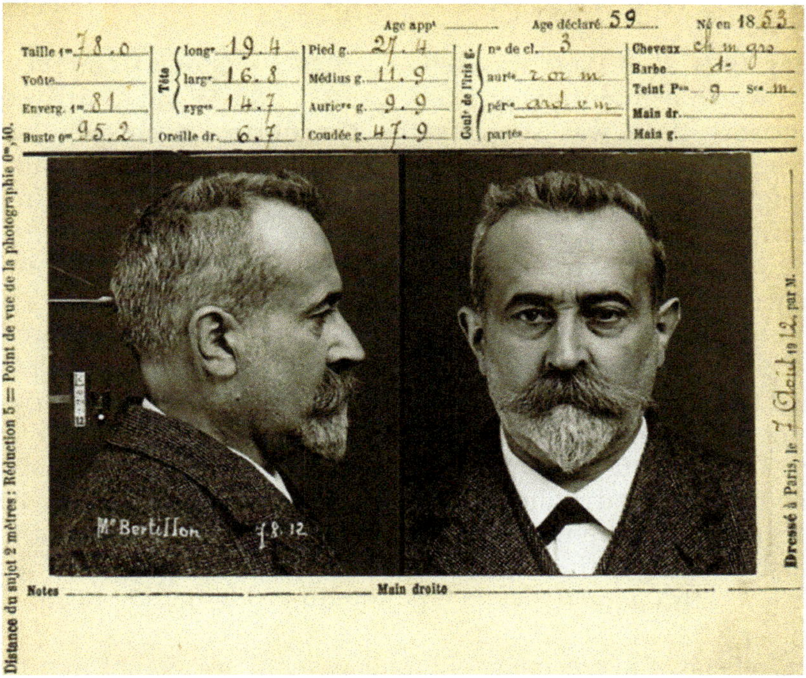

Figure 7.2 The inventor of the Bertillon system, Alphonse Bertillon's own anthropometric measurements. Until the introduction of fingerprints, this was the method used by police to identify and distinguish between individuals. Taille = waist, enverge. = left shoulder to right middle finger, buste = chest, tête = head, oreille = ear, pied = foot, médius = middle finger, auric^{re} = little finger, coudée = elbow to middle finger, coul^{r} de l'iris = eye colour, cheveux = hair, barbe = beard, and teint peu. = skin colour. Figure is in the Public Domain.

it was necessary to devise a system to classify fingerprints so that they could be filed, searched manually and compared with a suspect's prints at a later date.

In the early 1900s, the Commissioner of Police for the Metropolis [of London], Sir Edward Henry developed a formula to classify fingerprints; it included numbers and letters representing the fingerprint pattern (see Section 7.9 *Analysing Fingerprints (Dactyloscopy)*). The values obtained were unique to a particular fingerprint and could be filed (*i.e.*, a database) and retrieved and compared later. Over the years, the classification system was refined so that it became easier to retrieve fingerprints that were more likely to match a suspect's print.

7.4.1 The Will West Case – USA 1903

In May 1903, Will West was sentenced to a custodial sentence in Leavenworth Penitentiary (prison), Kansas, USA. When he arrived at the prison, West was subjected to the normal admissions procedure, which involved front and side view photography and recording a series of anthropometric measurements to include in his identification profile.

The prison clerks found that Will West's anthropometric profile exactly matched that of William West, who had a previous conviction for murder. They assumed that Will West and William West were the same person. Even the two men's photographs were identical. Will denied that he had a previous murder conviction. The prison officials did not believe him.

The Leavenworth Penitentiary officials looked further (in fact not much further!) and found that they already had a William West incarcerated in Leavenworth. They were shocked (I suspect this is an understatement) to find two men deemed identical using their gold standard anthropometry, were now standing side by side in front of them!

The officials fingerprinted Will and William and found that their fingerprints were very different. From that point on Leavenworth never used anthropometric identification again; instead they used far more reliable fingerprints as their means of identification. This began the shift toward fingerprints worldwide.

The FBI's official record of the event states:

> *The fallibility of the three systems of personal identification – names, photographs and Bertillon measurements – was demonstrated by this one case. On the other hand, the value of fingerprints as a positive means of identifying people was dramatically shown.*

From that time, fingerprints replaced the somewhat dubious anthropometric identification. However, even today elements of anthropometry remain in criminal records, including a photograph, height, hair and eye colour.

7.5 What Are Fingerprints?

This might seem a blatantly obvious question with an equally obvious answer. At the simplest level, they are the marks produced when a finger comes in contact with an object. But why do fingers have unique surface patterns that leave unique prints? To answer this, a quick dermatological anatomy lesson is in order.

The skin is made up of two main layers: the dermis (bottom) and epidermis (top) (Figure 7.3). The dermis is above the subcutaneous (*e.g.*, adipose) tissue; it houses the nerves (*e.g.*, touch sensitive), sweat (eccrine) glands with their ducts, which exit through the epidermis as pores (and from where we sweat) and the hair follicles that form hairs, which also exit through the epidermis. Hair follicles are associated with apocrine (sebaceous) glands, which secrete sebum (an oily fluid);

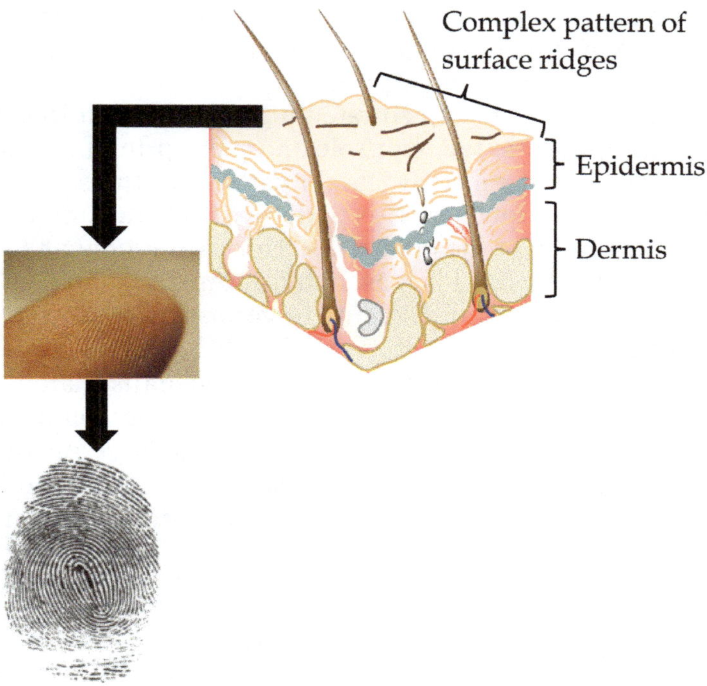

Figure 7.3 The anatomy of the skin (*e.g.*, fingertips) showing the epidermal ridges which form fingerprints (top) and the corresponding complex fingerprint pattern (lower left). Hairs and hair follicles are shown here for completeness but are not present in the skin of the fingertips, palms, underside of the toes or soles of the feet and thus are not relevant in the context of prints. Reproduced from ref. 4, https://commons.wikimedia.org/wiki/File:Fingerprintforcriminologystubs2.png, under the terms of the CC BY-SA 3.0 license https://creativecommons.org/licenses/by-sa/3.0/, and ref. 5, https://commons.wikimedia.org/wiki/File:Fingerprint_detail_on_male_finger_in_T%C5%99eb%C3%AD%C4%8D,_T%C5%99eb%C3%AD%C4%8D_District.jpg, under the terms of the CC BY 3.0 license https://creativecommons.org/licenses/by/3.0/.

however, they are absent from finger tips, palms, the underside of toes and soles of the feet.

The surface of the epidermis has a characteristic and complex array of ridges in lines and whorls, which form the unique surface structure of the epidermis of the fingertips. This characteristic surface structure is, in part, genetically determined, but its major determinants are random pressures, movement and orientations of the foetus, which modify the growth patterns of the genetically determined ridges. This characteristic epidermal pattern forms equally characterisitc fingerprints on surfaces that the finger touches.

Not only do fingers have a characteristic epidermal pattern of ridges; but the palms of the hands and the toes and soles of the feet also exhibit equally characteristic patterns and are used in forensic casework in the same way as fingerprints.

The fingerprint (I will concentrate on fingerprints in this chapter, but the same applies to toe, foot sole and palm prints) is made from skin exudates (*e.g.*, sweat) that leave a pattern on the surface (*e.g.*, a glass) that the finger touches. This pattern reflects the unique pattern of ridges and whorls of the epidermis. This skin exudate-based print is very difficult to see; for this reason, it is termed a latent (*i.e.*, hidden or concealed) fingerprint. Latent fingerprints must be treated (developed) to make them visible.

Before we look at the methods used to visualise latent fingerprints, it is important to know a little about the composition of sweat and other skin exudates that make up a fingerprint. This is important because both the physical (powder methods) and chemical (colour reactions) rely on the chemistry of the latent fingerprint to work.

7.6 The Chemistry of Fingerprints

Latent fingerprints include two components: sweat excreted from the sweat glands and possibly excretions from the sebaceous glands. Sweat glands are present in the dermis of most tissues. Their ducts exit on the surface of the epidermis (Figure 7.4), which means that the epidermis is 'moistened' with sweat – the amount depends on many factors, including temperature. On the other hand, sebaceous glands secrete sebum into hair follicles (Figure 7.4) and so it is only a major component of hairy skin surfaces (*e.g.*, the scalp). The palms of the hands and soles of the feet do not secrete sebum because they do not have hair follicles. However, sebum is readily transferred to the

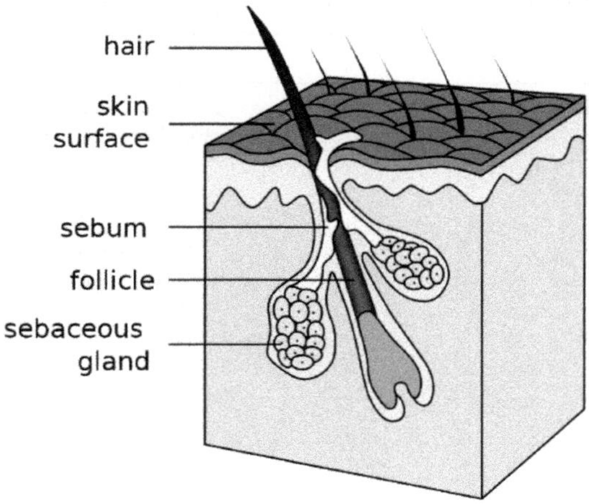

Figure 7.4 Cross-section of 'hairy skin' showing the sebaceous glands (in the dermis) that secrete sebum onto the surface of the epidermis (skin) *via* hair follicles.

fingertips if they have touched skin surfaces (*e.g.*, the head) that secrete sebum – just think how many times you touch your forehead and head!

The composition of these exudates is used to visualise latent fingerprints.

7.6.1 The Composition of Sweat

The components of sweat vary considerably from person to person due to both interindividual variations in biochemistry and health status. Much of this variability is quantitative and will have minimal impact on latent fingerprint visualisation. Some diseases lead to biochemical changes in sweat composition, but again these are unlikely to have a significant effect on fingerprints. Table 7.1 shows the basic biochemical components of sweat.

7.6.2 The Composition of Sebaceous Gland Secretions

The secretion from the sebaceous glands is termed sebum. As outlined above, it is only produced by 'hairy skin' because the sebaceous glands are associated with hair follicles (Figure 7.4); however, it can be transferred to fingers from the scalp and forehead. Sebum's composition is complex, consisting mainly of fats and oils, including triglyderides (about 40%), wax esters (about 25%), squalene (about 10%) and free fatty acids (about 15%), all are useful in latent fingerprint visualisation.

Table 7.1 Some of the basic components of sweat.

Group	Examples
Ions	Sodium – Na^+ Potassium – K^+ Chloride – Cl^- Calcium – Ca^{2+} Iron – Fe^{2+} and Fe^{3+} Phosphate – $PO_4{}^{3-}$ Sulphate – $SO_4{}^{2-}$
Proteins	Albumin Clusterin Caspase-14 Carboxypeptidase E
Amino acids	Alanine Arginine Glycine Histidine
Biochemical intermediates	Lactic acid Pyruvic acid Glycogen
Excretory products	Urea Creatinine Uric acid
Fats	Fatty acids Triglycerides Sterols

7.7 Visualising Latent Fingerprints

There are three ways to visualise latent fingerprints:

1. Using a fine powder (*e.g.*, aluminium) that sticks to the skin exudate.
2. Using a coloured chemical (*e.g.*, iodine) that dissolves in the skin exudate.
3. Using a colour reaction that detects a component of the skin exudate (*e.g.*, ninhydrin reacts with amino acids, giving a purple product).

7.7.1 Powder Methods

A very fine powder is applied to a surface (*e.g.*, glass) with a fine brush and gently dusted across the surface (termed 'dusting'). The powder sticks to the latent fingerprint, making it visible. The particles of the dusting powder must be very small (either spheres of approx. 1 μm diameter or flakes of approx. 10 μm diameter) to ensure that they adhere to the latent fingerprint and make the complex convolutions of the fingerprint clearly visible.

The composition of dusting powders varies considerably (different powders produce different fingerprint colours), including metals (*e.g.*, aluminium - grey) and their oxides (*e.g.*, manganese dioxide - black, iron oxide - red) and formulations containing rosin (a plant-derived resin - red). The dusting powder colour is chosen to give the best contrast (and thus make the latent fingerprint visible) based on the surface being dusted (*e.g.*, light coloured dusting powder used for a dark surface).

In simple terms, the powder is very gently dusted across the surface of the item (*e.g.*, a table) being studied using a very fine (*e.g.*, squirrel hair) brush (Figure 7.5). The excess powder is dusted off to reveal any fingerprints. A fingerprint visualised in this way is reasonably stable, but for records and digital comparison techniques, the print is either 'lifted' using clear gel lifts adhesive tape and/or photographed. This is also, of course, important if the fingerprint is on a surface that cannot be removed from the scene.

Figure 7.5 Dusting with fine powder to visualise latent fingerprints. In this case, a dark dusting powder is used to give good contrast on a light background. Reproduced from ref. 2, https://commons.wikimedia.org/wiki/File:Dacty_poederen.JPG, under the terms of the CC BY-SA 3.0 license https://creativecommons.org/licenses/by-sa/3.0/.

7.7.2 Iodine Visualisation (Fuming)

Elemental iodine (I_2) is soluble in fats and therefore since both sweat and sebum contain fats, it is soluble in latent fingerprints. Dissolved iodine is brownish, so latent fingerprints treated with iodine show up brown.

Iodine has an interesting property that is utilised in its application to visualise latent fingerprints: it transitions from its solid to its gaseous state without going through the liquid phase. In addition, solid iodine is quite volatile; indeed, a sealed container of solid iodine (I_2 (s)) always has a purple atmosphere of iodine gas (I_2 (g); Figure 7.6). I_2 (g) readily dissolves in fats (or other hydrophobic solvents), giving a characteristic brown colour. Visualising latent fingerprints using iodine depends on these properties of iodine: the item to be studied is placed in an atmosphere of I_2 (g) (created in a sealed vessel with I_2 (s) at the bottom) or using an iodine fuming gun which blows I_2 (g) over the test surface. The I_2 (g) dissolves in the fat in the latent fingerprints, staining them brown (Figure 7.6). When the iodine-stained fingerprint is removed from the I_2 (g) atmosphere, the I_2 evaporates from its fat solution to re-form purple I_2 (g) and thus the fingerprint loses its colour and 'disappears'. The fingerprint can be re-visualised by repeating the iodine fuming process, which can be repeated many times. Because of the instability of the iodine stain, iodine-visualised fingerprints must be photographed for the records and digital comparison.

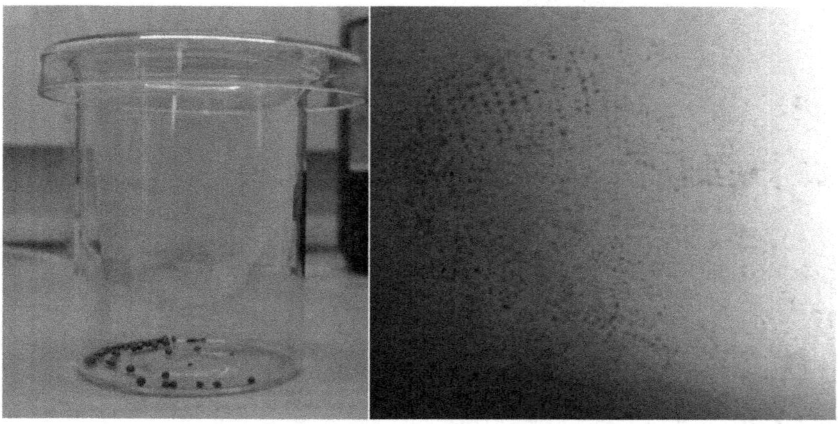

Figure 7.6 Left: a beaker with solid iodine (I_2) pellets at the bottom; the I_2 transitions straight from solid (metallic) to the gas (purple) phase. Also in the beaker is a piece of paper that the author had touched. The paper was left in the I_2 atmosphere for 15 min to visualise any fingerprints. Right: an enlarged section of the I_2 gas-exposed paper showing a fingerprint.

7.7.3 Ninhydrin Visualisation

Amino acids in sweat (and therefore in latent fingerprints) react with ninhydrin, forming a purple compound (Figure 7.7). Ninhydrin is sprayed onto the test item and left for a few minutes to allow the purple colour to appear (Figure 7.7).

As with other fingerprint visualisation methods, the ninhydrin colour is not very stable. It is possible to treat with other chemicals before the ninhydrin spray to prolong the life of the purple colour; however, photography is usually used to record the print and facilitate later comparisons.

7.7.4 Indanedione Visualisation

This is another commonly used method that relies on the presence of amino acids in the latent fingerprint. The chemistry (Figure 7.8) is very similar to that of the ninhydrin reaction, but in this case, the product formed fluoresces under UV light; so, after treating the

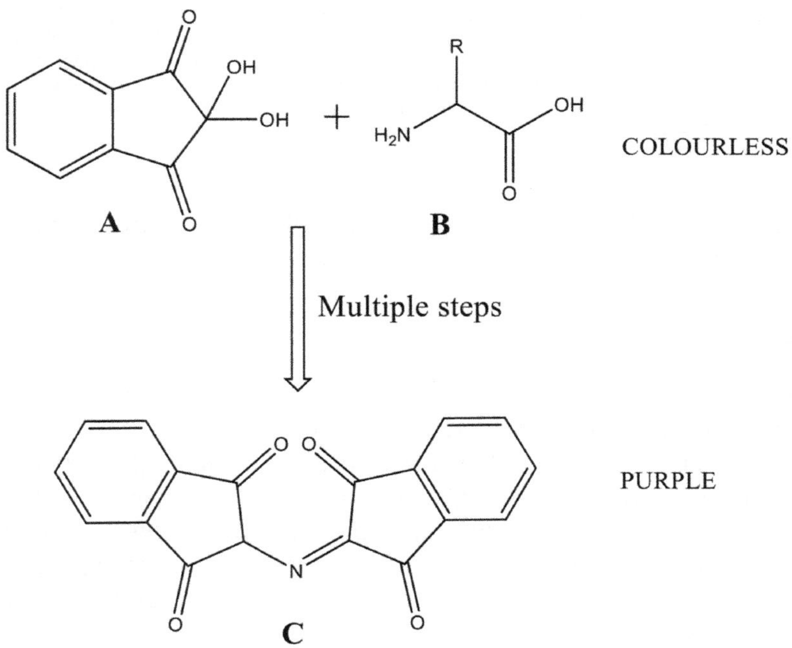

Figure 7.7 Ninhydrin fingerprint visualisation: ninhydrin (A) reacts with amino acids (B) *via* a complex series of reactions to form a purple compound (C). R = the chemical group that defines the identity of a particular amino acid; *e.g.*, glycine R = H, alanine R = CH_3, and cysteine R = CH_2CH_2SH.

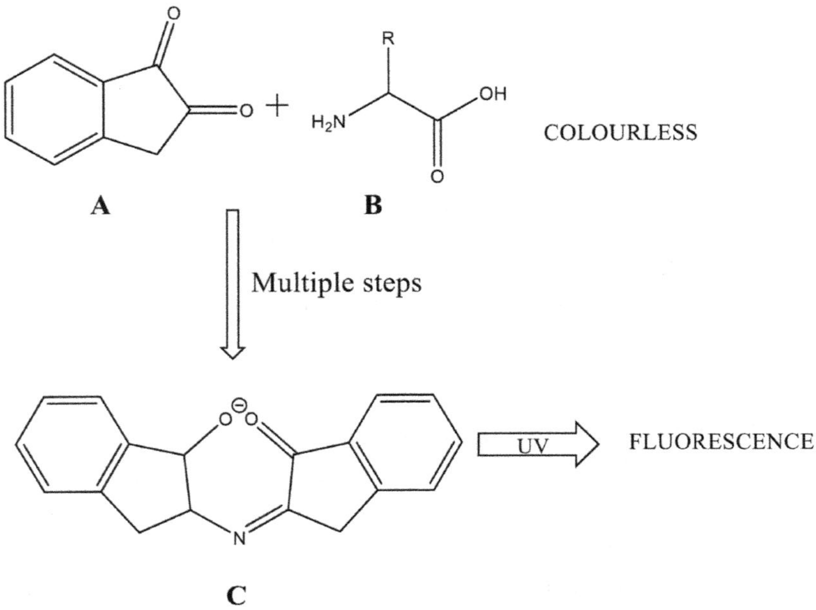

Figure 7.8 Indanedione fingerprint visualisation: 1,2-indanedione (A) reacts with amino acids (B) to form a compound (C), which when illuminated with UV light, fluoresces yellowish-orange. R = the chemical group that defines the identity of a particular amino acid; *e.g.*, glycine R = H, alanine R = CH_3, and cysteine R = CH_2CH_2SH.

test material with 1,2-indanedione, the field is illuminated with UV light and the fingerprints fluoresce yellowish-orange.

7.7.5 Silver Nitrate Visualisation

Silver nitrate solution reacts with chloride ions to form insoluble cream-coloured silver chloride. Silver chloride is light sensitive and slowly changes to black as silver metal is released (Figure 7.9). Chloride ions in latent fingerprints react with silver nitrate, producing a black silver precipitate (*via* silver chloride) which makes the fingerprint visible. Interestingly, this reaction is the basis (silver bromide – AgBr – is used) for black and white photography.

The visualisation of fingerprints using silver nitrate has the advantage of long-term stability, similar to that of a photograph, with which its chemistry shares similarities.

$$AgNO_3\ (aq) + Cl^-\ (aq) \rightarrow AgCl\ (s) + NO_3^-\ (aq)$$

$$\downarrow \text{Light}$$

$$Ag\ (s) + \tfrac{1}{2}Cl_2\ (g)$$

Figure 7.9 The chemistry of silver nitrate visualisation of fingerprints: silver nitrate ($AgNO_3$) in aqueous (aq) solution reacts with chloride ions (Cl^-) in fingerprints to form silver chloride solid (AgCl (s)) plus nitrate ions (NO_3^-). The AgCl (s) is light-sensitive and decomposes to form black silver solid (Ag (s)) plus chlorine gas (Cl_2 (g)). The precipitated silver makes the fingerprints black.

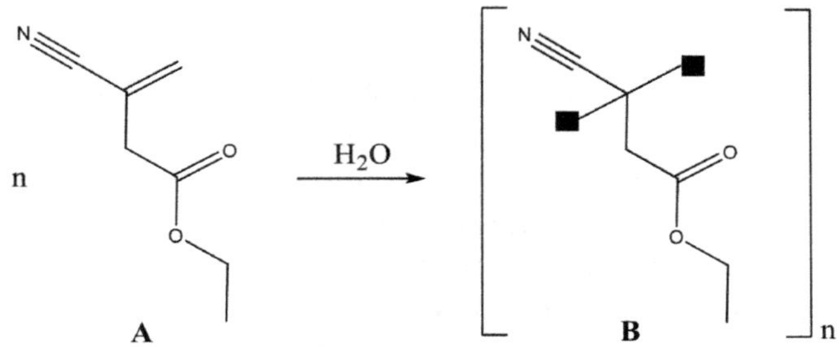

Figure 7.10 Ethylcyanoacrylate (A) fumes react with water in latent fingerprints, forming white polyethylcyanoacrylate (B). The chemical bonds between the monomer units in the polymer are shown by ■.

7.7.6 Cyanoacrylate Fuming

Cyanoacrylate is the monomer used in Superglue; it is volatile and when fingerprints are exposed to its vapour, it reacts with water in the fingerprints resulting in its polymerisation to form a white polycyanoacrylate (Figure 7.10). The white polycyanoacrylate allows visualisation of the latent fingerprint, but it is often treated further to enhance the print's visibility. The secondary treatment often involves dusting (see Section 7.7.1 *Powder Methods*) with a contrasting-coloured powder (*e.g.*, red Acid Fuchsin) or soaking the print in a fluorescent dye and observing under UV.

7.8 Non-latent Fingerprints

Non-latent fingerprints are those that can be seen without using any of the development methods described above. For example, a print in blood or perhaps on a freshly painted surface would be clearly visible.

However, methods are often used to enhance the visibility of non-latent prints.

7.8.1 Fingerprints in Blood

Stains that react with components of blood (*e.g.*, haemoglobin and proteins) are used to enhance blood prints. Amido Black is commonly used to stain blood (it reacts with proteins) prints on light surfaces and Acid (quinoline) Yellow is used for dark surfaces. Viewed under light at a wavelength of 416 nm, Acid Yellow-stained prints fluoresce, further enhancing their visibility.

7.9 Analysing Fingerprints (Dactyloscopy)

Forensic science could be regarded as the science of comparison (as is obvious from the methods discussed so far); fingerprints are no exception to this generalisation. Fingerprints found at the scene of a crime are compared with those from a suspect to determine whether s/he might have been at the scene, or a fingerprint database could be used to identify a suspect. There are two basic levels of comparison: simple visual comparison carried out by a fingerprint expert or digital comparison following digitalisation of the fingerprints and computer-aided comparison.

To facilitate dialogue about fingerprints, fingerprint characteristics nomenclature has been developed. At the basic level, the characteristic features of fingerprints are observed at three levels.

Level 1: The overall flow of the ridges form characteristic patterns termed loops, whorls or arches (Figure 7.11).

Level 2: Interruptions in the flow of ridges – where they stop, start or join.

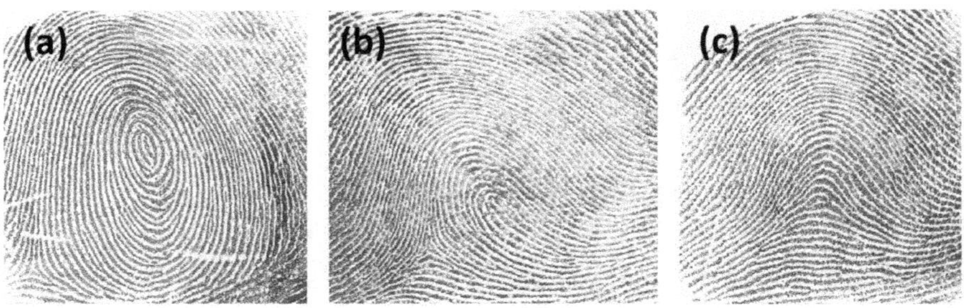

Figure 7.11 The three fingerprint characteristic patterns: whorl (a), loop (b) and arch (c). Reproduced from ref. 3 with permission from the Royal Society of Chemistry.

Level 3: Minute details such as differences in the thickness of ridges and size and shape of pores.

Levels 1 and 2 are the key observations used by fingerprint experts to compare prints.

Before the advent of computer systems for digitizing fingerprints (from images) and pattern comparisons, a complex system of classifying fingerprints was developed based on the number of whorls, arches and loops on each finger (thumb, index, middle, ring and little) on each hand (left and right) to give a numerical value that could be compared (*e.g.*, comparing the value from a suspect with the value from a fingerprint found at the crime scene). This has now largely been superseded by computerised image processing and pattern recognition techniques; they are accurate, reliable and allow comparisons of prints from the scene with a huge fingerprint database.

The fingerprint data stored in a database allow the print to be classified (*i.e.*, digitally), and a print found at the scene of a crime can be similarly classified, enabling the comparison of the 'unknown' print by electronically searching the vast number of prints in the database. This will generate a small number of similar candidate prints – they are scored according to the degree of fit with the unknown print. The fingerprint expert then compares the prints selected from the database with the print from the scene using manual methods and sheer expertise. The computer does not determine a positive identification or match; the result is the opinion of an expert.

7.10 Permanence and Clinical Absence of Fingerprints

The basic fingerprint structure of an individual does not change over their lifetime. However, injuries that leave scars might impart characteristic changes that are superimposed on the underlying 'unchanged' fingerprint pattern. This makes them an even more valuable tool in forensic investigations.

An extremely rare genetic disorder, adermatoglyphia, results in the sufferer having smooth fingertips, palms of their hands and soles of their feet; therefore, they have no fingerprints.

Some criminals may take extreme measures to eliminate the possibility of being convicted based on fingerprint evidence; indeed, Alvin Karpis (1907–1979) – also known as Creepy Karpis, a naturalised American gang leader originally from Canada, went to the extent of removing his fingerprints surgically (in 1934) to prevent detection through prints left at the scenes. Other criminals simply wear gloves!

7.11 Early Fingerprint Case Examples

7.11.1 Rojas Infanticide, Argentina 1892 – The First Use of Fingerprints in a Criminal Case Worldwide

On 29th June, 1892, two children, 6-year-old Ponchiano and his 4-year-old sister Feliza, were found brutally murdered and their mother, 27-year-old Francisca Rojas, was found with her throat cut, but still alive in their home in the Buenos Aires province of Necochea, Argentina. Rojas blamed her neighbour, Pedro Ramón Velázquez, for the attack.

The local police were at a loss; they simply could not find any evidence that pointed to the perpetrator of the terrible crime. News of the incident reached the provincial capital, La Plata, on 8th July. Police Inspector Alvarez was sent to Necochea to investigate the case.

Inspector Alvarez questioned the prime suspect, Pedro Velázquez. He vehemently denied attacking the Rojas family and provided a good alibi for his whereabouts at the time of the attack. Further investigations revealed that Rojas' boyfriend had been overheard saying that he would marry Rojas 'Except for those two brats.' If Rojas wanted to marry her boyfriend, she had a motive to dispose of her children.

During his examination of the scene, Inspector Alvarez found a brown mark on the bedroom door. He later identified this as a bloody fingerprint (Figure 7.12). He took the part of the door bearing the fingerprint back to his La Plata headquarters.

Inspector Alvares had been trained by Juan Vucetich (see Section 7.1 *The History of Fingerprints*) and he very successfully applied Vucetich's methods to the Rojas case. He took fingerprints from Rojas and found a match with the bloody print on the door. When confronted with this evidence, Rojas admitted that she had murdered her children and that she had cut her own throat to fool the police into believing that she and her children had been attacked by someone else.

The conviction of Francisca Rojas made Argentina the first country to use fingerprint evidence for a conviction and the first country to move from anthropometry to fingerprint identification (see Section 7.4 *The Transition from Anthropometry to Fingerprints*).

Figure 7.12 The bloody fingerprint from the bedroom door of the Rojas infanticide (Buenos Aires, Argentina, 1892), which was found to be identical to the murdered children's mother prints. This led to her conviction for murder, making it the first case in the world to secure conviction based on fingerprint evidence. Figure is in the Public Domain.

7.11.2 Farrow Murder, UK 1905 – The First Fingerprint Conviction in the UK

Thomas Farrow and his wife Ann were shop assistants in Chapman's Oil and Colour Shop in Deptford, south London, in the early 1900s. Thomas was found dead in the shop on 27th March 1905 and his wife was found upstairs in their flat with seriously injuries – she died several days later. The discovery of two masks by police at the scene suggested the presence of more than one intruder. A fingerprint was found on the store's cash box that did not match Thomas' or Ann's prints; this suggested the print belonged to one of the intruders.

The local milkman, Henry Jennings, had witnessed the break-in and gave descriptions of two men that matched those of the well known burglars, Albert and Alfred Stratton. The Stratton brothers were arrested and taken to the Tower Bridge police station.

Detective Inspector Charles Collins, a founding member of the Scotland Yard Fingerprint Branch (set up in 1901), worked at the Tower Bridge police station; he took the fingerprints of the Stratton brothers using an ink pad and found that the print on the cash box from the shop exactly matched Albert Stratton's right thumbprint.

Based on this evidence and Henry Jennings' description, the Stratton Brothers were charged with murder. This was the first case of convicting a murderer based on fingerprint evidence.

Alfred and Albert Stratton were hanged at Wandsworth Prison, London, on 23rd May 1905.

7.11.3 Hiller Murder, USA 1910 – The First American Murder Trial to Use Fingerprint Evidence

Clarence Hiller, his wife and their four children lived in Chicago, USA. Clarence was awakened at 2.00 am on the morning of 19th September 1910 by the screams of his wife and daughter, Clarice. He went to investigate and found an intruder on the landing. Mr Hillier fought the intruder and they both fell down the stairs. Two shots were fired (later recalled by Clarice), resulting in the death of Clarence.

The intruder fled, but did not go far. Thomas Jennings was found half a mile from the Hiller's home wearing a blood-stained jacket and carrying a revolver. He was arrested by the police.

The intruder had left a fingerprint on a newly painted rail in the porch through which he had escaped. The fingerprint matched Jennings' prints.

The court case relied heavily on the fingerprint comparison. The defence argued that *'such a new-fangled and little-understood technique could [not] be admitted in court'*. Despite these protestations, Jennings was convicted of Clarence Hiller's murder and sentenced to death.

Thomas Jennings was hanged on 16th February, 1912.

7.12 21st Century Fingerprint Case – Dubai, United Arab Emirates 2016

On 26th April 2016, a security guard's body was found gagged and secured with adhesive tape to the floor of a construction site in Dubai. There was a bottle of water and a piece of paper covered in a red powder beside the body.

The red powder was found to be capsaicin – a highly irritant pepper extract used in 'pepper spray'. Presumably, the perpetrators had used capsaicin to disable their victim before binding and securing his body with tape. The case came to be known as the 'Dubai Capsaicin Murder'.

Ten fingerprints were found on the adhesive tape (cyanoacrylate method; see Section 7.7.6 *Cyanoacrylate Fuming*) used to secure the body to the floor, three fingerprints were found on the water bottle (cyanoacrylate method) and two were found on the paper (ninhydrin method; see Section 7.7.3 *Ninhydrin Visualisation*) found next to the body – other fingerprints (total = 29) were also found. None of the prints belonged to the deceased, but all matched the 4 suspects in the case, who were apparently attempting to steel electrical cables.

The strong fingerprint findings led to convictions of the suspects and prison sentences of between three and seven years. Following an appeal (2021), their prison sentences were increased to between 5 and 10 years.

7.13 Fingerprint Databases

Most countries have huge fingerprint databases; for example, the UK police database includes more than 7 million fingerprints and the FBI (USA) holds some 30 million prints in its database. The records are kept digitally, which means that digital comparisons can be made with fingerprints found at a crime scene to identify potential suspects.

The UK police now uses mobile electronic fingerprint scanners to take fingerprints on the streets. If a potential match is found, the system will provide the police with the person's name, address and other details – this is useful if the police are suspicious about an individual's declared identity when they have no other means of assessing who they are. In addition, the system might be used to show whether the prints may match any unidentified fingerprints found at a previous crime scene.

The use of fingerprint databases in conjunction with DNA databases has significantly increased the probability of associating an individual with a crime scene as they can both have high evidential weight in court.

Further Reading

1. G. S. Sodhi and J. Kaur, Powder method for detecting latent fingerprints: a review, *Forensic Science International*, 2001, **120**, 172–176.
2. S. A. Cole, *Suspect Identities – A History of Fingerprinting and Criminal Identification*, Harvard University Press, Cambridge, Massachusetts, USA, 2001.
3. D. Maltoni, D. Maio, A. K. Jain and S. Prabhakar, *Handbook of Fingerprint Recognition*, Springer-Verlag, New York, USA, 2003.

References

1. J. E. Hoover, "fingerprint". Encyclopedia Britannica, 6 Oct. 2023, https://www.britannica.com/topic/fingerprint [accessed 27 November 2023].
2. File: Dacty poederen.JPG [Wiki]. Available from: https://commons.wikimedia.org/wiki/File:Dacty_poederen.JPG [accessed 4 September 2023].
3. P. C. White, *Crime Scene to Court: The Essentials of Forensic Science*, The Royal Society of Chemistry, 6th edn, 2016.
4. https://commons.wikimedia.org/wiki/File:Fingerprintforcriminologystubs2.png.
5. https://commons.wikimedia.org/wiki/File:Fingerprint_detail_on_male_finger_in_T%C5%99eb%C3%AD%C4%8D,_T%C5%99eb%C3%AD%C4%8D_District.jpg.

8 Forensic Pathology

Forensic pathology is a branch of pathology that focuses on determining the cause of death through a detailed internal and external examination of a dead body. The primary responsibility of a forensic pathologist is to conduct the *post-mortem* examination, during which the pathologist looks for signs that might help in determining the cause of death. These signs might be macroscopic (*e.g.*, bruising or gunshot wounds) or microscopic (changes in the morphology of tissues – histopathology). In addition, the forensic pathologist might recover important evidence from the corpse (*e.g.*, fibres under the fingernails) that might be important in the police investigation of the case. The *post-mortem* examination involves taking samples from the corpse for biochemical (*e.g.*, determining enzymes in blood that might indicate disease), toxicological (*e.g.*, measuring drug concentrations in the liver: see Chapter 9 *Forensic Toxicology*) or histopathological (*e.g.*, to determine structural change in the liver that might indicate disease) analysis. The forensic pathologist looks not only for signs of foul play, but also indicators of disease (*e.g.*, coronary heart disease), which might suggest that the death was due to natural causes.

8.1 What Will This Chapter Cover?

As with much of the subject matter in this book, this chapter is not intended to be a treatise on forensic pathology; this is a huge subject area that I am not going to attempt to cover it here in anything but cursory detail. The purpose of this chapter is to give you a 'feel' for

Forensic Science: The Science Behind the Truth
By Ian Shaw with Anna Sandiford

Published by the Royal Society of Chemistry, www.rsc.org

forensic pathology and the work performed by forensic pathologists. Professor Bernard Knight CBE, arguably the most eminent living forensic pathologist, co-authored (with Professor Pekka Saukko) *Knight's Forensic Pathology* (see *Further Reading*): this is the resource to consult for the detailed information that I will not give in this chapter.

8.2 What Is Forensic Pathology?

Before we can delve into forensic pathology, we need to understand what pathology *per se* is. Pathology is the science of disease (Greek *pathos* = suffering; *logia* = study of); it includes clinical pathology (chemical/biochemical pathology) and anatomical pathology. Clinical pathology involves the analysis of bodily fluids to help diagnose diseases; for example, increased activity of the enzyme serum glutamate pyruvate aminotransferase (SGPT) indicates liver disease or liver damage. Anatomical pathology is further subdivided into macroscopic (gross anatomy) and microscopic (histopathology and cytology). Macroscopic anatomical pathology covers the gross structures of tissues and organs, while microscopic anatomical pathology is the study of tissue anatomy (histology) involving taking samples from tissues, cutting thin sections, staining them and observing them under a microscope for any abnormalities that might indicate a particular disease. For example, characteristic histological changes in heart muscle might indicate myocardial infarction (heart attack) (Figure 8.1).

Forensic pathology is a branch of pathology spanning both clinical and anatomical pathologies that focuses on determining the cause of a death that occurs under suspicious circumstances or that was not expected (*i.e.*, the deceased was not known to be suffering from a disease or a disorder that might cause death). Depending on the jurisdiction, deaths that occur under these circumstances would be referred to a coroner for the determination of the cause of death. The coroner would order a *post-mortem* be carried out by either a 'general' pathologist or forensic pathologist, depending on the circumstances of the death.

8.3 The Role of the Coroner

The specific role and qualifications of a coroner vary from country to country, but basically their role is to determine the cause of death and issue recommendations to prevent a similar occurrence in the future.

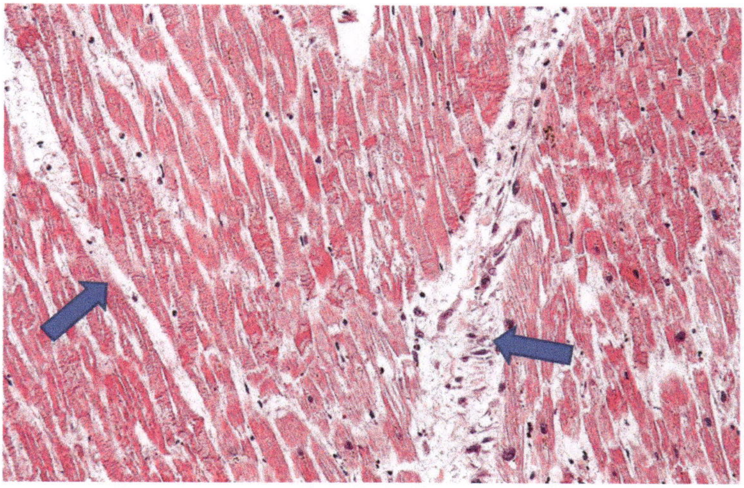

Figure 8.1 Heart section stained with haematoxylin and eosin from a patient who died of myocardial infarction (heart attack) ×400. Contraction band necrosis resulting in an inflammatory response with lymphocyte and monocyte infiltration is characteristic of myocardial infarction - marked with arrows. Reproduced from ref. 1 https://commons.wikimedia.org/wiki/File:MI_with_contraction_bands_high_mag.jpg, under the terms of CC BY-SA 3.0 license, https://creativecommons.org/licenses/by-sa/3.0/deed.en.

The coroner, depending on the jurisdiction, may be a medical doctor or a lawyer or perhaps both. They preside over the coroner's court, where the circumstances of death are explored and a verdict is drawn - this process is termed an inquest. The results of the *post-mortem* examination are crucially important in the decision-making process in the inquest.

The coroner is akin to a judge and, as such, has the authority to subpoena witnesses and appoint a jury. The verdicts of the coroner include:

- Death by natural causes
- Death by misadventure
- Accidental death
- Lawful killing (*e.g.*, self-defence)
- Suicide
- Unlawful killing (*e.g.*, murder)
- Occupational disease
- Drug dependence
- Non-dependent drug abuse

- Attempted abortion
- Self-induced abortion
- Still birth
- Self-neglect
- Lack of care/neglect
- Open verdict
- Disaster (but only if it has been the subject of a public enquiry)
- An open conclusion
- A narrative conclusion

The operation and responsibility of the coroner's court are too vast a subject (as indicated by the list of verdicts above) to cover fully here, but it is important to have an appreciation of the coroner's role because forensic cases are often under the auspices of a coroner.

8.4 The Forensic Pathologist

Pathology is a branch of medicine, but not all pathologists are medically qualified. Anatomical pathologists are usually medics, but clinical pathologists are often biochemists or members of the allied scientific disciplines (*e.g.*, microbiology and haematology). There are professional bodies in the field of pathology (*e.g.*, in the UK the Royal College of Pathologists) that recognise and accredit qualified members of the profession (*e.g.*, in the UK with Fellowship of the Royal College of Pathologists; FRCPath).

Forensic pathologists carry out *post-mortem* examinations on people who died suddenly, unexpectedly, or as a result of trauma or poisoning. Forensic pathologists usually undergo medical training before specialising in forensic pathology.

8.5 *Post-mortem* Examination (Autopsy, Necropsy)

In most countries, a *post* (Latin = after)-*mortem* (Latin = death) examination (also called autopsy from Greek *autos* = self and *opsis* = a sight, or necropsy from Greek *necros* = death, corpse) can only be carried out by an anatomical pathologist, who observes anatomical signs during his/her examination and sends samples to laboratory-based clinical pathologists or other specialists for further investigation (*e.g.*, biochemistry, histopathology). Usually the forensic (anatomical) pathologist, sometimes in consultation with clinical pathologist(s) or other experts such as neuropathologists, would determine the cause of death.

The forensic *post-mortem* is carried out in two parts; the external examination followed by an internal examination. Both examinations adhere to a protocol, even if an apparent cause of death is found early in the examination. This is because there might be other factors that contributed to the underlying cause of death. For example, the forensic pathologist might find a significant stab wound to the chest during his/her external examination. During the internal examination, the wound might be observed to extend to the heart. This might be the cause of death. However, additional investigation might reveal heart histology consistent with myocardial infarction (Figure 8.1). This could mean that the person had a heart attack that would have occurred irrespective of the assault and stabbing, or it might be that the heart attack was caused by the stress of the assault (although not consistent with the histology shown in Figure 8.1, indicating longstanding disease). Clearly, these mechanisms are very different. A stab to the heart will generally be fatal and result in a manslaughter or murder verdict, regardless of whether there was a pre-existing condition.

The objectives of the forensic *post-mortem* examination are summed up perfectly by Professors Bernard Knight and Pekka Saukko in their seminal text *Forensic Pathology* (see *Further Reading*):

- To make a positive identification of the body and to assess the size, physique and nourishment.
- To determine the cause of death or, in the newborn, whether live birth occurred.
- To determine the mode of dying and time of death, when necessary and possible.
- To demonstrate all external and internal abnormalities, malformations and diseases.
- To detect, describe and measure any external and internal injuries.
- To obtain samples for chemical, toxicological or general analysis, microbiological and histological examination, and any other necessary investigations.
- To retain relevant organs and tissues as evidence.
- To obtain photographs and videos for evidential and educational purposes.
- To provide a full written report of the autopsy findings.
- To offer an expert interpretation of those findings.
- To restore the body to the best possible cosmetic condition before release to the relatives.

8.5.1 Estimation of Time Since Death

The use of forensic entomology – particularly the larval stages of the blow fly – has already been discussed as a means of estimating time since death (see Chapter 5 *Forensic Biology*).

There are other techniques that are also used independently or in combination to enable the pathologist to estimate the time of death. I will outline just three of them here.

8.5.1.1 Body Temperature

Body temperature is important because theoretically it declines at a known rate *post-mortem* (based on Newton's Law of Cooling), which means that taking the temperature (*e.g.*, using an anal probe thermometer) of the body as soon as possible after it is found can be used to estimate time of death. The temperature of the environment where the body was found must be considered because, obviously, a body cools down more slowly in a warm environment than a cold environment. Also, a covered (*e.g.*, with clothing or a blanket) body cools down more slowly than a naked body, and an obese body cools more slowly than a lean body. All of these parameters are considered when a pathologist estimates the time of death. Professor John Glaister (1856–1932; see Chapter 2 *The History of Forensic Science*) estimated that a body loses 1.5 °C h^{-1} (Glaister worked in °F; I have converted to today's unit of temperature °C) until it reaches ambient temperature; he proposed the following equation (the Glaister Equation) to determine time since death:

$$\text{Time since death (h)} = \frac{37.5 - (\text{Rectal temperature }^{\circ}\text{C})}{1.5}$$

All methods used to estimate time since death only give a rough guide. The eminent forensic pathologist, Professor Bernard Knight, summed the time of death uncertainty up in his comment, '*The person died between the time they were last seen alive and the time they were found dead.*'

8.5.1.2 Rigor Mortis

Immediately after death, the body becomes flaccid (approx. 3 h after death) for a short time (to approx. 6 h after death) after which it begins to stiffen (approx. 6–12 h after death): this is *rigor* (Latin = stiffness) *mortis* (Latin = death). As autolysis sets in, *rigor mortis* subsides (approx. 12 h after death); the relatively brief periods of change from flaccid to rigor and then back to flaccid can be used as a rough guide of time since death. As a very general guide:[6]

- If the body feels warm and no rigor is present, it suggests that death occurred within the last 3 h.
- If the body feels warm and stiff, it suggests that death occurred 3–8 h earlier.
- If the body feels cold and stiff, it suggests that death occurred 8–36 h earlier.
- If the body is cold and not stiff, it suggests that death occurred more than 36 h earlier.

In addition, the spread of *rigor mortis* throughout the body, starting with muscles of the jaw and progressing to elbows, wrists, hip, knees and then ankles, can be used to roughly fine tune time since death. The conditions in which the body is placed (*e.g.*, ambient temperature) affect these timescales and thus must be taken into account when using this approach to estimate the time since death. This method of estimating the time since death is unreliable because of the many variables (*e.g.*, ambient temperature) involved; however; it can give a rough indication of the time since death when the pathologist first examines the body – this might be useful for the investigating police officer.

The biochemistry underpinning the flaccidity and rigor progression is relatively simple: under normal life conditions, nerves constantly send stimuli (impulses), inducing muscle contractions (*e.g.*, if you want to move your arms, your brain sends signals *via* nervous impulses to the muscles in your arms, causing the muscles to contact). This stiffens the muscles (by shortening the muscle fibres) and results in arm movement. When the impulse stops, the muscle fibres relax and the muscles extend again and your arm returns to its 'usual' relaxed position. The process is mediated by neurotransmitters (*e.g.*, acetylcholine), which carry the nerve impulse from one nerve to another nerve. These neurotransmitters are removed immediately after use to prevent a single signal from causing multiple muscle contractions (termed tetany, hence 'tetanus' which is a disease caused by a bacterium which produces a toxin that inhibits the process). The removal of neurotransmitters occurs either through uptake into nerve cells or by enzymes (*e.g.*, acetylcholinesterase) which break down the neurotransmitter. Immediately after death, neurotransmission stops, leading to body flaccidity because muscles are no longer being stimulated to contract. As time passes, the neurotransmitter removal processes (*e.g.*, enzymes like acetylcholinesterase) become inactive and so any remaining neurotransmitter builds up and causes muscle contraction – this is *rigor mortis*. As more time passes, dead cells begin to lyse and release their contents, including enzymes that break down proteins (proteases); these proteases begin to break down muscle fibres (autolysis), resulting in the restoration of flaccidity.

8.5.1.3 Vitreous Humour Potassium Concentration

Measuring potassium (K^+) concentration in the vitreous humour (a transparent gelatinous material that fills the eyeball) can also be used to estimate the time of death. K^+ diffuses from the retina (the light-sensitive tissue at the back of the eye) into the vitreous humour as the retinal cell membranes become permeable after death. Therefore, the K^+ concentration in the vitreous humour rises *post-mortem*, serving as an indicator of the time since death. I include this method here because it is still relatively novel and the subject of much research (and controversy), but it shows the biochemical ingenuity that can be applied to address important questions like the time since death.

8.5.2 External Examination

The body is usually transported in a body bag (Figure 8.2) to prevent contamination after it has been removed from the scene – this is akin

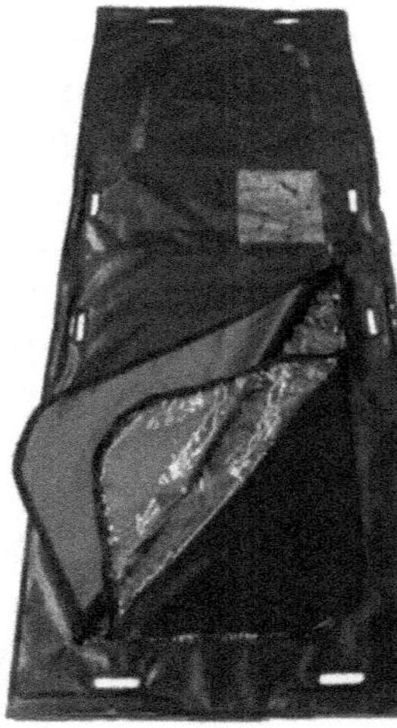

Figure 8.2 A body bag. The body bag prevents contamination of the body after it is removed from the crime scene. Courtesy of Bodybags.com.

to bagging smaller samples taken from the crime scene (see Chapter 3 *Crime Scene Investigation*).

When the body arrives at the mortuary, it is placed on a cadaver dissection table for both external and internal examinations (it might be stored in a refrigerator prior to *post-mortem* examination). There are many different designs of dissection tables, but modern dissecting tables are usually made up of stainless steel (Figure 8.3) and older 'tables' or 'slabs' as they were called were often made of vitreous china. It is important that these tables can contain bodily fluids, *etc.* during the examination process, and that they can be effectively cleaned between *post-mortem* examinations.

Before starting the examination, the cadaver is photographed in any clothes that it may have worn upon arrival at the mortuary. Detailed notes are taken regarding the clothes and any rips, tears, *etc.* that might be visible. The clothes are then removed for investigation in the lab and the naked cadaver is photographed to record all external features, particularly any abrasions, bruises or wounds. This is important because these details might be destroyed during the internal examination process. Again, detailed notes and measurements of any abrasions, bruises or wounds should be made.

Post-mortem hypostasis is also observed; this is the movement of blood after death under the influence of gravity – the blood settles

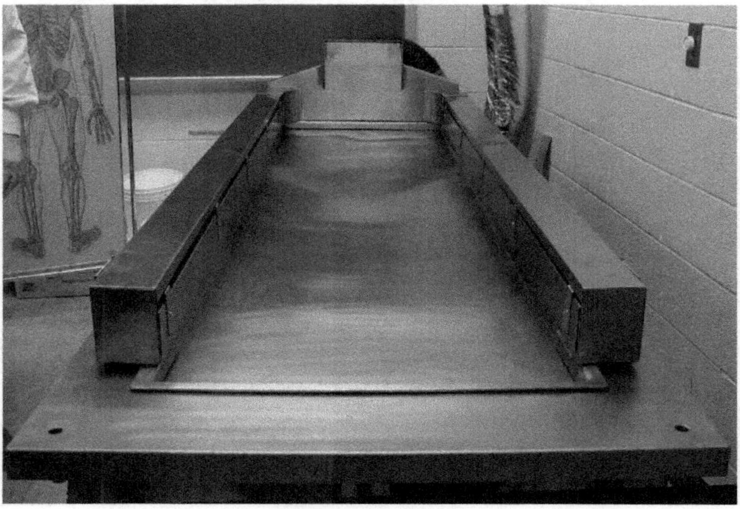

Figure 8.3 A typical modern cadaver dissecting table made of stainless steel. Reproduced from ref. 2, https://commons.wikimedia.org/w/index.php?curid=170499, under the terms of the CC BY-SA 3.0 License https://creativecommons.org/licenses/by-sa/3.0/deed.en.

in the lowest parts of the body making them red in appearance. Hypostasis indicates the position or orientation of the body after death. For example, if the person died on their back on a flat surface and remained there after death, there will be significant hypostasis at the lower points of the body (*e.g.*, shoulders, small of back, and rear of thighs) with a distinct absence of reddening at the pressure points (*e.g.*, buttocks) (Figure 8.4). On the other hand, if a person died by hanging, the hypostasis will be in the lower limbs because these are the lowest parts of the body in an upright hanging position.

The body may then be painstakingly searched for any residues adhering to it; flakes of paint, plant material, *etc.* are noted and photographed *in situ* before being removed, bagged and sent to the laboratory for detailed examination. Of particular importance are the fingernails; materials caught under the nails are noted and very carefully removed for later examination. Skin under the fingernails might indicate a struggle and might help, *via* DNA, to identify an assailant. UV light is sometimes used to look for residues adhering to the body and under fingernails because some of these materials fluoresce under UV and so are more easily visible.

Sometimes, the cadaver is X-rayed or put through a CT scanner to check for fractured bones, wound tracks or perhaps foreign bodies in the cadaver's tissues (*e.g.*, a bullet lodged in the brain; Figure 8.5).

Samples of hair and fingernails may be taken for closer examination later.

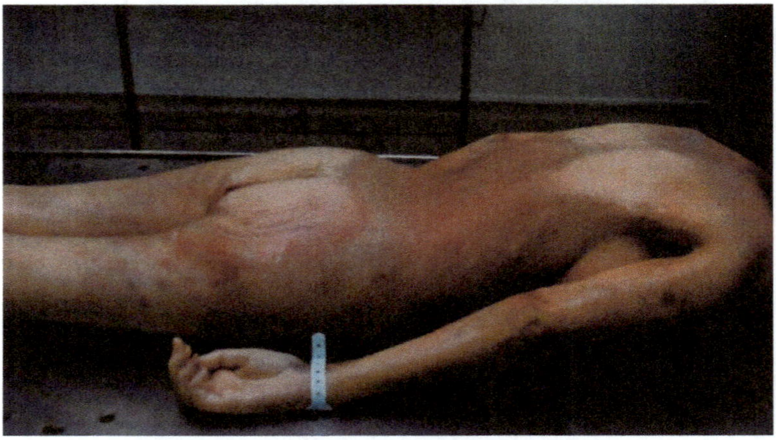

Figure 8.4 Hypostasis in the body of a person who died on their back and remained in that position for a significant time *post-mortem*. The reddened areas are where the blood has settled and the light areas denote the pressure points (*e.g.*, buttocks). Image: Courtesy of Dr Paul Fitzmaurice.

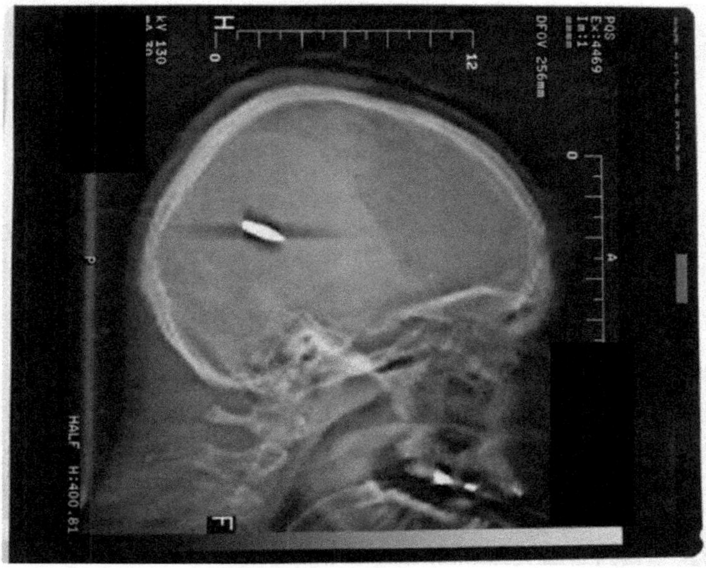

Figure 8.5 X-ray of the skull of a victim of a firearm homicide. A bullet can clearly be seen lodged in the brain. The two bright areas in the upper jaw are tooth fillings. Image: Copyright Dr Ghaffarpasand.

Upon completion of the external examination, the cadaver is weighed and measured (*e.g.*, height) and any distinguishing features (*e.g.*, birthmarks and tattoos) are described, noted and photographed. The cadaver's ethnic group, hair colour and length, eye colour, sex and approximate age are all recorded before the internal examination begins.

8.5.3 Internal Macroscopic Examination

Before the internal examination begins, a 'head block' (a solid plastic or rubber brick-like block) is placed under the cadaver's shoulders to raise and angle the head backwards and push the chest upwards to make the dissection easier.

The internal examination begins with a Y-shaped incision through the skin and muscle layers of the thorax, consisting of two diagonal cuts from the shoulders to the sternum (breast bone), followed by an incision to the end of the sternum (xiphoid process). Then, the incision from the xiphoid process is extended into the abdomen to the pubic region (pubis) (Figure 8.6) – the incision avoids the navel (umbilicus). The skin and tissues are pulled back to expose the breast bone (sternum) and ribs, along with the contents of the thorax

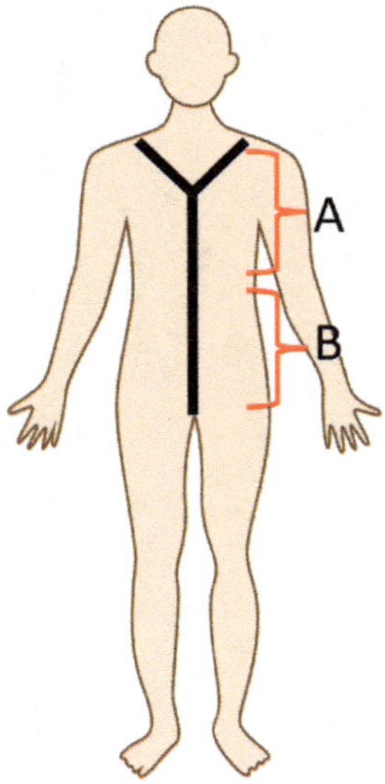

Figure 8.6 The Y-incision made to begin the internal *post-mortem* examination. A = thorax; B = abdomen.

(heart and lungs), and then the abdomen (digestive and sex organs). Different incision techniques are used by different pathologists – the Y-incision being a commonly used example.

Next, the rib cage is cut along either side of the sternum using *post-mortem* shears and the sternum is removed. This exposes the heart and lungs, which are then usually removed 'as one' (Figure 8.7) and investigated separately. The evisceration process extends into the abdomen, from where the digestive organs, *etc.* (*e.g.*, stomach, intestines, liver, and bladder) are removed for further examination. Sometimes, the thoracic cavity plus abdominal cavity organs are removed 'en masse' (Figure 8.7).

The organs are observed 'en masse', then separated, weighed (organ weight relative to body weight is an indicator of organ health; Table 8.1), dissected, and samples taken for microscopical examination (histology; see Section 8.5.4.1 *Histopathology*).

Next the skull is opened to expose the brain and facilitate its removal and further investigation later. A deep (to the bone) incision is made across the scalp (approximately from the temple above the nose to the nape of the neck) and the skin is pulled back (towards the ears) to expose the skull, which is then cut around its circumference using a specially designed *post-mortem* circular saw, revealing the brain (Figure 8.8).

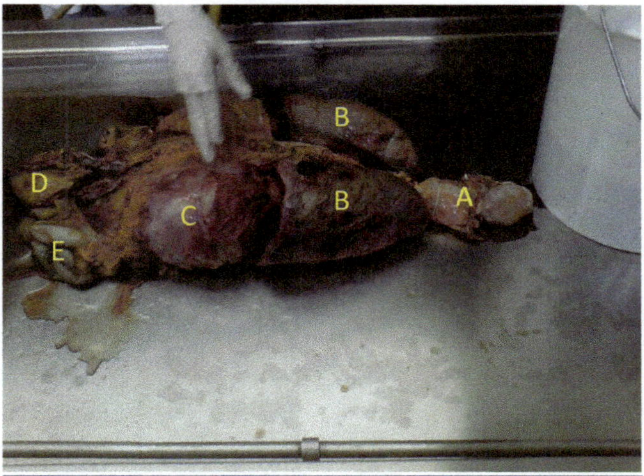

Figure 8.7 The thoracic and abdominal organs removed from a cadaver 'en masse' for further investigation. A = larynx (voice box) leading to the trachea (wind pipe); B = lungs; C = stomach; D = bladder; E = small intestine. Reproduced from ref. 3, https://commons.wikimedia.org/wiki/File:Human_dissection_of_the_abdominal_and_toraxic_organs.jpg, under the terms of the CC BY-SA 3.0 license https://creativecommons.org/licenses/by-sa/3.0/deed.en.

Table 8.1 Weights of hearts ± standard deviation from cadavers with cardiovascular disease and cancer compared with 'normal' heart weight. Reproduced from ref. 4 with permission from Elsevier, Copyright 2014.

	Mean heart weight (g)	
Cadaver group	Male[b]	Female[c]
'Normal'[a]	426 ± 108	351 ± 76
Cardiovascular disease	550 ± 140	430 ± 105
Cancer	392 ± 89	309 ± 66

[a] 'Normal' denotes a heart from a person who died of natural causes.
[b] $n = 805$; mean age = 66.5 years.
[c] $n = 605$; mean age = 70.6 years.

The brain is observed *in situ* to see if there are any signs of damage (*e.g.*, bruising following blunt trauma). It is then removed, weighed, dissected and samples are taken for histology (see Section 8.5.4.1 *Histopathology*).

The limbs are rarely dissected unless there are clear indications that something is amiss (*e.g.*, gunshot wounds, signs of injection sites, and bruising of interest).

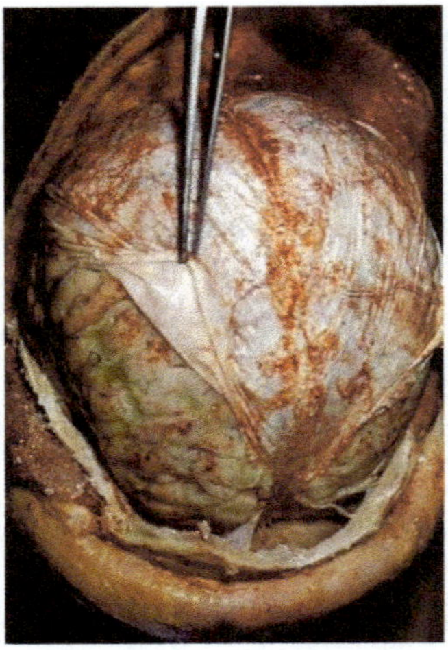

Figure 8.8 The brain is exposed following the removal of the top of the skull using a *post-mortem* circular saw. The pathologist is removing the dura mata (membrane covering the brain) with forceps. Figure is in the Public Domain.

8.5.3.1 Restoring the Body Before Its Release to Relatives

At the end of the *post-mortem* examination, the organs are placed back into the body cavity, which is then filled with (for example) cotton wool so that the mortuary technician can reconstruct the body, sew up the cavities, reattach the top of the skull and pull the scalp skin across and sew up, ensuring that there is little or no evidence of the procedure once the body is clad in a shroud. This is done out of respect for the relatives who might want to view their loved one.

8.5.4 Collecting Samples for Analysis

The organs are dissected (usually by slicing with a large *post-mortem* knife to observe their internal structures). Many disease processes or poisoning agents result in characteristic effects on an organ's morphology and/or colour. For example, paracetamol (Panadol, Tylenol, acetaminophen; see Chapter 9 *Forensic Toxicology*) results in a characteristic yellowing of the liver in distinct foci, and carbon monoxide (*e.g.*, from a car exhaust) results in the formation of bright orange/red carboxyhaemoglobin, which results in a characteristic bright red liver. These changes are all part of the observations that the forensic pathologist puts together to determine the cause of death. In addition, damage to the heart due to a stabbing wound or, perhaps, a bullet found lodged in the brain will be important physical information used to assist in determining the cause of death.

Bodily fluid samples might also be taken during the internal examination. The most commonly taken samples are blood, urine and bile; they are particularly important in forensic toxicology cases because toxic chemicals (including drugs) are usually present in the blood and often excreted in bile and/or urine. The concentrations of toxic chemicals in these fluids are important in determining whether they might have resulted in death (see Chapter 9 *Forensic Toxicology*) or to assess a person's level of drug or alcohol intoxication at the time around their death. In addition, biochemical markers can be measured, which indicate organ-associated diseases (*e.g.*, high blood activity of the enzyme glutamate pyruvate aminotransferase – SGPT – indicates liver damage). It must, of course, be remembered that after death biochemical parameters change significantly (see Section 8.5.4.2 *Biochemistry*).

8.5.4.1 Histopathology

Histopathology (Greek *histos* = tissue) is the study of changes in tissues caused by disease. Even though the forensic pathologist is often looking for non-disease causes of death, s/he still has to determine whether disease might be a cause or a contributor to death. In addition, toxic chemicals (*e.g.*, in poisoning cases) often induce histopathological changes that serve as diagnostic indicators of intoxication; for example, death caused by an overdose of paracetamol causes a very characteristic histopathological change in the liver (Figure 8.9; see Chapter 9 *Forensic Toxicology*).

In order to visualise histopathological changes, samples of tissues (*e.g.*, liver) are dehydrated (with ethanol), embedded in wax and cut into very thin (approx. 5 μm thick) sections using a microtome (Figure 8.10). The wax is removed using a solvent (*e.g.*, xylene) and the remaining tissue section is stained (*e.g.*, with haematoxylin and eosin – H&E) before being mounted on a glass microscope slide, allowing the cells and cell organelles to be observed under a light microscope.

Figure 8.1 shows a section of human heart stained with H&E. The eosin stains the cell cytoplasm pink and the haematoxylin stains the cell nuclei dark blue. There are many other stains used by histopathologists; each stain (or combination of stains) shows different cell features in different colours. Stains are chosen to highlight particular facets of cells to aid in specific diagnoses. Examining a stained tissue section under the microscope gives the pathologist a wealth of

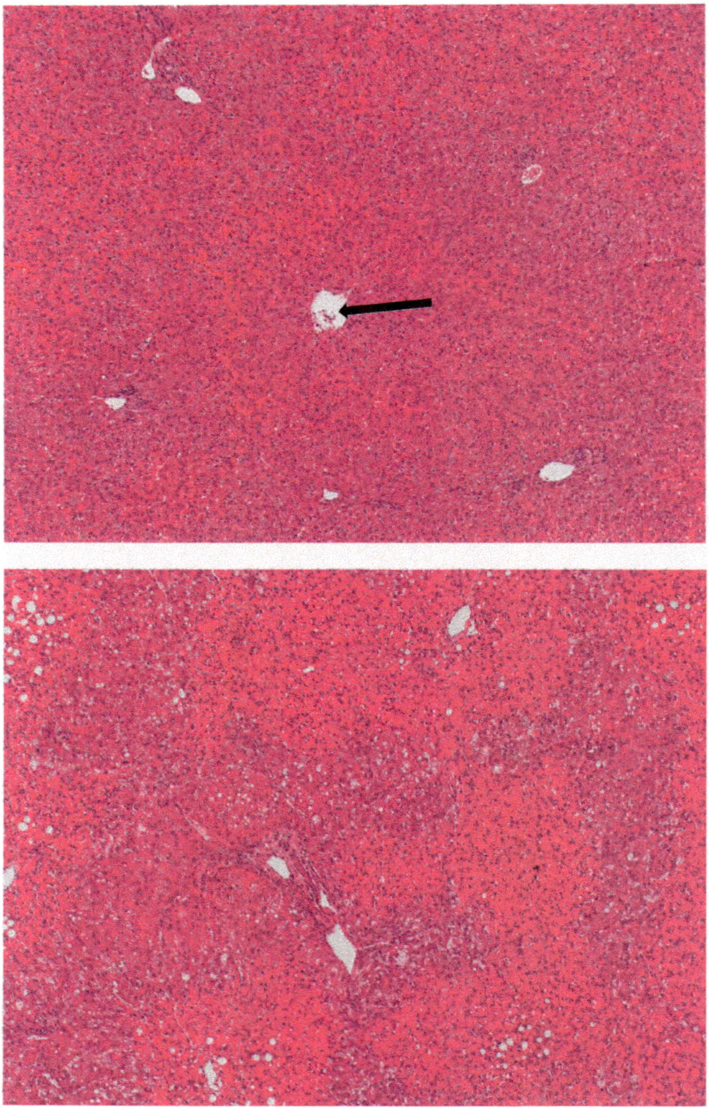

Figure 8.9 Sections of the liver stained with haematoxylin and eosin (H&E). Images: courtesy of Dr Rosa Miquel, Institute of Liver Studies, King's College Hospital, London. Above: normal liver showing normal cell organisation, structure and staining. The central vein is marked with an arrow. Below: liver from a paracetamol overdose fatality; there are clear differences between the normal and paracetamol liver sections which show the centrilobular necrosis characteristic of paracetamol poisoning. You can see important diagnostic changes around the central vein – the tissue structure is quite different from the normal liver and there is an increased haematoxylin (purple) staining. Magnification approx. ×400.

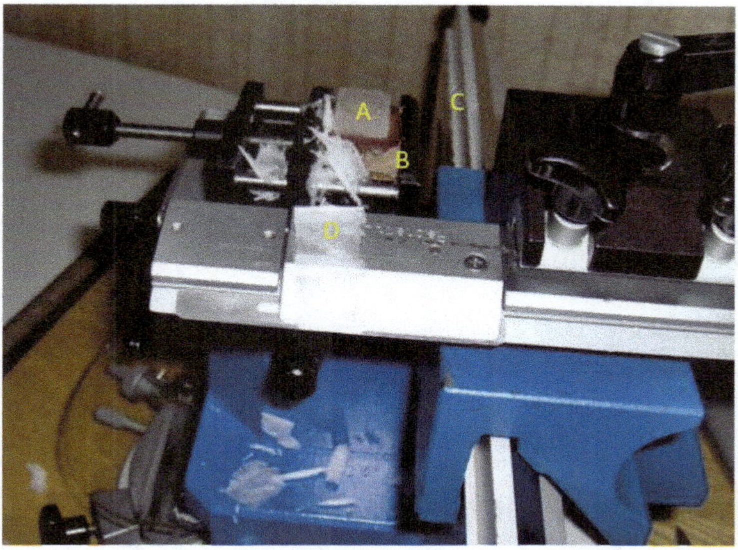

Figure 8.10 Cutting thin tissue sections with a microtome: the dehydrated tissues are mounted in wax blocks (A), a wax-mounted tissue block is held in a vice (B) and a fine, extremely sharp blade (C) moves up and down cutting approximately 5 μm-thick sections (D) from the block, which are gently peeled off onto a microscope slide (not shown). Reproduced from ref. 5, https://commons.wikimedia.org/wiki/File:Sledge_microtome.jpg, under the terms of the CC BY-SA 3.0 license https://creativecommons.org/licenses/by-sa/3.0/deed.en.

information about the health of the organ from which the tissue was taken. This information is often crucial in determining the cause of death.

8.5.4.2 Biochemistry

Bodily fluids, including blood, urine and stomach contents samples are often taken during the internal *post-mortem* examination. They can be analysed quantitatively for biochemicals, poisons (*e.g.*, metals) or drugs. The presence of toxic concentrations of drugs or poisons in these samples might establish the cause of death; the presence of, for example, pills or food in the stomach contents might help identify what the person had ingested before death. In addition, the concentrations of particular biochemicals might indicate disease or perhaps organ damage caused by the ingestion of toxic chemicals (*e.g.*, the presence of carboxyhaemoglobin following carbon monoxide poisoning); again, this might be important in determining the cause of death – this will be discussed further in Chapter 9 *Forensic Toxicology*.

Further Reading

1. P. Saukko and B. Knight, *Knight's Forensic Pathology*, CRC Press, Boco Ratan, London, New York, 2016.
2. R. B. Dettmeyer, *Forensic Histopathology – Fundamentals & Perspectives*, Springer-Verlag, London, Berlin, New York, 2011.
3. B. Madea, Methods of determining time of death, *Forensic Sci., Med., Pathol.*, 2016, **12**, 451–485.

References

1. File:MI with contraction bands high mag.jpg. Wikimedia Commons. https://commons.wikimedia.org/wiki/File:MI_with_contraction_bands_high_mag.jpg.
2. File:Cadaver dissection table – long shot.jpg. Wikimedia Commons. https://commons.wikimedia.org/w/index.php?curid=170499.
3. File:Human dissection of the abdominal and toraxic organs.jpg. Wikimedia Commons. https://commons.wikimedia.org/wiki/File:Human_dissection_of_the_abdominal_and_toraxic_organs.jpg.
4. N. T. Kumar, *et al.*, *Cardiovasc. Pathol.*, 2014, **23**, 5–11.
5. File:Sledge microtome.jpg. Wikimedia Commons. https://commons.wikimedia.org/wiki/File:Sledge_microtome.jpg.
6. Death and Kinetics [Internet]. Fsu.edu. 2019. Available from: https://www.chem.fsu.edu/chemlab/chm1020c/Lecture%208/02.php.

9 Forensic Toxicology

Toxicology is the scientific study of poisons (including drugs); it covers the underlying biochemical mechanisms by which poisons cause harm and death, the processes that occur when a poison is ingested (toxicokinetics), the doses/blood concentrations that result in toxic effects and death, and the methods of extraction and analysis used to determine the poison in the body and the effects of the detected concentration (*i.e.*, would it cause harm or death?). Forensic toxicologists focus on the latter: their role involves identifying and determining tissue and fluid concentrations of ingested poisons and giving an expert opinion on whether the concentration found is likely to have caused harm or death. While forensic toxicologists concentrate on impairment from drug use or death by poisoning, clinical toxicologists are more concerned with sub-lethal doses of poisons and the treatment of poisoning.

Illicit drug investigations are also covered in this chapter, although in most countries they would be conducted by forensic chemists rather than forensic toxicologists. However, the methods used are akin to those used by forensic toxicologists, which is why they are included in this chapter.

9.1 Toxicology

Toxicology (Latin *toxicus* = poisoned; Greek *toxicon* = arrow poison; Greek *logia* = discourse, doctrine, theory, science) is the study of poisons in its broadest sense. It includes absorption, distribution (around the body) and excretion (toxicokinetics), effects of toxic

Forensic Science: The Science Behind the Truth
By Ian Shaw with Anna Sandiford

Published by the Royal Society of Chemistry, www.rsc.org

chemicals on animals (including humans), the biochemical mechanisms underlying toxicity, and the doses or exposure concentrations that cause harm or death. In addition, regulatory toxicologists apply toxicological thinking to risk assessments of new chemicals (*e.g.*, medicines, pesticides), and clinical toxicologists diagnose poisoning and treat poisoned patients. Toxicologists are usually qualified in biochemistry or perhaps chemistry (or an allied subject) and/or medicine (clinical toxicologists are usually qualified in medicine).

Forensic toxicology is a specialized branch of toxicology, which focuses on human (and sometimes veterinary) toxicology. In this chapter, I will focus on fatal poisoning cases. It is important to note that practising forensic toxicologists carry out a broader array of investigations, including the effects of drugs and alcohol on people involved in criminal incidents as well as cases related to drink and drug driving and drug facilitated sexual assault (see Chapter 4 *Forensic Chemistry* for blood alcohol investigations).

9.2 Forensic Toxicology in Poisoning Cases

Forensic toxicology is the application of toxicological principles to support medicolegal investigations. It relies very heavily on analytical chemistry (see Chapter 4 *Forensic Chemistry*) because in order to determine whether a toxic chemical (*e.g.*, a drug) caused death, the concentration of the chemical in blood or perhaps organs (*e.g.*, the liver) must be determined. The concentration found in blood is compared with the concentration range known to be fatal to determine whether or not the drug or toxic chemical might have caused the death under investigation.

The simple fact that a chemical is present in the body does not mean it caused death. For example, paracetamol (also known as Panadol, acetaminophen and Tylenol; Figure 9.1) is a very commonly used over-the-counter treatment for mild fever and pain and is the most commonly encountered self-poisoning substance worldwide; if it is found in blood at 10 mg L^{-1}, this indicates therapeutic use; however, if the blood concentration is 180 mg L^{-1}, a paracetamol overdose is very likely to have been the cause of death (Table 9.1). It is also important to bear in mind that more than one poisoning agent might be involved because this could significantly reduce the toxic concentration of the individual substances – I will discuss paracetamol poisoning in greater detail later in this chapter (see Section 9.2.1 *Case 1: Paracetamol Poisoning*; see also Chapter 8 *Forensic Pathology*).

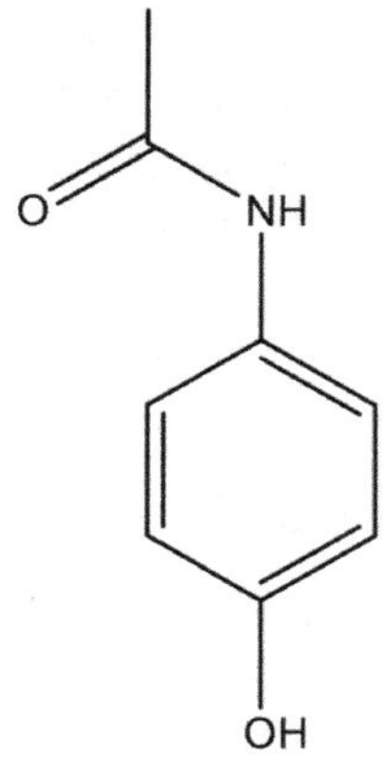

Figure 9.1 Paracetamol (also known as Panadol, acetaminophen or Tylenol): N-acetyl-4-aminophenol or more simply *p*-acetylaminophenol (*p* denotes the para- or 4-position on the phenol ring), which is contracted to paracetamol.

Table 9.1 Therapeutic, toxic and lethal blood concentrations of paracetamol. A paracetamol overdose is not necessarily fatal, but there is a significant overlap between non-fatal overdose and acute fatal dose blood concentrations due to inter-individual variations in susceptibility.

	Paracetamol concentration	
Effect	Blood (mg L^{-1})	Liver (mg kg^{-1})
Therapeutic	<30[a]	
Overdose	30–300[a,b]	>60[b]
Acute fatality	50–400[a]	

[a] Data reproduced from ref. 1 with permission from Elsevier, Copyright 2013.
[b] Data reproduced from ref. 2 with permission from Oxford University Press, Copyright 1981.

Forensic toxicologists use sophisticated analytical techniques (see Chapter 4 *Forensic Chemistry*) to measure concentrations of drugs and other toxic chemicals in *post-mortem* samples (usually blood), and then they compare the concentrations found with data from other cases with known outcomes (*e.g.*, death) to determine whether the concentration found is likely to have caused death. Not only do forensic toxicologists determine concentrations in tissues, they also identify unknown substances either in tissues or that might be otherwise associated with a death (*e.g.*, residues in a glass from which the deceased drank). Their role is to identify unknown substances and decide whether they might have caused or contributed to the death. In addition, forensic toxicologists may be called upon to identify drugs of abuse both in suspected overdose cases and suspected illicit substance trade (*e.g.*, seized at a port). However, the responsibility for illicit drug investigations varies from country to country, but is usually undertaken by forensic chemists. I have included it here because the methods used are very similar to those used by forensic toxicologists.

The work of a forensic toxicologist is best illustrated by case examples.

9.2.1 Case 1: Paracetamol Poisoning (See Ref. 10)

A 30-year-old woman suffering from a psychiatric disorder was admitted to hospital after apparently ingesting at least 50 paracetamol tablets (*i.e.*, 25 g paracetamol; the lethal dose in humans is 5–20 g). The woman died 4 days later. The *post-mortem* examination showed acute liver necrosis (see Chapter 8 *Forensic Pathology*), which is consistent with paracetamol toxicity. As part of her treatment regimen, the woman had been prescribed nortriptylene (an antidepressant) and barbiturates (used to treat anxiety).

Nortriptylene, barbiturates and their metabolites were found in blood at therapeutic concentrations. In addition, 4-aminophenol (a minor metabolite of paracetamol) was found at a concentration of 224 mg L^{-1} in blood – this is equivalent to 310 mg L^{-1} paracetamol and is consistent with acute fatality (Table 9.1). The liver histology (showing necrosis) and blood concentration of 4-aminophenol prompted the forensic toxicologist to conclude that death was caused by paracetamol poisoning.

9.2.1.1 Toxicology of Paracetamol

Paracetamol is readily absorbed from the gastrointestinal tract, resulting in a peak blood concentration (C_{max}) 20 min after dosing. It is metabolised in the liver, producing a series of metabolites, most of which are quickly excreted in urine (Figure 9.2). At therapeutic doses (*i.e.*, 1 g every 4 h), the rapid metabolism (detoxification) and excretion of paracetamol and its metabolites in urine make toxicity very unlikely. However, there is a very minor, but highly toxic metabolite produced (*N*-acetyl-4-benzoquinone imine; NAPQI – remember the 4-position on a ring structure is also termed *para* or *p*, hence 'P' in the abbreviated name); at therapeutic doses of paracetamol, the minimal amount of NAPQI produced is quickly neutralized by the cell protective molecule, glutathione (a tripeptide comprising glycine: cysteine:γ-glutamate), preventing NAPQI from causing harm to liver cells. If the paracetamol dose is high, there is insufficient glutathione

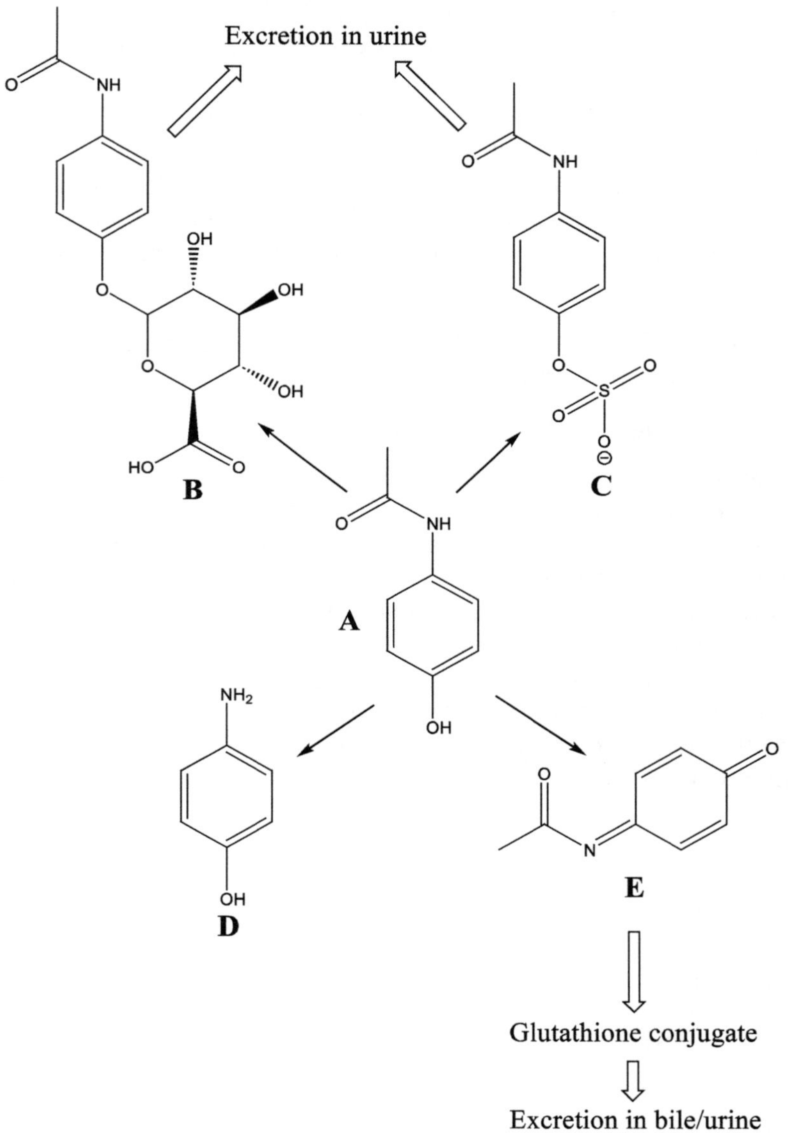

Figure 9.2 Metabolism of paracetamol in the liver: paracetamol (A) forms highly water-soluble glucuronide (B) and sulfate (C) conjugates which are excreted in urine – this is the major (~80%) detoxification and excretion route. A small proportion (~2%) of the paracetamol dose is deacetylated to form 4-aminophenol (D) or oxidised (5–15%) to form highly hepatotoxic NAPQI (E); the latter is detoxified by conjugation with glutathione and excreted in bile or urine. If the paracetamol dose is very large, the glutathione pathway becomes saturated, leaving free NAPQI, which results in hepatotoxicity, liver necrosis and death.

to neutralize the NAPQI, resulting in serious liver cell damage. The liver cell damage (hepatotoxicity) leads to liver necrosis and eventually liver failure and death.

After a high dose of paracetamol, the produced NAPQI begins the sequence of events that leads to hepatotoxicity, liver failure and death. The toxic effects on the liver cells are fast, but the knock-on effects leading to liver necrosis and liver failure are slower; therefore, it is usually a period of 4–12 days between a paracetamol overdose and death.

9.2.2 Case 2: Death Associated with a Synthetic Cannabinoid (See Ref. 11)

9.2.2.1 Synthetic Cannabinoids

In recent years, a number of synthetic cannabinoids have appeared on the illicit market. Many of them are far more toxic than their 'natural' cannabinoid counterparts (*e.g.*, tetrahydrocannabinol – THC) found in cannabis (dried herbage from *Cannabis sativa*). They have molecular structures akin to THC (Figure 9.3) and interact with the same nervous system receptors, but they often exhibit greater pharmacological effects. *N*-(1-amino-3,3-dimethyl-1-oxobutan-2-yl)-1-(4-fluorylbenzyl)-1*H*-indazole-3-carboxamide (fortunately it is abbreviated as ADB-FUBINACA! Figure 9.3) is an example of a synthetic cannabinoid that has been associated with the death of a woman in the USA.

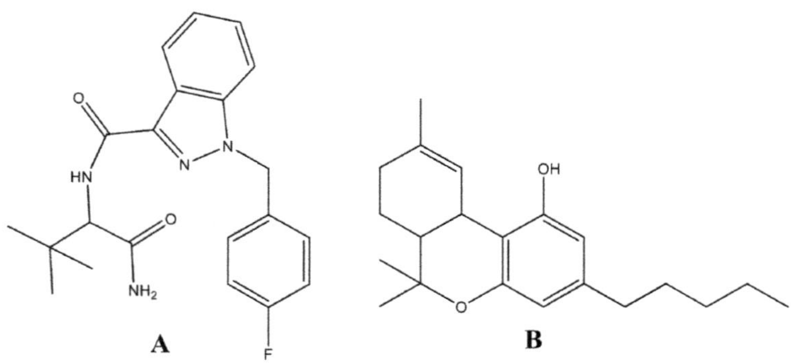

Figure 9.3 Molecular structures of ADB-FUBINACA (A) and the natural cannabinoid from cannabis, THC (B). They have molecular similarities which means that they both interact with the same central nervous system (CB_1) and peripheral nerve receptor (CB_2), and thus have cannabinoid effects, but ADB-FUBINACA's molecular structure is sufficiently different from that of THC, resulting in a different and more severe toxicity profile. CB_1 is responsible for the drugs' psychoactive effects and CB_2 is involved in pain reception (this is why cannabis and THC are sometimes used to treat chronic pain).

Interestingly, ADB-FUBINACA was originally developed by a pharmaceutical company as a medicine but was later dropped from the company's development programme.

9.2.2.2 The Case and Its Investigation

A 41-year-old woman smoked 'Mojo', a synthetic cannabinoid-containing product, at her home in the USA. Soon after she had smoked the product, she became violent and aggressive. Her children restrained her to prevent her from injuring them. During the time they restrained her, she became unconscious and unresponsive. Emergency services were called, and the woman was certified dead shortly after their arrival.

A *post-mortem* examination was carried out, revealing the following:

- Vascular congestion [blocked blood vessels].
- Thrombic occlusion of the lumen of the left anterior descending coronary artery – this was due to disruption of coronary arterial plaque caused by haemorrhage [plaque had fallen off the artery wall and blocked the coronary artery].
- Ischaemia of the left ventricular myocardium [lack of oxygen supply to the heart muscle, probably due to the blocked coronary artery].

Clearly, the woman had pre-existing heart issues (*e.g.*, plaque – thickening of the blood vessel walls often due to a build-up of cholesterol).

Post-mortem blood samples were screened for a broad array of drugs. The screen was positive for cannabinoids and negative for opioids/opiates (*e.g.*, morphine and heroin). HPLC and GLC assays were carried out to check for other drugs or toxic chemicals that might be present; both assays detected nothing of interest.

The positive cannabinoids screen warranted further investigation by HPLC-MS (see Chapter 4 *Forensic Chemistry*). Both THC and 11-nor-9-carboxy-9-tetrahydrocannabinol (carboxy-THC, a metabolite of THC) were present at 1.1 ng mL^{-1} and 4.7 ng mL^{-1}, respectively (3.3–25 ng mL^{-1} THC is the typical range observed 1 h after smoking cannabis; therefore, the result is consistent with the woman having recently smoked cannabis). Furthermore, HPLC-MS studies showed the presence of ADB-FUBINACA at 7.3 ng mL^{-1}. As discussed in Chapter 4 *Forensic Chemistry,* specific ion transitions in the mass spectrometer are used to verify a molecule's structure; an example of an ion transition used to identify ADB-FUBINACA in this case is shown in Figure 9.4.

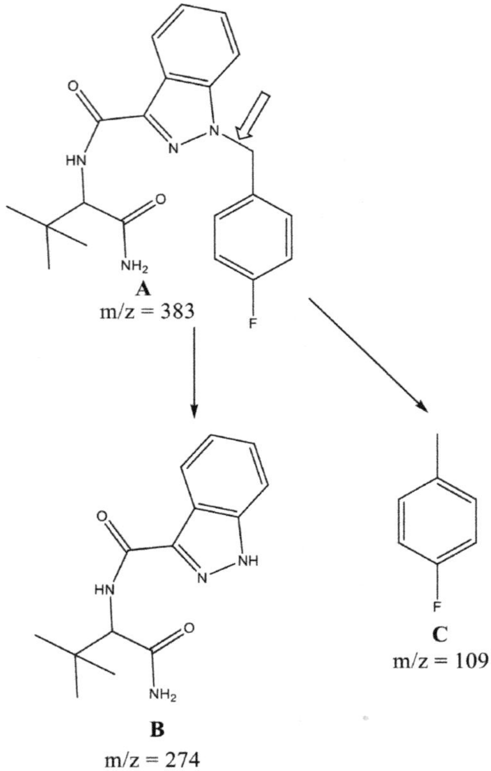

Figure 9.4 One of the ion transitions (*m/z* 383 → 109) used to verify the identity of ADB-FUBINACA (A) in the deceased woman's blood. The arrow shows the ADB-FUBINACA (A) fragmentation position which leads to the generation of two fragments B and C. All three ions include an extra proton (not shown).

These results showed that the woman had most likely smoked cannabis mixed with ADB-FUBINACA. From these results, it appears that the Mojo the woman smoked contained both cannabis and ADB-FUBINACA (often Mojo is a non-cannabis plant material coated with a synthetic cannabinoid, but in this case it appears to have been cannabis herb mixed with a synthetic cannabinoid).

Combining all the scientific findings it was thought that the woman's death resulted from pre-existing coronary heart disease exacerbated by the ingestion of synthetic cannabinoids. A possible scenario is that the ADB-FUBINACA induced a change in her mood, triggering aggression and a rapid increase in her heart rate; this coupled with her underlying coronary disease led to dysrhythmia (also termed arrhythmia: an abnormal beating pattern of the heart) which caused her death.

9.2.2.3 Determining the Cause of Death in Synthetic Cannabinoid-related Cases

It is often difficult to determine the 'exact' cause of death in a synthetic cannabinoid case – as was illustrated in the ADB-FUBINACA case discussed above. Often the cannabinoid is not the sole cause of death, but rather it is a combination of risk factors (including the synthetic cannabinoid and underlying heart disease as in the case above) converging and resulting in death (see *Further Reading*).

9.3 Investigating a Forensic Toxicology Case Involving a Death

The case examples outlined above illustrate the principles of a forensic toxicology investigation. These cases highlight the need to identify any toxic chemicals present in the body and to determine their blood concentrations to decide whether the amount present would have likely caused death. Using modern chromatographic techniques coupled with mass spectrometry (see Chapter 4 *Forensic Chemistry*), it is often possible to identify and quantify toxic chemicals in blood directly; however, sometimes extraction is necessary to concentrate the test chemical, and sometimes other *post-mortem* tissue (*e.g.*, liver) samples are analysed if it is difficult to obtain sufficient blood (*e.g.*, in cases involving putrefied bodies) (Figure 9.2).

9.3.1 Sampling and Analysing *Post-mortem* Blood

After death, the blood pools in the lower portions of the body (this depends on how the body is lying – see Chapter 8 *Forensic Pathology*) and usually clots about 3–15 min after death. Therefore, taking a blood sample during the *post-mortem* examination is not as easy as it usually is when a person is alive. For quantitative analytical work, it is common to extract blood from the peripheral circulatory system (often the femoral vein) because this blood provides a better reflection of the pharmaco(toxico)kinetics of the drug or toxic chemical. Sometimes, the vein has to be dissected to get a blood sample because of clots preventing its sampling with a hypodermic needle (see *Further Reading*).

The redistribution of blood in the body after death (*e.g.*, as mentioned above, it pools in the lower portions of the body) means that drugs and toxic chemicals in the blood redistribute too. For this reason, drug or toxic chemical concentrations in *post-mortem* blood might not reflect their concentrations in blood from a particular part of the body before death – this might have implications for the interpretation of blood analysis results and this is why fluid and tissue samples should be collected from different places in the body and caution used when interpreting results.

Blood analysis is divided into two phases: screening and quantification. Screening determines the drugs or toxic chemicals present. If the screening results show the presence of a drug or toxic chemical, quantification would usually be carried out. This enables the toxicologist to determine whether the concentration of a drug found in the blood corresponds to a sub-therapeutic, therapeutic, toxic or fatal dose (*e.g.*, paracetamol, see Table 9.1).

9.3.1.1 Blood Drug Screen

The specifics of blood drug screening vary from country to country and lab to lab; therefore, I have given a generalised analysis approach here, giving you a 'feel' for the process rather than an understanding of the specific methods used.

Blood samples are treated with a chemical (*e.g.*, formic acid) to precipitate proteins (*e.g.*, blood albumin) and centrifuged to separate the serum from blood cells and protein – blood cells and proteins interfere with chromatography. Alternatively, the blood sample might be passed through a solid phase extraction column, where the compounds in the blood bind to the column and are later eluted with a solvent. The serum or solid-phase extraction column eluate is analysed by an LC–MS method (see Chapter 4 *Forensic Chemistry*) – usually LC–MSMS or LC–time of flight (TOF)MS because these methods are precise and very sensitive. Using authentic analytical standards (*e.g.*, specific drugs) and databases, it is often possible to identify drugs and other toxic chemicals in the blood sample. If compounds are identified that are relevant to the case circumstances, further analysis is carried out to quantify the compound(s) and determine if a sufficient amount was present to cause death.

9.3.1.2 Quantification of Drugs and Toxic Chemicals in Post-mortem *Blood Samples*

The methods used to quantify drugs and toxic chemicals in *post-mortem* blood samples are often based on LC–MS. However, sometimes other methods (*e.g.*, GC–MS: see Chapter 4 *Forensic Chemistry*) are used for particular compounds. In addition, clean up methods might be used to either concentrate the analyte or remove interfering compounds.

The important difference between screening and quantification methods is that the latter must be validated to make sure that the results (particularly quantitative results) are accurate and reliable. Validation includes the following parameters:

- Selectivity – indicates whether the signals obtained from the analytical method (*e.g.*, peaks on a chromatogram) are specific to a particular analyte (*e.g.*, whether the peak retention time relates to only the target analyte).
- Linearity – indicates whether the graph of concentration *vs* signal magnitude is a straight line, or over what concentration range it is a straight line.
- Limit of detection (LOD) – is the smallest amount an analytical method can detect. This is important because even if an analytical method does not detect a particular compound, it does not mean that it is absent; rather it could be present in an amount below the LOD.
- Limit of quantification (LOQ) – is the minimum amount that a method can reliably quantify.
- Recovery – is the percentage of the compound present in the sample that is detected by the analytical method. It is determined by adding a known amount of compound to a blood sample and analyzing the 'spiked' sample using the analytical method: $\% \text{ Recovery} = \dfrac{\text{Amount found}}{\text{Amount added}} \times 100$
- Accuracy – refers to how close an analytical reading is to the true value. For LC analysis, this involves assessing how close the compound's V_R is to the authentic standard's V_R (see Chapter 4 *Forensic Chemistry*), *i.e.*, how sure can we be that we have identified the compound correctly if its V_R is not identical to the authentic standard's V_R?
- Precision – refers to how closely two separate measurements of the same sample align with each other; can we distinguish two closely related compounds from each other? For LC, this is influenced by the width of the peaks on the chromatogram. Broader peaks increase the likelihood of overlap with other compounds, resulting in reduced analytical precision.
- Stability – refers to the stability of the analyte in the analytical matrix (*e.g.*, blood). If it is unstable, the analytical result will not reflect the compound's concentration in the sample at the time of *post-mortem* sampling because in between sampling and analysis, some (or all) of the compound might degrade.

Having values for each of these validation parameters means that the toxicologist can interpret the results properly and with confidence. For example, if the compound detected in the blood sample was unstable with a decay half-life in blood at +4 °C of 6 h and the sample had been stored in a refrigerator (*i.e.*, +4 °C), and its recovery after solvent extraction and LC–MSMS was 70%, the toxicologist should consider these validation parameters when establishing the 'true' concentration of the compound in the blood sample. This is, of course, very important when determining whether the blood concentration found might have been fatal. There are differing views about how to account for recovery and stability, but such details are beyond the scope of this book.

9.3.1.3 Other Post-mortem Samples Analysed by Forensic Toxicologists

As already discussed, blood is by far the best sample for modern forensic toxicology investigations. However, sometimes it is not possible to get a blood sample *post-mortem*. This might be the case if the body has been dead for an extended period and has begun to putrefy or the body was severely traumatized and lost a lot of blood. In such cases, tissue samples (Table 9.2) may be taken for analysis if they are in a suitable condition – often samples taken from the liver are subjected to analysis. A measured amount of liver (usually 10 g) is taken for analysis; it is enzymatically digested in a buffer. The enzyme hydrolysate is then extracted (the pH of the homogenate is often changed to aid in extraction, but I will not go into detail here) with a solvent (*e.g.*, diethyl ether) or using a solid phase extraction method (see Chapter 4 *Forensic Chemistry*).

Stomach contents might be analysed in drug overdose cases when there are often very high concentrations of the drug present. Tablets or capsules might also be recovered; they can be identified by their shape, colour and markings.

Urine might also be taken directly from the bladder during the *post-mortem* examination. Its analysis is useful to show the presence of drugs or toxins, but it is often difficult to extrapolate this information back to blood concentrations and thus therapeutic or toxic doses.

The extract (*e.g.*, from blood) or the sample (*e.g.*, urine) is usually analysed by chromatography (HPLC or GLC) coupled to MS or MS–MS (see Chapter 4 *Forensic Chemistry*) to identify and quantify the drugs or toxins present.

Table 9.2 *Post-mortem* samples analysed in forensic toxicology.[a]

Sample	Analytical process	Notes
Blood	Solvent extraction → HPLC/MS, GLC/MS	Blood clots *post-mortem*, therefore blood samples are often difficult to obtain. Both toxic chemicals and their metabolites might be present.
Stomach contents	Evidence of pills or capsules? Solvent extraction → HPLC/MS, GLC/MS	Recovery of pills/capsules, followed by analysis. Only parent toxic chemical is usually present.
Urine	Solvent extraction → HPLC/MS, GLC/MS	Parent toxic chemical and metabolites often present.
Faeces	Solvent extraction → HPLC/MS, GLC/MS	Rarely collected. Unabsorbed and biliary excreted parent toxic chemicals and metabolites present.
Organs (usually liver)	Solvent extraction → HPLC/MS, GLC/MS	Parent toxic chemicals and metabolites present.
Bile	Solvent extraction → HPLC/MS, GLC/MS	Parent toxic chemicals and metabolites present – not all drugs or their metabolites are excreted in bile.

[a] Note: HPLC is often referred to as LC and GLC is referred to as GC. When these methods are coupled to MS, this often involves tandem MS which in the case of LC is referred to as LC/MS–MS (see Chapter 4 *Forensic Chemistry*).

9.3.2 Drugs and Toxins in Hair

Many compounds (including drugs) found in blood are deposited as residues in the hair medulla *via* the capillary blood supply to the hair root (Figure 9.5).

Since the presence of a compound in the hair medulla is related to its blood kinetics (see Section 9.3.2.1 *Kinetics of Drug and Toxic Chemical Deposition in Hair*), it is possible to determine the timescale of the presence of the compound in blood and estimate the dosing time of the compound based on the hair growth rate – the human hair growth rate is about 10 mm per month at the vertex (crown) of the head. This has been used extensively in forensic casework to estimate the approximate date range of administration of a drug or toxic compound in poisoning or illicit drug cases.

The mechanism of drug deposition in hair involves passive diffusion of the drug (or its metabolites; I will use 'drug' to mean either drug and/or drug metabolites) from the circulatory system *via* the root

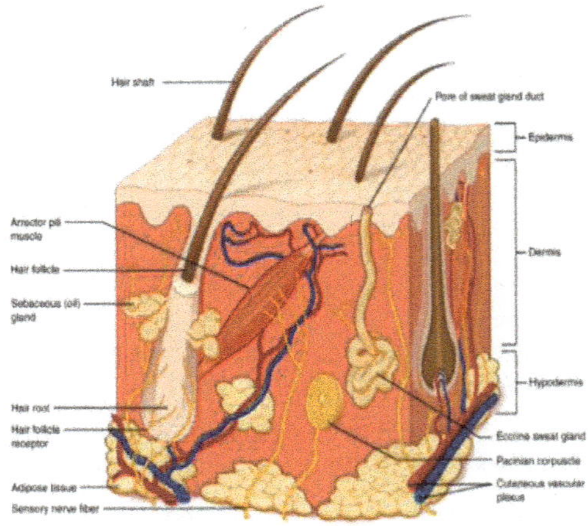

Figure 9.5 The anatomy of skin showing the hair root, follicle and shaft. Compounds (including drugs) enter the hair medulla *via* the capillary supply – shown as red (veins) and blue (arteries) – at the hair root. Image created with Bing Image.

hair capillary supply (Figure 9.5) into the growing cells at the base of the hair follicle. As the hair grows, the group of cells into which the drug has diffused is pushed along the developing hair by the dividing cells behind them. The blood kinetics of the drug determine how long the growing hair cells are exposed to and absorb the drug, thus influencing the length of hair containing drug residues. These residues are found *in* the hair medulla (*i.e.*, incorporated) rather than *on* the outside of the hair.

Drug might also be present on the outside of the hair *via* various mechanisms:

- Secretion onto the hair surface *via* the sebaceous gland.
- Contamination *via* drugs on the skin and environment after the hair has grown out of the epidermis (*e.g.*, *via* the exocrine sweat glands).
- Transfer from specific tissues (*e.g.*, hypodermal adipose tissue or pools of drug in the tissues surrounding the growing hair).

Only the incorporated drug residues can be used to determine the time range of exposure. The external residues (*e.g.*, from sebaceous secretions) are likely to be smeared along the surface of the hair as it emerges from the epidermis. These residues might be redistributed on or between hairs. For this reason, they are washed off before hairs are analysed for forensic investigations (see Section 9.3.2.2 *Hair Analysis for Drugs and Toxic Chemicals*).

9.3.2.1 *Kinetics of Drug and Toxic Chemical Deposition in Hair*

Since drugs (and other toxic chemicals) are transferred to and incorporated into hair by passive diffusion their appearance in hair follows (with a slight delay) the drug's blood kinetics (Figure 9.6). So, the peak concentration in hair will roughly correspond to the peak concentration in blood; however, usually only the week of administration can be estimated from hair drug residues analysis.

The peak concentration of a drug in hair corresponds (with a slight delay to account for diffusion) to the peak concentration of the drug in plasma (C_{max}) and to the time of C_{max} (*i.e.*, t_{max}) and the hair concentration profile will follow the plasma drug concentration changes. If a drug is repeatedly administered, the blood kinetics will show repeated peaks in plasma drug concentration, resulting in a prolonged presence of residues in the hair.

9.3.2.2 *Hair Analysis for Drugs and Toxic Chemicals*

In order to determine whether a drug had been administered historically, hair can be analysed. If the hair has not been cut after the dose of a drug, it can provide a long-term 'record' of a drug's presence in the circulatory system. This, of course, requires the drug to be absorbed (*e.g.*, from an oral dose) into the circulatory system and for it

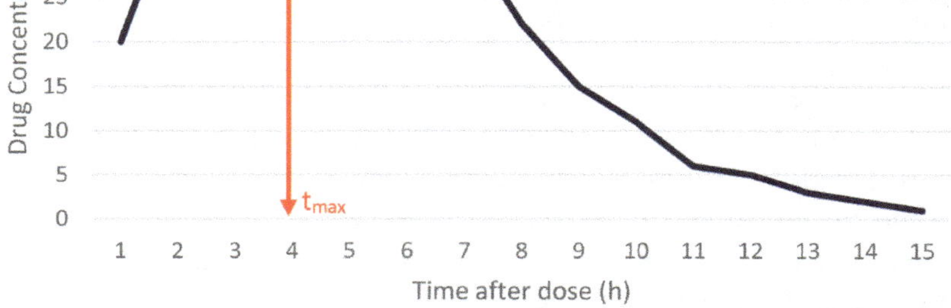

Figure 9.6 A hypothetical plasma drug concentration *vs* time graph for an orally administered drug showing C_{max} (45 ng mL^{-1}) and t_{max} (4 h).

to diffuse into hair (this is very likely to be the case for any drug present in the circulatory system).

Hair (usually head hair) samples taken for analysis are cut as close to the skin (scalp) as possible; the sample is then cut into sections and each section is quantitatively analysed separately; this gives a timeline for the drug administration – the period of this timeline is only limited by the length of the hair analysed. When the hair is analysed, it is important that only the incorporated drug is included in the analysis because the drug present on the outside of the hair (*e.g.*, from sebaceous secretion) can give a misleading drug administration timeline. It is for this reason that when hair is analysed for drug incorporation, the hair is first washed thoroughly with detergent to remove any drug present on the outside of the hair.

There are numerous published methods for hair analysis for drugs, but all of them follow the same general protocol (Figure 9.7), involving taking the hair sample, washing it with detergent to remove drugs contaminating the outside of the hair, digesting the hair to make a solution, extracting the digest, and analysing the extract for drugs and other toxic chemicals using standard analytical methods (*e.g.*, HPLC-MS; see Chapter 4 *Forensic Chemistry*).

9.3.2.3 Hair Analysis Case Example

In July 1990 in Miami, USA, A man became very ill after drinking a beverage imported from Colombia. The man suspected that he had been poisoned; he said the drink tasted bad and that it made his mouth and tongue numb. The man fell into a coma and was rushed to hospital, where a urine analysis identified cocaine. The man died 24 days later. During this time, it was estimated that his hair grew 10 to 15 mm.

Forensic investigations revealed cocaine in the residues of the beverage remaining in the bottle. Bottles from the same batch of beverage were recalled and their contents were analysed: approximately 30 g of cocaine per bottle was found. Product tampering was suspected.

The man was suspected of being a cocaine addict and therefore it was necessary to determine whether he was a cocaine user, whether he had overdosed on abused cocaine or whether the cocaine in the beverage he had consumed led to his death.

A *post-mortem* hair sample was taken. The hair was cut into 5 mm segments, and the samples were washed with detergent; the washed segments were digested, solvent extracted and the extract was analysed for cocaine.

1. Hair samples taken.

↓

2. Hairs cut into segments.

↓

3. Hair sections washed with 1 g/L sodium dodecylsulfate.

↓

4. Hair sample pulverised (using a bench ball grinder).

↓

5. Aliquot (100 mg) of pulverised hair digested with 1M sodium hydroxide (aq) at 95°C for 15 min.

↓

6. Digest neutralised with 1M hydrochloric acid (aq) and pH adjusted to 7.6 with Sørensen buffer.

↓

7. pH-adjusted digest extracted with diethyl ether (x3 – extracts combined). The diethyl ether is evaporated and the residue dissolved in phosphate buffer (5 mL, pH 7.0).

↓

8. Phosphate buffer extract put through a 3 mL C18 solid-phase extraction cartridge and eluted with methanol.

↓

9. Methanol eluate analysed by GLC-MS.

Figure 9.7 A typical protocol for hair analysis for incorporated drugs. Notes: Step 3: Sodium dodecylsulfate is a detergent. Step 8: A C18 solid-phase extraction cartridge is a small cartridge packed with octadecylsilane (C18)-coated micro-beads (see Chapter 4 *Forensic Chemistry*). The octadecylsilane binds any fat-soluble (hydrophobic) chemicals (*e.g.*, drugs), thus concentrating them on the beads. The bound chemicals are washed off with a small volume of a solvent (*e.g.*, methanol) yielding a concentrated extract for analysis. Method reproduced from ref. 3 with permission from Elsevier, Copyright 2007.

It is clear from the man's hair cocaine profile that he had been taking cocaine before drinking the contaminated beverage – the 20 and 25 mm hair segments (Figure 9.8) already had incorporated cocaine (approx. 58 and 54 ng mg^{-1} respectively), but the 15 mm segment had a significantly higher cocaine concentration (approx. 98 ng mg^{-1}). This suggests that the cocaine concentration was highest in the section of hair that included the date the man drank the allegedly laced beverage, and the older sections of hair (*i.e.*, at 20 mm and 25 mm from the head)

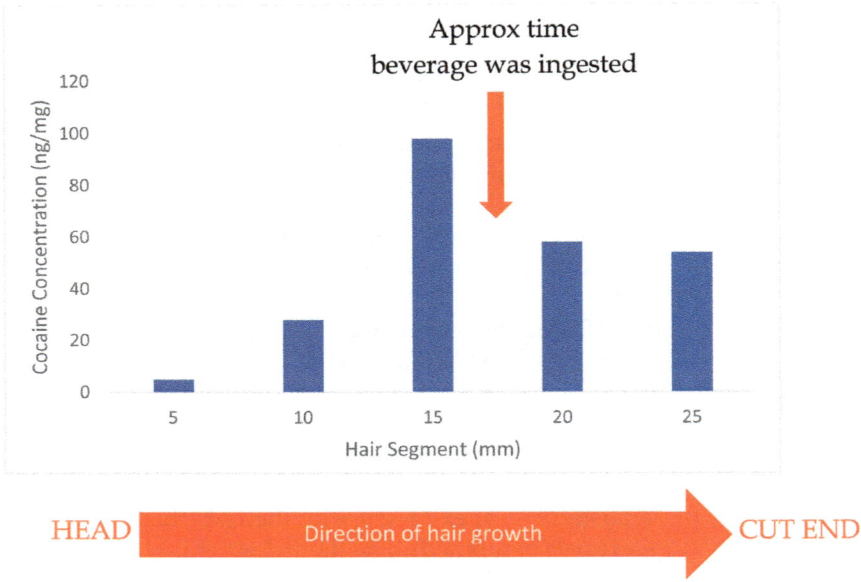

Figure 9.8 Cocaine profile in hair from a cocaine abuser who drank a beverage contaminated with a very high concentration of cocaine. Reproduced from ref. 4 with permission from NIH.

indicated he had been taking cocaine in the time leading up to the alleged incident and that the cocaine concentration declined in the time after he was admitted to hospital and up to his death. Therefore, it is likely that the cocaine in the beverage boosted the man's blood cocaine above the lethal concentration.

9.3.3 Metabolism of Toxic Chemicals

This is not the place for a treatise on drug and toxic chemical metabolism, but it is important to know a little about the processes in the body that detoxify and prepare toxic chemicals (including drugs) for excretion.

When a substance is taken orally, it goes to the stomach *via* the esophagus, and then it is absorbed (not all substances are absorbed) in either the stomach or the small intestine (this is the usual route of absorption; some substances are absorbed in the mouth). The absorbed substance is transported to the liver *via* the hepatic portal vein. Most substances are metabolised to some extent in the liver (this is termed the first pass effect) before the parent compound plus metabolites are released from the liver into the circulatory system. A proportion of the substance or its metabolites might be excreted in bile into the small intestine from where it might be reabsorbed (this is termed enterohepatic recirculation) or excreted in the faeces (Figure 9.9).

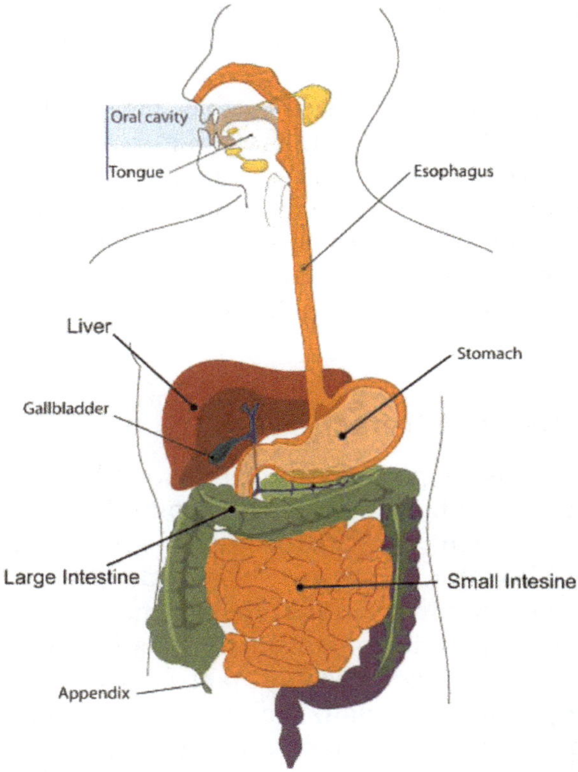

Figure 9.9 The human digestive system. A substance taken orally goes from the mouth (oral cavity) *via* the esophagus to the stomach. Some substances are absorbed in the stomach, but most progress to the small intestine where they are absorbed and transported *via* the hepatic portal vein (not shown) to the liver. Most substances are metabolised in the liver. A proportion of the substance and/or its metabolites may be excreted in the bile (this depends on their molecular characteristics), but most are released into the circulatory system. From the circulatory system, the substance/metabolites go to various organs and cause pharmacological or toxic effects. The substance/metabolites excreted in bile is/are stored in the gall bladder and periodically released *via* the bile duct (shown in blue) into the small intestine, from where it can be reabsorbed or excreted in faeces *via* the large intestine. Image created with Bing Image.

The metabolic process aims to detoxify and prepare substances for excretion by making them more water soluble because the body's excretory processes (*e.g.*, urine *via* the kidney) are water-based. The metabolism and excretion of paracetamol (Figure 9.2) illustrates this very well. Paracetamol's two main metabolites, paracetamol sulfate and paracetamol glucuronide, are highly water soluble and are efficiently excreted in urine *via* the kidneys. Paracetamol, on the other hand is much less water soluble and is not well 'designed' for excretion until it is metabolised.

The metabolism and excretion process means that some *post-mortem* samples will contain only the parent compound (*e.g.*, stomach contents), whereas others (*e.g.*, blood and liver) are likely to contain both the parent compound and its metabolites. It is important to consider metabolites when assessing analytical toxicology findings, particularly because some substances are extensively metabolised and therefore the only indicators of their presence are their metabolites.

9.3.4 Effects of Putrefaction

Sometimes, victims of poisoning are not found for days (or even longer) after their death. During this time, putrefaction occurs (if the remains are not frozen). The microorganisms responsible for putrefaction might also 'attack' any toxic chemicals present in the body; this will almost certainly mean that the substance *per se* disappears and is replaced with microorganism degradation products. Knowledge of putrefaction biodegradation pathways might help identify substances (*e.g.*, drugs) originally present in a putrefied body. This is rarely used in modern forensic toxicology investigations, but might be used on rare occasions when a body is badly putrefied and no parent compounds can be found. Extrapolating the concentration of a putrefaction product back to the original dose of the parent drug is nearly impossible, and so these methods can only provide a qualitative indication of the drugs or toxins that might have been present before putrefaction.

9.3.5 Inorganic Toxicants

I have focused on organic toxicants in this chapter because they are by far the most commonly encountered substances in modern forensic toxicology cases. This was definitely not the case a hundred years ago when inorganic poisons (*e.g.*, arsenic) predominated. However, a forensic toxicologist must always have an open mind and not forget the range of potential toxicants that might be involved in a poisoning case. Carbon monoxide (CO) poisoning is, perhaps, an exception to this generalisation, because it is commonly encountered in suicide and accidental poisoning cases.

9.3.5.1 Carbon Monoxide Poisoning

CO is the product of incomplete combustion of carbon or its compounds (*e.g.*, domestic gas) and is encountered in both accidental and intentional (*e.g.*, suicide) poisoning cases. Accidental CO poisoning might occur if a person sleeps in a room with a gas heater with inefficient air supply. This will mean that instead of the gas (*e.g.*, propane) being completely combusted to carbon dioxide (CO_2) it will be, in part, combusted to CO, which can lead to toxicity. This is particularly the case if someone is sleeping in a CO-contaminated atmosphere because they will not notice the CO effects before they succumb to its toxicity. Suicide by CO poisoning is common; for example, the combustion products from a car exhaust (rich in CO) are directed *via* a pipe attached to the exhaust to the inside of the car, where the person intending to commit suicide sits and inhales the exhaust – the person quickly loses consciousness and dies.

CO is incredibly toxic because it combines with hemoglobin (Hb), the oxygen carrying red pigment in red blood cells, to form carboxyhemoglobin (HbCO) which does not carry oxygen, thus starving the tissues of their oxygen source. If the brain is deprived of oxygen, unconsciousness and death quickly result. Similarly, if the heart muscle (myocardium) is deprived of oxygen, the heart stops beating and death occurs.

Determining whether a poisoning case is due to CO relies on the fact that HbCO is bright red (often described as 'cherry red') with a characteristic absorption spectrum. During the *post-mortem* examination, the liver will exhibit a bright red colour because of the presence of HbCO. Usually, a blood sample or liver homogenate is subjected to spectrophotometric analysis to see if the characteristic HbCO spectrum is present, which would indicate CO poisoning. The amount of HbCO (and thus the dose of CO) is determined by the absorbance readings at specific wavelengths (*e.g.*, 532 nm and 558 nm).

9.3.5.2 A CO Poisoning Case (See Ref. 5)

A 35-year-old mechanic was found unconscious in a commercial (automobile) garage in Delhi, India. He died in hospital later the same day.

External *post-mortem* examination revealed cherry red *post-mortem* lividity (see Chapter 8 *Forensic Pathology*) which suggested CO poisoning. Internal *post-mortem* examination showed cherry red muscles in the chest and abdominal wall, again suggesting CO poisoning. The heart and brain showed small surface (petechial) haemorrhages consistent with asphyxiation – this would be the case with CO poisoning because the tissues were deprived of oxygen (hypoxia), the condition also observed in asphyxia.

Analysis of blood and organs confirmed the presence of HbCO.

It was concluded that CO emissions from a car in the enclosed garage led to a toxic atmospheric CO concentration, which resulted in the man losing consciousness and breathingin further CO, ultimately causing his death.

9.3.5.3 Metals

Nowadays, toxic metals are rarely encountered in forensic toxicology cases. However, accidental fatal poisoning cases occur from time to time due to the use of toxic metals (*e.g.*, mercury and arsenic) in industry and as pesticides (*e.g.*, methyl mercury) in some countries.

The methods of investigating suspected metal poisoning cases are very similar to the scheme shown in Figure 9.7 for organic poisons; however, the analytical methods are often different. For example, for mercury, atomic absorption spectroscopy (determines the characteristic absorption spectrum of a metal in its gas form) is used, whereas for organic mercury (*e.g.*, methyl mercury; CH_3Hg^+), conventional HPLC with UV detection is often used (see Chapter 4 *Forensic Chemistry*).

I will illustrate the process of investigating metal poisoning with a mercury poisoning case reported in Portugal in 2009.

9.3.5.4 A Mercury Poisoning Case (See Ref. 6)

A 39-year-old pharmacist consumed a large quantity (~50 g) of mercury chloride (it is not specified whether this was HgCl or $HgCl_2$; both

Table 9.3 Tissue mercury concentrations in a mercury chloride suicide case compared with normal and fatal mercury concentrations (reproduced from ref. 6 with permission from Elsevier, Copyright 2008). These data clearly show that mercury toxicity was the cause of death.

	Mercury concentration ($\mu g\ g^{-1}$)		
Tissue	Suicide case	Normal range	Fatal
Brain	0.33	0.04–0.23	1.1
Liver	49.9	0.16–1.3	33
Blood	6,750[a]	0.02–0.11[a]	0.8[a]

[a] $\mu g\ mL^{-1}$.

are highly toxic) to commit suicide. He was treated in hospital, but died 42 days later.

Blood (taken 24 h after exposure) and *post-mortem* brain and liver samples were analysed for mercury by atomic absorption spectroscopy. The mercury concentrations found were consistent with mercury toxicity causing the death (Table 9.3).

9.3.5.5 Cyanide

No chapter on forensic toxicology poisoning cases would be complete without mention of cyanide (CN^-). Cyanide, usually in the form of potassium cyanide (KCN), sodium cyanide (NaCN) or hydrocyanic acid (HCN), is most commonly encountered in poisoning cases – such cases are rare nowadays. All compounds that readily liberate CN^- are extremely toxic; for example, 200–300 mg of KCN would kill a human (*i.e.*, lethal dose ~ 3–4 mg per kg body weight). Cyanide is toxic because it inhibits cellular respiration by interfering with a crucial component of electron transport in the mitochondrial membrane, namely, the cytochrome C oxidase complex. This enzyme complex transfers electrons, creating a gradient from which the energy needed to drive cellular metabolism is derived. Without cytochrome C oxidase, mitochondria cannot generate energy and the cell dies. The victim of cyanide poisoning usually dies of cardiac arrest because the heart muscle can no longer function.

Cyanide poisoning cases are much less common now than in previous years. Most poisoning cases are suicides (*e.g.*, in Korea, approximately 97% of 255 fatal cyanide poisoning cases over 5 years were suicides) and involved men (*e.g.*, 78% in Korea) who worked in places where cyanide was used and so was accessible to them.

The *post-mortem* signs of cyanide poisoning include cyanosis (*i.e.*, blue appearance) of the lips and skin under the finger and toe nails, and there is often a smell of almonds emanating from the body (cyanide smells like almonds – indeed it is cyanogenic alkaloids in almonds that impart their characteristic flavour and odour). The internal *post-mortem* examination often reveals brick-red coloured gastric mucosa, frothiness of the stomach contents and liquid within the bronchial system (*i.e.*, trachea, bronchi and lungs), the smell of almonds is more pronounced when the gastrointestinal tract is opened and the brain is dissected.

Cyanide is rapidly metabolised by microorganisms and therefore its measurement in *post-mortem* samples can yield misleading (*e.g.*, false negative) results. However fatal cases often exhibit very high cyanide concentrations in the stomach contents, indicating that the victim had ingested the poison and died quickly.

9.3.5.6 Cyanide Analysis

There are many analytical methods for cyanide; I will give a few examples here.

9.3.5.7 Prussian Blue Test

The Prussian blue test is a colorimetric test utilising the reaction between cyanide and ferrous ($FeSO_4$) and ferric sulfate ($Fe_2(SO_4)_3$) to give blue ferric ferricyanide (Prussian blue) (Figure 9.10).

9.3.5.8 Pyridine–Barbituric Acid Method

This is a relatively simple colour reaction for cyanide; however, the chemistry (the König reaction) underlying the colour formation is complex (Figure 9.11), but is specific for cyanide.

I $$6\,KCN\,(aq) + FeSO_4\,(aq) \rightarrow K_4[Fe(CN)_6]\,(aq) + K_2SO_4\,(aq)$$

II $$3\,K_4[Fe(CN)_6]\,(aq) + 2\,Fe_2(SO_4)_3\,(aq) \rightarrow \underset{\text{BLUE}}{Fe_4[Fe(CN)_6]_3\,(s)} + 6\,K_2SO_4\,(aq)$$

Figure 9.10 The chemistry of the Prussian blue test for cyanide. I: potassium cyanide (KCN) reacts with ferrous sulfate ($FeSO_4$), forming potassium ferricyanide ($K_4[Fe(CN)_6]$) plus potassium sulfate (K_2SO_4). II: potassium ferricyanide then reacts with ferric sulfate ($Fe_2(SO_4)_3$), forming a precipitate of Prussian Blue (ferric ferricyanide - $Fe_4[Fe(CN)_6]_3$) and more potassium sulfate. The amount of Prussian Blue formed corresponds to the amount of cyanide originally present. aq = aqeous solution and s = solid.

I $KCN + Br_2 \rightarrow BrCN + KBr$

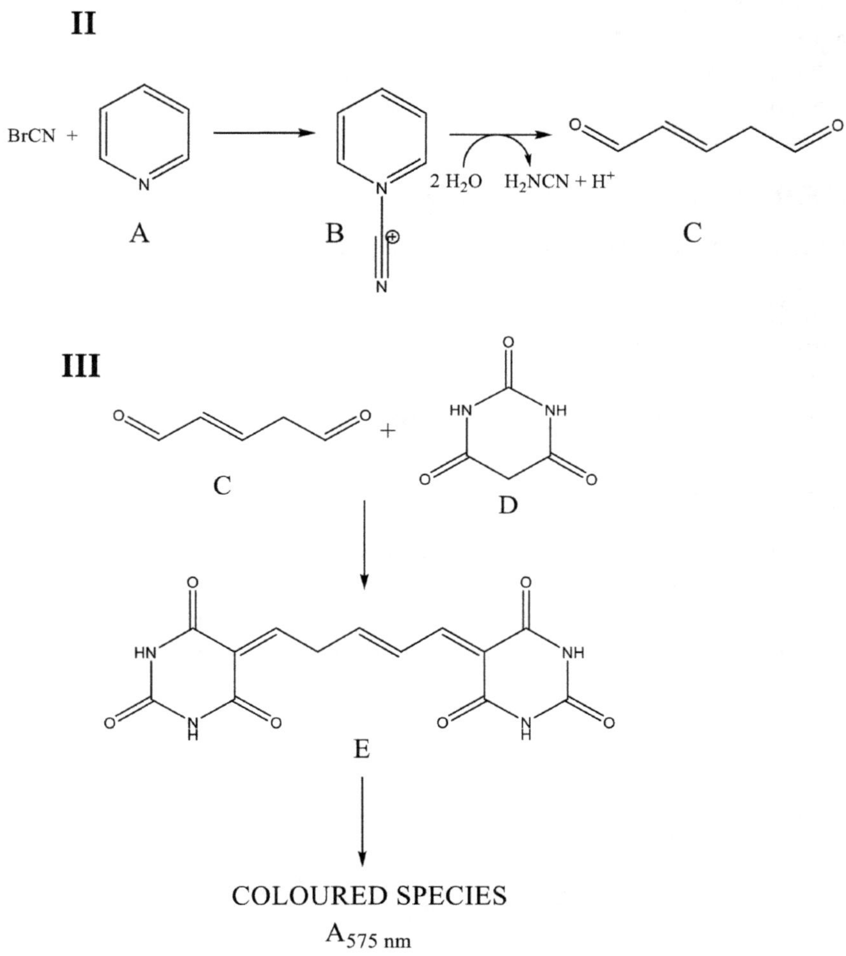

Figure 9.11 The complex chemistry of the barbiturate-pyridine-cyanide reaction used for the quantitative analysis of cyanide. I: the suspected cyanide (in this case potassium cyanide – KCN) is reacted with bromine (Br_2, or other oxidant) to form cyanogen bromide (BrCN) plus potassium bromide (KBr). II: BrCN is then reacted with pyridine (A) to form an unstable intermediate (B) which hydrolyses to form glutaconic aldehyde (C). III: the glutaconic aldehyde is then reacted with barbituric acid (D) to form a complex compound (E), which rearranges to form a coloured species that absorbs at 575 nm (*i.e.*, it appears blue/violet). The reactions are carried out in an aqueous solution. Reproduced from ref. 7 with permission from American Chemical Society, Copyright 1975.

The assay is carried out using a single reagent containing an oxidant (*e.g.*, bromine), pyridine and barbituric acid in an acid (usually hydrochloric acid – HCl) solution. The test sample is mixed with the reagent, incubated for about 15 min and its absorbance is determined at 575 nm ($A_{575\ nm}$) using a spectrophotometer. The $A_{575\ nm}$ value is proportional to the cyanide concentration in the sample.

9.3.5.9 Headspace GLC Analysis

Cyanide is also measured (particularly in blood) by headspace GLC (see Chapter 4 *Forensic Chemistry*) using a nitrogen phosphorus detector (NPD): remember cyanide (CN^-) has a nitrogen atom. The sample is acidified with sulfuric acid to liberate HCN from KCN or NaCN (Figure 9.12) because HCN is very volatile (boiling point 26 °C) and is thus ideal for headspace analysis.

9.3.5.10 HPLC

HPLC analysis is often used for cyanide analysis either using a NPD or by post-column derivatisation. One such method involves reacting cyanide eluting from the HPLC column with *o*-phthaldehyde to form a fluorescent derivative, which is easily detected (Figure 9.13).

$$2\,KCN\,(aq) + H_2SO_4\,(aq) \rightarrow 2\,HCN\,(aq) + K_2SO_4\,(aq)$$

Figure 9.12 GLC headspace analysis for cyanide relies on the volatility of hydrogen cyanide (HCN) which is formed when the sample (*e.g.*, blood) containing potassium cyanide (KCN), or other cyanide salt (*e.g.*, sodium cyanide, NaCN) is acidified with sulfuric acid (H_2SO_4). aq = aqueous solution.

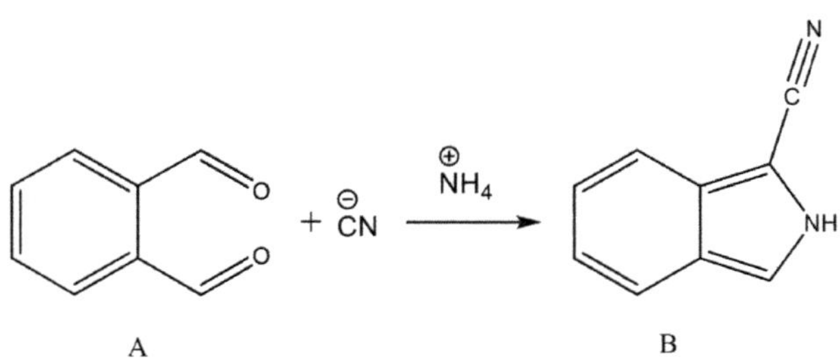

Figure 9.13 The HPLC post-column derivatisation for cyanide analysis: *o*-phthalaldehyde (A) reacts with cyanide (CN^-) in the presence of ammonium (NH_4^+) to form fluorescent 1-cyanoindole (B).

9.3.6 Cyanide Poisoning Cases

Case examples illustrate well how a forensic toxicologist may approach a suspected cyanide poisoning.

9.3.6.1 Cyanide Poisoning Case 1 – Suicide (Case Details from Ref. 8)

A 56 year-old man was found dead on the floor of his kitchen in a city in Brazil. The man was a goldsmith, who worked from his home workshop. A plastic bottle containing a crystalline substance (later found to be a cyanide salt: cyanide salts are used in gold plating and jewellery cleaning) was found next to his body and a suicide note was found in his workshop.

The external *post-mortem* examination showed severe cyanosis and epistaxis (nosebleed – this may not be related to cyanide; perhaps the man fell as he became unconscious although we would expect other impact injuries if that were the case), but no signs of violence were observed. Samples of blood, urine and stomach contents were taken during the internal *post-mortem* examination.

The *post-mortem* samples were acidified with sulfuric acid and analysed for cyanide through headspace GLC. The headspace oven temperature was 60 °C (remember HCN boils at 26 °C). NPD was used to quantify the HCN. The results of the analyses are shown in Table 9.4.

The analysis results in Table 9.4 show that the concentration of HCN in the stomach contents was very high, showing that the man had consumed cyanide. The blood cyanide concentration is approximately 10 times the concentration known to be fatal. The total blood volume in adults is about 5,000 mL; therefore, a blood concentration of 30.7 $\mu g\ mL^{-1}$ relates to a dose of at least 30.7 $\mu g\ mL^{-1} \times$ 5000 mL = 153 mg – the lethal dose of potassium cyanide is estimated to be 140 mg; therefore, the blood concentration found corresponds to a lethal dose. The urine concentration is low

Table 9.4 Cyanide concentrations in *post-mortem* samples from a goldsmith found dead near a bottle of cyanide and a suicide note.

Sample	Cyanide concentration ($\mu g\ mL^{-1}$)	Fatal concentration ($\mu g\ mL^{-1}$)
Blood	30.7[a]	>3[b]
Stomach contents	100.5[b]	—
Urine	0.1[a]	—

[a] Reproduced from ref. 8 with permission from Oxford University Press, Copyright 2017.
[b] Reproduced from ref. 9 with permission from Wolters Kluwer Health, Inc., Copyright 2013.

because the dose of cyanide was high enough to result in very rapid death (probably within minutes); therefore, there was no time for cyanide to be excreted in urine. It was concluded that this was a case of suicide through self-administration of a cyanide salt.

9.3.6.2 Cyanide Poisoning Case 2 – Murder (Case Details from Further Reading)

A 29-year-old laboratory technician in a metal processing factory was handed a cup of coffee from a vending machine by a colleague. She drank only a little coffee before experiencing seizures and then collapsing. She entered a comatose state before the emergency services arrived; she was immediately taken to hospital.

Blood and stomach contents samples were taken about 4 h after she was hospitalised. She died 4 days later. Doctors could find no pre-existing disease that might have led to her death, and so they suspected poisoning (perhaps suicide), since a number of toxic chemicals were used at her workplace, including cyanide, to which the deceased had access.

During the case investigation, it was reported that the deceased had commented that her coffee had '*a pungent putrid odour resembling ammonia and a bitter disgusting taste*'. This could be the reason why she only drank 'a few drops' of the coffee. Unfortunately, the coffee was not available for analysis.

The blood sample taken 4 h after she was admitted to hospital was analysed by the pyridine–barbituric acid method, revealing a cyanide concentration of 3 $\mu g\ mL^{-1}$ which indicates a fatal cyanide dose.

Further investigation of the man who brought the coffee to the deceased revealed that he was stealing gold from his employers – perhaps the deceased knew of this which could give him a motive for her murder. He was convicted of murder mainly based on circumstantial evidence, which relied on expert testimony about the smell and taste of cyanide-laced coffee, the deceased's lethal blood cyanide concentration, the availability of cyanide at the workplace, and his motive for killing his colleague.

Further Reading

1. E. G. C. Clarke and A. C. Moffat, *Clarke's Isolation and Identification of Drugs in Pharmaceuticals, Body Fluids, and Post-mortem Material*, Pharmaceutical Press, London, 1986.

2. R. J. Flanagan, G. Connally and J. M. Evans, Analytical toxicology – guidelines for sample collection postmortem, *Analytical Toxicology*, 2005, **24**, 63–71.
3. A. Girgetti, F. P. Busardo, R. Tittarelli, V. Auwärter and F. Giorgetti, Post-mortem toxicology: a systematic review of death cases involving synthetic cannabinoid receptor agonists, *Frontiers in Psychiatry*, 2020, **11**, 464.
4. P. Kintz, Value of hair analysis in postmortem toxicology, *Forensic Science International*, 2004, **142**, 127–134.
5. F. Musshoff, P. Schmidt, T. Dalrup and B. Madea, Cyanide fatalities: case studies of four suicides and one homicide, *The American Journal of Forensic Medicine and Pathology*, 2002, **23**, 315–320.
6. K. G. Shanks, W. Clarke and G. Behonick, Death associated with the use of the synthetic cannabinoid ADB-FUBINACA, *Journal of Analytical Toxicology*, 2016, **40**, 236–239.
7. H. M. Stevens, The Stability of Some Drugs and Poisons in Putrefying Human Liver Tissues, *Journal of the Forensic Science Society*, 1984, **24**, 577–589.
8. P. Triumfante, M. E. Soares, A. Santos, S. Tavares, H. Carmo and M. de L. Bastos, Mercury fatal intoxication: two case reports, *Forensic Science International*, 2009, **184**, e1–e6.
9. J. C. West, Rapid HPLC analysis of paracetamol (acetaminophen) in blood and postmortem viscera, *Journal of Analytical Toxicology*, 1981, **5**, 118–121.

References

1. J. Marx, R. Walls and R. Hockberger, *Rosen's Emergency Medicine*, Elsevier Health Sciences, 2013.
2. J. C. West, Rapid HPLC Analysis of Paracetamol (Acetaminophen) in Blood and Postmortem Viscera, *J. Anal. Toxicol.*, 1981, **5**, 118–121.
3. S. Boyer, *et al*, *J. Chromatogr. B*, 2007, **852**, 684–688.
4. Miller *et al.*, National Institute on Drug Abuse (NIDA), US Department of Health and Human Services. Monograph 167, 1997, pp. 146–160. https://archives.nida.nih.gov/sites/default/files/monograph167_0.pdf
5. M. Chand-Meena, *Int. J. Med. Toxicol. Forensic Med.*, 2014, **4**, 158–161.
6. P. Triumfante, M. E. Soares, A. Santos, S. Tavares, H. Carmo and M. de L Bastos, Mercury fatal intoxication: two case reports, *Forensic Sci. Int.*, 2009, **184**, e1–e6.
7. Lambert, *et al.*, *Anal. Chem.*, 1975, **47**, 916–918.
8. Peliçāo, *et al.*, *J. Toxicol. Anal.*, 2018, **1**, 1–3.
9. Anseeuw, *et al.*, *Eur. J. Emerg. Med.*, 2013, **20**, 2–9.
10. E. G. C. Clarke, *Isolation and Identification of Drugs*, 1978, vol. 1, p. 465.
11. K. G. Shanks, W. Clarke and G. Behonick, Death associated with the use of the synthetic cannabinoid ADB-FUBINACA, *J. Anal. Toxicol.*, 2016, **40**, 236–239.

10 Forensic Psychology

The meaning of forensic psychology has changed significantly in recent years. Originally, it was the application of psychology to help answer legal questions in a criminal justice setting, but its scope has now expanded significantly, encompassing the application of the scientific, technical, and specialized knowledge of psychologists to the field of law. Forensic psychologists help the police to understand a criminal's mind by examining patterns in their offences, predicting their potential strike locations, assessing their IQ (whether it is below average, average or above average), and evaluating their social and/or sexual competence. This is termed investigative psychology and it might help the police in tracking down a suspect and informing their questioning strategy during suspect interviews. At the other end of the psychology spectrum is rehabilitative psychology, where personal profiles are used to guide treatment. This will not be considered in this chapter.

This chapter is not intended to be a detailed, all-encompassing account of forensic psychology, but rather an overview to illustrate how forensic psychologists are involved in the process of criminal detection and investigation; *i.e.*, investigative psychology.

This chapter is based on an excellent lecture given by Dr Jacinta Cording to criminal justice students at the University of Canterbury, Christchurch, New Zealand. I am very grateful to Dr Cording for granting permission to use her lecture as the basis for this chapter.

10.1 Psychology

Psychology is the study of human behaviour, the science of the mind.[1] The word psychology is derived from the Greek words 'psyche',

Forensic Science: The Science Behind the Truth
By Ian Shaw with Anna Sandiford

Published by the Royal Society of Chemistry, www.rsc.org

meaning spirit or soul, and 'logia', meaning study or research. In its broadest sense, it covers the nature, functions and phenomena of the human soul and mind.[2] This encompasses a wide range, spanning from pure biochemistry (*e.g.*, the biochemistry of nervous transmission) to the social science of human behaviour. Forensic psychology emphasizes the social science aspects of psychology: understanding the mind and behaviour of criminals.

10.2 Forensic Psychology

Forensic psychologists are involved in a broad array of aspects of the criminal justice system, including:[3]

- Competency evaluations
- Sentencing recommendations
- Evaluations of the risk of reoffending
- Testimony as an expert witness
- Child custody evaluations
- Academic research on criminality
- Consulting with law enforcement agencies
- Treatment of criminal offenders
- Providing psychological services to inmates and offenders

In this chapter, I will only cover the forensic psychologist's work relating specifically to forensic case investigations, which encompasses helping the police to locate or identify suspects and determine whether or not a suspect might be linked to a particular crime. Assessing whether a suspect is telling the truth or lying is an important aspect of the forensic psychologist's work, especially when considering the possibility of a false confession. False confessions are surprisingly common: it has been estimated that for all crimes, false confessions are about 12% of confessions, but the false confession rate for homicides is close to 23% (see *Further Reading*). Arguably, this alone justifies the work of a forensic psychologist.

10.3 Delving into the Criminal Mind – Criminal Profiling

Understanding the criminal mind can provide some very useful information to help the police to identify suspects and perhaps determine whether or not the identified suspect might have committed the crime. This is termed criminal profiling (also known as offender profiling). It involves inferring the criminal's attributes from their behaviours gleaned from evidence at the crime scene and predicting where they might strike next based on previous patterns of offending (criminal profiling) and/or patterns of offending at the locations of the offences (geographical profiling).

10.3.1 Criminal Profiling

Psychological profiling is a '*method of suspect identification that aims to identify a person's mental, emotional, and personality characteristics based on actions or things left at the crime scene*' (dictionary.sensagent.com). Criminal profiling involves examining the characteristics of a crime and inferring aspects of the character or physical traits of the offender based on those characteristics. For example, this might involve close scrutiny of messages left at the crime scene and, perhaps, warning notes sent to individuals against whom the crime was directed. A look at the first case to utilise forensic psychology shows clearly how the psychologist used information from the crime to paint a psychological and, to some extent, a physical picture of the perpetrator. This case involved a series of bombings in New York, USA, between 1940 and 1956; the 'Mad Bomber of New York', as he was dubbed by the press, had a grudge against Consolidated (Con) Edison Inc. (now called ConEdison), a major energy supplier to New York. The Mad Bomber of New York carried out his crimes as a form of reprisal against Con Edison.

10.3.2 The Con Edison Bombings – New York, 1940–1956

On November 16, 1940, a bomb and a note were found at the West 64th Street power plant of Con Edison. The note read:

Con Edison crooks, this is for you.

It was signed, FP and written in neat block capitals. The device did not explode, but marked the beginning of 16 years during which similar devices were employed, all targeting Con Edison. The police were particularly interested in the bomb because they were still investigating an explosion that had occurred four months earlier at the

New York World Fair which had killed two police officers. Perhaps the incidents were connected.

Almost a year later (September 1941), another explosive device was found at another Con Edison site, but there was no message this time. Police suspicions shifted towards the possibility that a Nazi sympathiser might be responsible because they had previously received a letter, again in block letters signed FP:

> *I will make no more bomb units for the duration of the war – my patriotic feelings have made me decide this – later I will bring the Con Edison to justice.*

In March 1951 (after the end of World War II, which complied with the bomber's patriotic ideals), the police received another letter warning them that a bomb would explode at Grand Central [Station], but they found and defused the bomb, thus averting a potential disaster. A month later (April 1951), a bomb exploded in front of the New York Public Library, but miraculously it caused no injuries, and in August of the same year, a device exploded outside Grand Central Station, but again it caused no injuries.

In October 1951, the New York Herald Tribune newspaper received a letter written in FP's characteristic block capitals:

> *Bombs will continue until [Con Ed] is brought to justice for their dastardly acts against me.*

Bombs were found around New York city for the next 3 years, although there were surprisingly few serious injuries, New Yorkers were understandably on edge – in total, 33 bombs were planted over 16-years of which 22 exploded injuring 15 people. The police investigation to track down the elusive Mad Bomber of New York was unsuccessful for a long time.

Eventually, the New York Police Department (NYPD) took an unprecedented step of seeking the help of a psychiatrist, Dr James Brussel (1905–1982), who had been involved in counter-intelligence profiling during the Korean War (1950–1953). They asked Dr Brussel to apply his wartime profiling techniques to the Mad Bomber of New York case. As a result, he came up with the first criminal profile (Table 10.1).

Eventually, the Mad Bomber of New York was identified by police as George Metesky (1903–1994) (Figure 10.1). Metesky's arrest was a result of traditional police work rather than Dr Brussel's profile, and

Table 10.1 Dr Brussel's profile of the Mad Bomber of New York.

Inferred Characteristic	Reasoning
Male	Most criminals were male.
Foreign, possibly Eastern European descent	Some of the wording in his notes was considered 'un-American'; *e.g.*, the use of the phrase 'dastardly acts...'. Eastern European came from the predominance of anarchism, particularly in the Slavic nations at the time
Unmarried and lived with a woman (perhaps a relative) who filled a maternal role in his life	This is derived from Freudian thinking (not necessarily still held) applied to his letters and the way he planted some of his bombs in slits cut in cinema seats. This pointed to a hatred of male dominance and a preoccupation with the female anatomy.
Clean-shaven and neatly dressed	His writing style was fastidious and it was thought that this trait would extend to his personal habits.
Athletic build	It had been shown in the 1950s that people with an athletic build are more likely to develop paranoid schizophrenia than other body types.
Suffered from progressive paranoia	His traits were typical of paranoid schizophrenia – key symptoms are prominent delusional thoughts about persecution or conspiracy: he held grudges (against Con Edison), felt that people were plotting against him (in one of his letters he refers to Con Edison's 'dastardly acts against him'), and that he would not let go of these views (he continued his bombing campaign for 16-years).
Middle-aged	In the 1950s, it was thought that the symptoms of paranoid schizophrenia usually did not manifest until the age of approx. 30, and since his bombing exploits continued for a decade, it was inferred that he was approx. 40 years old.
Lived in Connecticut	Many of the letters were posted in White Plains, NY, which is between New York City and the Connecticut border.
He would be wearing a double-breasted suit with the buttons done up when arrested.	Pure guesswork! But it does relate to the prediction of neatness.

when the police caught up with Metesky at his apartment in Conecticut one January night in 1957, they found him wearing his pyjamas and not a double-breasted suit. However, Metesky requested the opportunity to change before being taken to the police station and

Figure 10.1 George Metesky after his arrest in January 1957. Figure is in the Public Domain.

he emerged from his bedroom wearing a double-breasted suit with all of the buttons done up!

George Metesky was 44 years old when he was arrested (*i.e.*, middle aged), and he was of Lithuanian (*i.e.*, East European) descent and of medium build. He was unmarried (indeed he had never had a girlfriend) and lived with his two sisters. He was described by his neighbours as fastidious and his bedroom was said to be 'creepily neat'. He had a reputation for petty disputes. He had worked for Con Edison, where he experienced an industrial accident which left him disabled and unable to work, and so he was forced to retire from the company. He had a history of disputes with his employers, which were brought to a head at the time of his forced retirement.

His guilt was clinched when samples of his handwriting matched (see Chapter 12 *Forensic Document Examination*) the handwriting on the notes sent to the police and those left with his bombs.

At his trial, a psychiatric expert concluded that Metesky was suffering from paranoid schizophrenia (Dr Brussel was right again!) and the judge declared him '*hopeless and incurable both mentally and physically*' and declared him 'legally insane', rendering him unable to stand trial. On April 18, 1957, Judge Liebowitz committed Metesky to Matteawan Hospital for the Criminally Insane, New York. Sixteen years later in 1973, following a USA law change preventing detention of criminals in psychiatric hospitals, it was decided that on account of his exemplary behaviour during his time at Matteawan Hospital that Metesky should be transferred to a psychiatric hospital outside the criminal justice system. He was later declared harmless and released on 13 December, 1973 – he returned to Connecticut to live with his sisters.

Even though psychological profiling was not used to catch or prosecute Metesky, the case shows clearly that the psychologically-derived profile of the Mad Bomber of New York was remarkably accurate, even though the methods used to derive it were somewhat dubious!

10.4 Development of Criminal Profiling by the FBI

The Con Edison case illustrates the potential of criminal profiling in criminal detection even though Metesky's profile was not used by the police to track him down. In the 1970s, the USA's Federal Bureau of Investigation (FBI) explored the use of criminal profiling as a means of helping them with their enquiries. They interviewed 36 serial killers to explore their personalities, behaviours and motives, and to see if there were features of their psychological profiles that might be evident in the crimes they had committed; additionally, could these traits have guided the police to the perpetrators of the crimes? Underlying their investigations was a plan to develop a method of criminal profiling that could be used to inform investigations of similar crimes.

10.4.1 *Modus Operandi Versus* Criminal Signature

There are philosophies involved in criminal profiling: *modus operandi* and criminal signature. *Modus operandi* refers to the ways in which a criminal carries out their crime (*e.g.*, killing their victims by stabbing). This might change with subsequent offences as the criminal adapts to minimise the risk of identification and arrest. For example, the infamous Jack the Ripper murdered 5 (possibly 6) female sex workers in Whitechapel in the east end of London, UK (1888–1891); when he began his crimes, his victims were killed in the streets; however, he narrowly evaded capture during one of his murders and changed his *modus operandi*, opting to kill his victims inside to reduce his risk of being captured by police. Criminal signature is a particular aspect (or aspects) of the way a criminal carries out their crime; it is not necessary for the crime, but represents a flair or foible characteristic of the criminal. For example, the murderer Donald Neilson wrote notes to the police using a Dymo label maker; this was not necessary for his murderous activities, but was (to him) a form of amusement – a game (see Chapter 1 *The Whittle Case – 1976*). If the police encounter another case involving Dymo label messages, they would likely put Nielson at the top of their suspect list because of the criminal signature of using Dymo label messages. The criminal signature is unlikely to change over time because it is ingrained in the criminal's nature and personality.

10.4.2 Organised *Versus* Disorganised Offenders

Another important facet of the FBI's criminal profiling is to distinguish between organised and disorganised offenders because the different character and behavioural traits of these two categories of criminals might help in their detection (Table 10.2). For example, organised offenders take pleasure in the attention their criminal activities receive and they are likely to follow the police investigations closely (*e.g.*, *via* the newspapers). The Lesley Whittle murderer Donald Neilson is an excellent example of an organised offender. When he was arrested and the police visited his home, they found a control room in his attic where the tools of his illicit trade were stored (including the Dymo label maker), and they found many newspaper clippings reporting his activities pinned to the walls of his room (see Chapter 1 *The Whittle Case – 1976*).

The particular traits that distinguish organised from disorganised offenders might be used by the police to catch the criminal. For example, an organised offender's desire to follow the investigation of their activities can be used to trap them. To achieve this, the police might organise a community meeting regarding the crime; an organised offender might attend such a meeting because they would get a buzz from hearing the police talk about them and their crime. If the police document the meeting attendees, one of them might be the offender – all they have to do is find out which one! In addition, the police might use the character profile of the offender type to direct

Table 10.2 Traits of organised *versus* disorganised offenders.

Organised offenders	Disorganised offenders
Average to high IQ[a]	Low IQ[a], often unskilled and unemployed
Socially and sexually competent	Socially and sexually incompetent
Live with partner and children	Live alone
Plan their offences	No planning – their victims are selected randomly on the spur of the moment. They are opportunists
Conceal evidence (*e.g.*, hide weapons)	Limited effort to hide evidence
Follow the police investigation (*e.g.*, collect newspaper articles)	
Alcohol associated with their crime	
Possibly psychopathic[b]	Personal history of sexual or physical abuse. They engage in deviant sexual behaviours (*e.g.*, necrophilia[c])
	Carry out crimes close to their home
	Change job or become religious after their crime

[a] Intelligence quotient: an assessment of human intelligence.
[b] Psychological disorder involving deficient emotional responses, lack of empathy, and poor behavioural controls.
[c] Sex with or attraction to a corpse (from Greek *nekros*, death or corpse).

their interrogation strategy. For example, disorganised offenders often have personal histories of being sexually abused and so the police show empathy during their questioning to engender trust and, perhaps, secure a confession.

Criminal profiling might give important clues about a criminal's personality, nature and traits, which might be used to identify them and guide the interrogation process. However, it is important to note that criminal profiling is not definitive (there is little hard scientific evidence to support it), but indicative. It is possible to see a profile and erroneously fit it to a particular suspect when, in fact, the profile relates to a different person – this is the Barnum Effect: the tendency to accept as true types of information such as character assessments or horoscopes, even when the information is so vague as to be worthless. There have been notable 'misses' employing criminal profiling; for example, the Washington Snipers.

10.4.3 The Washington Snipers – When Criminal Profiling Failed

Over a 3-week period beginning February 2002, 10 people were shot dead and 3 were seriously injured in and around Washington DC,

USA. Based on the characteristics of the shootings, the FBI generated a criminal profile which included the following:

White male
30–40 years old
A loner
Weapons training background

They eventually arrested two black men for the crime – Allen Muhammad (age 41) and Lee Malvo (age 17) – both of whom had no weapons training. Clearly, criminal profiling had failed on this occasion. In September 2003, Muhammad was convicted and sentenced to death (he was killed by lethal injection in November 2009), and in October 2003, Malvo was convicted and sentenced to life in prison without parole.

10.5 Further Developments of Criminal Profiling – Professor David Canter

David Canter (born 1944), a psychologist and Emeritus Professor at the University of Liverpool, UK, refined criminal profiling by designing a new approach to criminal profiling that is driven by data, which was different to the FBI profiling approach which was largely based on interviews with sexual serial murderers. This gave his technique (sometimes termed statistical profiling) a much greater chance of success. He examined data for different crimes and the corresponding characteristics of the criminals. He plotted the data and found clusters of traits corresponding to particular crimes; this meant that certain crimes were associated with particular criminal's traits, making it possible to predict a criminal's personal traits from their crime – this, of course, is useful when tracking down the offender. The difference between Canter's method and the techniques described above is that Canter's method was underpinned by significant data, whereas the previous methods were purely judgement-based. The probability of Canter's criminal profiling accurately reflecting the character and behaviours of the perpetrator of a crime is much higher than the previous methods because it is evidence-based. The association between offender traits and their crimes is illustrated well by sex offenders (Figure 10.2). It is clear that particular offender traits are associated with specific offender types; for example, individuals characterized as pedophiles/hebophiles (sex offenders with a

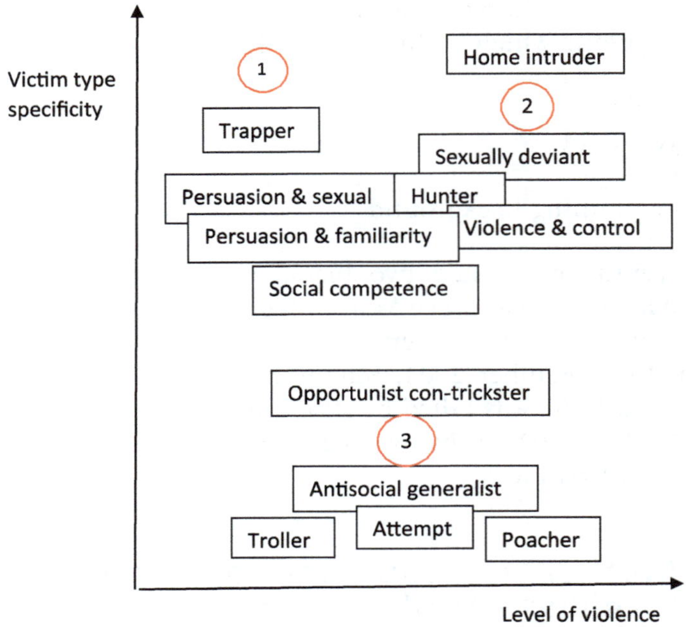

Figure 10.2 Schematic representation of a graph of victim types of sex offenders *versus* the levels of violence used by sex offenders. The three different classes of sex offenders are shown as red circles: 1 = pedophiles/hebophiles; 2 = teleiophiles; and 3 = non-specific offenders. The different approaches employed in carrying out their sex offences tend to cluster around the different classes of sex offenders. This shows that different types of sex offenders use different, but consistent means to carry out their crimes. Data from ref. 4. Glossary: *Pedophiles* are sex offenders who predate children; *hebophiles* predate children on the cusp of puberty; *teleiophiles* have a sexual preference for young adults <17 years old; and *non-specific offenders* do not show a preference for any age group.

preference for children/children on the cusp of puberty) have tendencies of persuasion (both to achieve familiarity with their victims and to cajole them into illegal sexual activity); they are socially competent (this also helps them to attract their victims), and trap their victims (after luring them with their social abilities and powers of persuasion). On the other hand, teleiophiles (sex offenders with a preference for young adults less than 17 years old) have quite different character traits erring towards control, violence, and a tendency to break into places (*e.g.*, the victim's home) to find and secure their prey. In addition, they display sexually deviant activities. These different traits would be useful not only in tracking down a suspect but also in determining the questioning strategy for the interrogation of the suspect.

10.6 The Final Word on Criminal Profiling...

In a UK study (see *Further Reading*) in which 184 police officers were asked to assess the usefulness of criminal profiling in their case investigations, 83% said that criminal profiling information was useful, although mainly to confirm what they already knew. Just 14% said that profiling assisted in solving the case, and only 3% had successfully used criminal profiling to identify an offender.

Criminal profiling is therefore like the proverbial curate's egg: partly good and partly bad. Clearly, the police cannot rely on a criminal profile to track down a suspect, but on the other hand they cannot dismiss its potentially useful information. In most cases, where the expertise of a forensic psychologist is utilized, the resulting criminal profile becomes part of the overall information gathered during the case investigation and might be used to help build a picture of the case and its perpetrator, even if it does not definitively identify the offender.

10.6.1 Geographical Profiling

Analyzing the location and timing of crimes gives information about a criminal's behavioural pattern. The pattern might be used to pinpoint *where* the criminal might strike next and perhaps *when* they might strike next.

Geographical profiling was introduced by Dr (later Professor) David Canter, a psychologist at the University of Surrey (he later moved to Liverpool University where he directed the International Research Centre for Investigative Psychology: see Section 10.5 *Further Developments of Criminal Profiling – Professor David Canter*). Dr Canter was called in by the police to help with the Railway Murder cases (London, 1982–1986: see Section 10.6.3 *The Railway Murders*). Professor Canter's thesis is that '*criminals reveal who they are and where they live not only through the way they commit crimes, but also through the locations they choose*'. He applied this thinking to the Railway Murders and presented compelling psychological evidence that supported the convictions – of the 17 perpetrator's character traits that Dr Canter identified while investigating the case, 13 were identified in the suspects when they were arrested. This was the first time that geographical profiling had been used in a criminal case and because its application to the Railway Murders was so successful, geographical profiling became part of the police's forensic armoury.

Geographic profiling is complex, but there are several facets that I will use to illustrate the principle here. As is evident from the distribution of the rapes and murders in the Railway Murders Cases (Figure 10.3), they all occurred in the same geographical region of London; moreover, all of the initial 19 cases were closely clustered in the Camden area. Criminals often commit their crimes in close proximity to their homes – simplistically, it is easier to travel a short distance to commit a crime, especially if there is a good means of transport (in this case the London Underground). So, in geographic profiling terms, it would be predicted that the offender lived near to the initial cluster of cases: spot on! One of the offenders (there were two, see Section 10.6.3 *The Railway Murders*) lived at Kilburn, north London, just a few underground stops from the murder scenes (Figure 10.3).

Geographical profiling often shows that as a crime progresses, the cases move further away from the offender's home. The Railway Murders example illustrates this perfectly. The initial 19 rapes

Figure 10.3 The geographical spread of the Railway Murders across greater London and beyond. Approximate locations of rapes (red circles showing number of cases) and rape/murders (red stars) and John Duffy's home (green filled circle).

(1982–1986) were clustered close to the offender's home, while the next group of 4 rapes (1985–1986) occurred further away, but remained connected to the offender's home by the London Underground; the final 3 murders (1985–1986) occurred much farther away – this time they were connected to the offender's home by the mainline railway system. It is clear that this kind of information would be useful to the police to help prioritise suspects so that investigative resources can be directed towards those suspects who might offer valuable leads.

There are two types of criminals in the geographical profiling system: 'marauders' and 'commuters'.

10.6.2 Marauders and Commuters

Marauders (about 85% of criminals) live close to the locations where they commit their crimes while commuters (about 15% of criminals) travel to commit their crimes (Figure 10.4). The railway murders began as marauders and progressed to commuters. Since there were two offenders working together in these cases, it is possible that one was a marauder and the other was a commuter and the dominant offender changed as the crimes progressed.

Defining a criminal in this way allows the police to direct their resources within a particular geographical area.

The Railway Murders (1982–1986) committed in and around London were the first cases to be investigated using geographical profiling and are a perfect example of the usefulness of this important psychological approach to criminology.

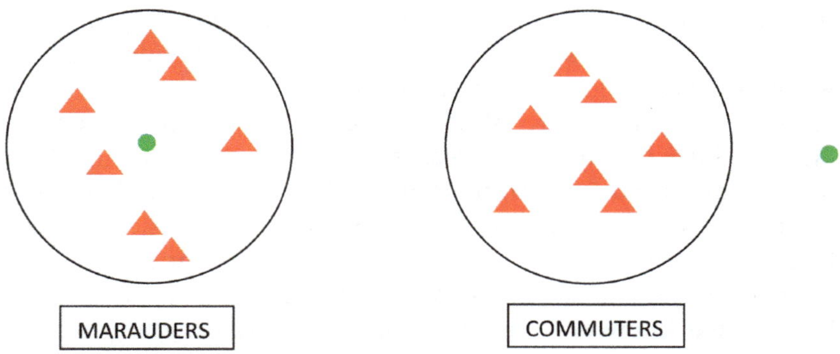

Figure 10.4 The two geographic profiling classes of criminals: marauders and commuters showing their homes (green circles) and locations of their crimes (red triangles).

10.6.3 The Railway Murders

In 1982, a woman was raped near Hampstead Heath Underground station in north London. An additional 18 similar offences were committed over the next year, all in close proximity to Underground stations and concentrated in or around north London. The press referred to the cases as the Railway Rapes.

The victims described the offenders as a 'ginger-haired man and a larger man'; since the descriptions were very similar, the police decided that the crimes had been committed by the same duo. DNA technology was very much in its infancy during the early to mid-1980s and therefore was not used to identify the perpetrators. However, biological fluid (blood, semen, and saliva) samples recovered from the crime scene showed that one of the men was a blood group A secretor – this means that the individual had blood group A (see Chapter 5 *Forensic Biology*) and that he secreted the A antigen into other biological fluids (*e.g.*, saliva). In addition, the same man was a specific, rare phosphoglucomutase (PGM) blood group. Although these blood groups could not be used to definitively identify the offender, the presence of both in blood from the same person provides valuable evidence that might contribute to the identification process. When caught, the ginger-haired man was indeed found to be a blood group A excreter and to have the same PGM group as the biological fluids found at the scene.

Three years later, on the evening of 29th December 1985, a 19-year-old woman was raped near Hackney Wick Underground station and taken to nearby wasteland and strangled to death. Initially, the police did not connect this rape/murder case with the earlier Railway Rapes and set up a separate investigation.

Later, two rapes/murders occurred: one of a 15-year-old girl on 17th April 1986 in West Horsley (Surrey: 45 km south of London), and the other of a 29-year-old woman at Brookman's Park railway station (Hertfordshire: 34 km north of London) on 18th May the same year.

The case investigations progressed slowly: in short, in the 1980s, the police identified John Duffy (born 1958) as a suspect because he had previously been charged with the rape of his estranged wife and he was known to have the appalling view that rape is a 'natural male instinct'. In addition, he was a former railway worker and had a detailed knowledge of the London area railway network; and furthermore a rare type of string the same as that used for the Brookman's Park railway murder was found at Duffy's parents' home. And he had ginger hair! These cicumstances raised suspicion among the police and they took him in for questioning. At this point, the connection between the Railway Rapes and the subsequent murders became clear. The press modified their epithet to the Railway Murders; this is now used to cover all of the rapes and murders committed by the deranged duo.

Shortly after Duffy's arrest on 23rd November 1986, police arrested Duffy's close friend, David Mulcahy (born 1959), but released him because they had no evidence linking him to the crimes. Mulcahy was re-arrested on 3rd February 1999 and with the help of DNA technology (not available when Mulcahy was first arrested; see Chapter 6 *DNA*), he was conclusively linked to the Railway Murders.

In February 1988 at the Central Criminal Courts (Old Bailey) in London, Duffy was convicted and sentenced (for several of the rapes/murders) to life in prison. He was re-tried in 1999 and pleaded guilty to 17 of the offences; he received additional life sentences for these extra convictions; however, this made no difference to the fact that he would spend the rest of his life in prison. On 1st February 2001, Mulcahy was sentenced to life imprisonment at the Old Bailey. Interestingly, Mulcahy's conviction was largely based on Duffy's evidence; this was the first time a category one prisoner had given evidence in court in the UK.

10.7 Forensic Psychology Is a Broad Discipline...

In this book, we are trying to include all of the forensic techniques that might be used to investigate and hopefully solve crime. Forensic psychology is an important tool in the forensic armoury. However, it is a much broader discipline that extends beyond providing tools for crime investigations. Forensic psychology also includes, among other things, predicting whether criminals might reoffend (risk assessment) or whether their confessions can be deemed reliable (false confessions). Detailed discussions of both are beyond the remit of this book, but for completeness, they will be outlined briefly here.

10.7.1 Risk Assessment

Risk assessment is about predicting whether – or assessing the risk that – an offender will reoffend. Statistics show that forensic psychologists' predictions are usually good (with a success rate comparable to the effectiveness of a mammogram detecting breast cancer).

However, there are some factors which underpin reoffence. One (which is rather obvious) is that the more offences a criminal commits, the greater the likelihood he or she will commit more. You would be right if you are thinking that you do not need to be a forensic psychologist to work that out! However, a closer analysis of the risk of reoffending does bring out some useful determinants which might be used to determine the risk of reoffending and (importantly) be used as part of offender 'treatment' regimens (*e.g.*, in a prison setting).

There are two classes of reoffending determinants, 'static' and 'dynamic': static determinants are those that cannot change (*e.g.*, number of offences committed), while dynamic determinants are changeable (*e.g.*, how well an offender can deal with his/her pride being hurt as a trigger for offending). It is clear how therapy relating to dynamic determinants might reduce reoffence.

Interestingly, the nature of a crime is not a particularly good determinant of the nature of future crimes. For example (New Zealand data), only 20% of sex offenders commit sex crimes in the future and only 7% of crimes involving a weapon predict the future use of weapons by the same offender. As you might expect, one of the greatest determinants (30%) of reoffending is drugs because addiction is involved, but even that is perhaps lower than might be expected. The best indicator of future crime is theft (42%), and interestingly, a significant proportion of reoffences by other offenders involves theft. For example, 24% of drug offenders commit thefts later (perhaps to fund their addiction), 26% of sex offenders and 20% of driving offenders also commit theft later in their criminal careers.

In short, reoffending risk assessment is of limited use in criminal detection, but more important in offender rehabilitation and to determine whether someone would be safe on a community sentence or to be released from prison.

10.7.2 False Confessions

This is when someone confesses to a crime they did not commit. It is clearly important in a forensic context because if it goes undiagnosed, the real offender remains at large and might reoffend. Why on earth would a person confess to a crime they did not commit? Often the answer involves some sort of psychological disorder or, perhaps, undue stress caused by the interrogation process.

An example of a false confession was Andrew Evans' confession to Judith Roberts' murder in the UK in 1972. Evans was convicted based on a dream that he had about Judith's murder and his belief that the dream was indeed true. Despite many inconsistencies in his evidence, Evans was sentenced to life in prison.

10.7.3 Judith Roberts' Murder - UK, 1972

Fourteen-year-old Judith Roberts left her home near Tamworth in Staffordshire, UK, on 7th June 1972 for a bike ride – she never returned. Her body was found naked from the waist down 3 days later under hedge clippings at the edge of a field with her bicycle nearby. She had 19 head wounds, one being severe enough to be the fatal blow.

A 17-year-old soldier, Andrew Evans had a dream in which he 'saw' Judith's murder. The dream became reality to him and he was convinced that he was the murderer. Four months after the murder, Evans confessed. The police interviewed him over a three-day period and despite many inconsistencies in his story (*e.g.*, he got the colour of Judith's underwear and the colour of her hair wrong), they accepted his confession. He was convicted of murder and sentenced to life in prison at Birmingham Crown Court in June 1973 – he spent the next 24 years in Verne Prison in Dorset in the south west of England.

Fast forward to 1994: the media became interested in Evans' conviction after a Greenpeace member, Steve Elsworth, had given a talk to Verne Prison inmates and met Evans. Elsworth returned to Verne Prison to interview Evans. He talked to a TV company about the interview which resulted in a TV programme as part of Central Television's *Crime Stalker* series. Later that year, Evans wrote to *Justice* (a human rights organisation) about his case which led to his Court of Appeal hearing in December 1997.

The Appeal judges found many transgressions that had occurred during Evans' interview by police. He was receiving medication for depression at the time, a doctor was not present at his interview, and he was not given his right to have a solicitor present. The judges concluded that if a doctor and a solicitor had been present, his confession would not have been accepted due to his state of mind. To corroborate this, a doctor at the Appeal told the court that Evans was suffering from 'false memory' due to anxiety and depression at the time of his confession. In addition, significant inconsistencies in the case's forensic investigations were noted. Not least, the fact that there was none of Judith's blood on Evans' clothing, which would be incredibly unlikely if he had bludgeoned her to death, and unknown fingerprints on Judith's bicycle did not match Evans'. The Court of Appeal quashed Evans' conviction and he was released from prison in 1997.

Later, in an interview, Andrew Evans commented, '*By confessing, I thought I would be able to rid myself of all the crap going on in my head.*' This is clear evidence of a psychological disorder-based false confession.

The police did not reopen the investigation and so Judith Roberts' killer remains at large, unless of course, the most natural form of justice – death – has caught up with him or her.

Further Reading

1. R. Volbert, L. May, J. Hausam and S. Lau, Confessions and denials when guilty and innocent: forensic patients self-reported behavior during police interviews, *Frontiers in Psychiatry*, 2019, **10**, 168.
2. A. M. Goodwill, R. J. B. Lehmann, E. Beauregard and A. Andrei, An action phase approach to offender profiling, *Legal and Criminological Psychology*, 2016, **21**, 229–250.
3. G. Copson, Coles to Newcastle – Police Use of Offender Profiling, *Police Research Group: Special Interest Series, Paper 7*, Home Office, UK, 1995.

References

1. T. C. Collocott and A. B. Dobson, *Chambers Dictionary of Science and Technology*, W & R Chambers Ltd., Edinburgh, 1974.
2. *The Shorter Oxford English Dictionary*, Oxford University Press (Book Club Associates), 1983.
3. K. Cherry, An Overview of Forensic Psychology. An Overview of Forensic Psychology (verywellmind.com), 2020, accessed 8th June 2022.
4. A. M. Goodwill, R. J. B. Lehmann, E. Beauregard and A. Andrei, An action phase approach to offender profiling, *Legal Criminol. Psychol.*, 2016, **21**, 229–250.

11 Firearms

Firearms are increasingly involved in criminal cases, either to threaten (*e.g.*, in aggravated burglary, approx. 9% of violent crimes in the USA involve firearms) or as a murder weapon (62% of murders involve firearms in the USA). Forensic firearms investigation is an important aspect of violent crime investigations. It involves highly specialised experts who identify projectiles (bullets) or firearms associated with a crime; they can link projectiles and cartridges to particular firearms; they can assess distances from where a shot was fired and, in some cases, assess firing angles and interpret blood and wound patterns. In addition, the explosion that generates the energy necessary to accelerate the projectile leaves residues that may be deposited on the hands or clothes of the shooter, as well as on surfaces and other people in the immediate vicinity. All of this information, and more, is part of the powerful armoury of forensic firearms investigators, although not all of these skills are consolidated within a single expert, contrary to what is shown on television.

11.1 Introduction

Of the approx. 5.5 million rapes, robberies and aggravated assaults in the USA in 2019, 8.1% involved firearms, and 62% of the approx. 6.5 thousand murders in the same year involved firearms (data calculated from the USA National Crime Victimization Survey, 2019). In New Zealand, 18% of murders and 13.6% of aggravated robberies carried out in 2022 involved firearms (data from Firearms Information Summary, NZ Police, 2022). Clearly, looking just at

Forensic Science: The Science Behind the Truth
By Ian Shaw with Anna Sandiford

Published by the Royal Society of Chemistry, www.rsc.org

these two very different countries indicates a significant number of crimes involving firearms occurring each year worldwide. This is why forensic firearms experts are key members of forensic investigation teams.

Firearms experts, obviously, must possess expertise in the science of firearms, including their projectiles, how they travel in flight and the damage they can cause at the point of impact: what happens when a projectile hits its target? Can a projectile be linked to a particular firearm? Can a firearm from which a projectile was fired be identified? The deliberations of the firearms expert require a substantial degree of professional judgement and experience (rather than necessarily hard scientific proof) and this expert professional judgement is generally accepted by the courts.

11.2 The Anatomy of Firearms

There is a vast array of firearms available, but this book is not the place to go into great detail about the ins and outs of that diversity. The most common types of firearms are handguns, shotguns, rifles, assault rifles, and machine guns. I will outline each firearm type later; however, since the basic principles are common to all firearms, understanding the anatomy of a firearm and its projectiles is important before discussing the different types and the forensic information they generate when they are fired.

I will use the handgun as an example to outline how a firearm works. Basically, a small amount of explosive (propellant) is detonated to provide the energy to eject the projectile (bullet) at high speed from the muzzle so that it has maximum impact on its target. To achieve this, the firearm is equipped with a receptacle for the projectile (chamber: Figure 11.1), an amount of propellant behind the projectile, a trigger to release the hammer (or striker) that strikes the primer (detonator), igniting the propellant and initiating an explosion that produces a large volume of gas that forces the projectile down the barrel and expels it *via* the muzzle towards its target (Figure 11.1). The speed at which projectiles exit the muzzle varies and depends on the type of firearm. A typical speed for a common 9 mm round could be around 1100 feet per second (approx. 3300 m s^{-1}).

The primer can be an electric impulse (used for large munitions not usually encountered in forensic investigations) or a shock-sensitive chemical mixture which is used for most small munitions that might be encountered by the forensic firearms expert.

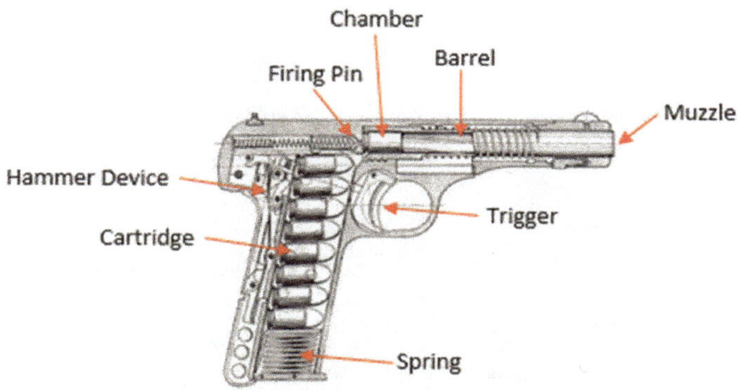

Figure 11.1 The anatomy of a handgun. After the cartridge enters the chamber, pulling the trigger causes the hammer device to release the firing pin, which strikes the primer at the base of the cartridge. This detonates the propellant in the cartridge base or rim, triggering the explosion that forces the projectile down the barrel at great speed. The spring-loaded reload system then pushes the next cartridge into the chamber, ready to be fired. Adapted from ref. 1.

11.2.1 Rifling

With the exception of most shotguns, most firearms have spiral grooves on the inner surface of the barrel – this is called 'rifling'. These spiral grooves impart spin to the projectile when it leaves the barrel *via* the muzzle, enhancing its stability in flight and thus potentially increasing the accuracy of the shot.

Shotguns generally do not have rifling as the projectiles they fire are housed within a larger cartridge, the projectiles are smaller than the bore/barrel and therefore rifling cannot influence individual projectiles. Instead, the term 'smooth bore' is used to indicate the lack of rifling.

11.2.2 Ammunition

Ammunition is described by the United Nations Office on Drugs and Crime as "*the complete round or its components, including cartridge cases, primers, propellant powder, bullets or projectiles, that are used in a firearm, provided that those components are themselves subject to authorization in the respective State Party*" (https://www.unodc.org/e4j/en/firearms/module-2/key-issues/ammunition.html).

The key components of this statement are the cartridge case, primer, propellant powder, bullets/projectiles (Figure 11.2). Cartridges can

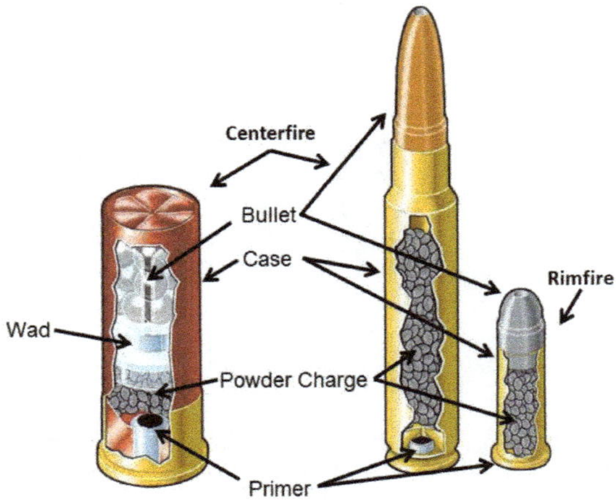

Figure 11.2 Cross-sections of a shotgun cartridge (left), a centrefire rifle cartridge (middle) and a rimfire cartridge (right). Created with Bing Image.

contain a single projectile or multiple projectiles (such as used in shotguns). There are two main classifications of cartridges: rimfire and centrefire. This distinction is based on whether the primer is located around the rim of the cartridge base or in the centre of the base and therefore whether the firing pin strikes the centre or the edge of the cartridge base in order to start the explosive reaction. Generally, rimfire cartridges are utilised in smaller cartridge sizes such as 0.22 or 0.17 calibre. Centrefire cartridges are preferred in calibres larger than these.

The calibre determines the size of the projectile the firearm can take. However, determining the meaning of 'calibre' is not at all straightforward. The description of ammunition does not follow a standard methodology, as it is based on both metric and imperial measurements and it has been developed by Americans, British, other Europeans and the rest of the world using different descriptive systems at different times. As a result, there is such a diverse range of descriptive terminologies that it is best left to individual case circumstances to describe ammunition relevant to a particular case. 'Calibre' can refer to projectile diameter or barrel diameter (either land or groove), which are not equivalent measurements. To quote from the first paragraph of the first page of the first chapter of Cartridges of the World (now up to the 17th edition), "*The subject [of caliber designation], regrettably, is full of inconsistencies and confusion*".

11.3 The Science of Detonation and Explosion

The principle of firing the projectile from a firearm involves two stages: detonation and explosion. As explained above, the shock-sensitive detonator is housed in the base of the cartridge; it is struck by the hammer released by the trigger. This action provides the energy necessary to ignite the propellant, which combusts to produce a large volume of gas that forces the projectile along the firearm barrel and expels it *via* the muzzle towards its target.

11.3.1 Detonation

This is achieved though a mixture of explosive chemicals plus a chemical that releases oxygen (this mixture is termed the primer). The explosion initiates the explosive combustion of the propellant. Primer mixtures vary from manufacturer to manufacturer, but they are usually based on a formulation patented in 1932 by James Burns (US Patent 1 880 235) (Table 11.1).

Table 11.1 A typical primer formulation used to initiate the explosion that expels the projectile from a firearm towards its target (Burns JE (1932) US Patent 1 880 235).

Chemical	Formula	Properties
Mercury fulminate	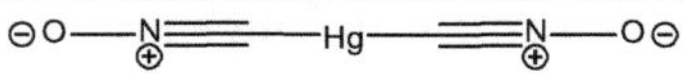	Explosive
Lead trinitroresorcinate (lead styphnate)	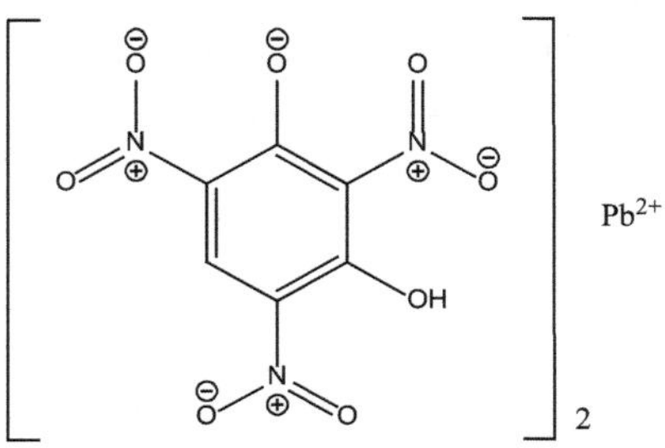	Explosive
Barium nitrate	$Ba(NO_3)_2$	Oxidiser – provides oxygen for the explosion
Lead thiocyanate (lead sulfocyanate)	$Pb(SCN)_2$	Explosive with nitrates
Abrasive		Helps to initiate the explosion when shocked
Binder		Holds the mixture together

11.3.2 Explosion

As outlined above, detonation involves a relatively small explosion which provides the energy to initiate the significantly larger propellant explosion that propels the projectile.

For example, traditional gunpowder contains potassium nitrate (saltpetre; KNO_3) a powerful oxidising agent, sulfur and carbon (the 'fuels' for the explosive reaction). The complex chemical reaction between the gunpowder components results in the release of a large volume of carbon dioxide (CO_2) and nitrogen (N_2) gases (Figure 11.3) – for 10 g of KNO_3, approx. 2.4 L of gas would be produced very quickly. This results in a huge increase in pressure behind the projectile, forcing it out of the barrel at high speed.

A complete industry is dedicated to the manufacturing of primers and propellants and as a result, there is no standard compound; to add other complexities to the situation, people also engage in 'home loading', which involves loading their own cartridges with their own combinations of primers, propellants and projectiles.

The explosion not only results in the production of a huge volume of gas but also leads to the generation of a fine dust containing, among other particulate matter, small amounts of the primer and propellant components and large amounts of the explosion products (Figure 11.3). This dust settles on surfaces, including people, near the fired firearm. These gunshot residues can be analysed and used to assess potential sources of ammunition, determining whether or not they could have been deposited by the purported ammunition, proximity of people, places or items to a fired firearm, handling of a firearm or transfer of the same material onto other surfaces. I will deal with this later.

11.4 Firearm Types

This is not a treatise on firearms but is intended to give you a brief idea of the most common firearm types because this is important for understanding how a forensic firearms expert interprets findings related to firearms.

$$10\ KNO_3 + 3\ S + 8\ C \rightarrow K_2CO_3 + K_2SO_4 + 6\ CO_2 + 5\ N_2$$

Figure 11.3 The chemistry of a traditional gunpowder explosion: potassium nitrate (KNO_3) reacts explosively (when detonated) with carbon (C) and sulfur (S) to form potassium carbonate (K_2CO_3), potassium sulphate (K_2SO_4) and a large volume of CO_2 and N_2 gases, which increases the pressure behind the projectile forcing it out of the firearm barrel at high speed.

11.4.1 Handguns

Handguns can be fired with one hand, hence their name. The barrel is short and generally contains rifling.

The terms 'pistol' and 'handgun' are often erroneously used interchangeably. A pistol is a type of handgun with its chamber incorporated into the barrel (Figure 11.4). Pistols have a magazine that is loaded with cartridges, which is then inserted into the base of the firearm, usually in the handgrip. A new cartridge is fed into the chamber by a spring-loaded system (Figure 11.1) after each firing.

11.4.2 Revolvers

Revolvers (Figure 11.5) are handguns that have a revolving cylinder containing multiple chambers, each holding a single cartridge. When the trigger is pulled, the cylinder rotates to the next chamber, aligning the next cartridge with the firing pin and the barrel (Figure 11.1).

11.4.3 Rifles

Rifles are larger than handguns. The stock (Figure 11.6) is positioned against the shoulder to provide stability during firing. Rifles have long barrels that are often, but not always, rifled.

The speed at which a projectile leaves the muzzle of a rifle is generally faster compared to handguns.

Figure 11.4 A pistol. Figure is in the Public Domain.

Figure 11.5 A revolver showing the revolving cylinder which aligns the next cartridge with the firing pin whenever the trigger is pulled. Figure is in the Public Domain.

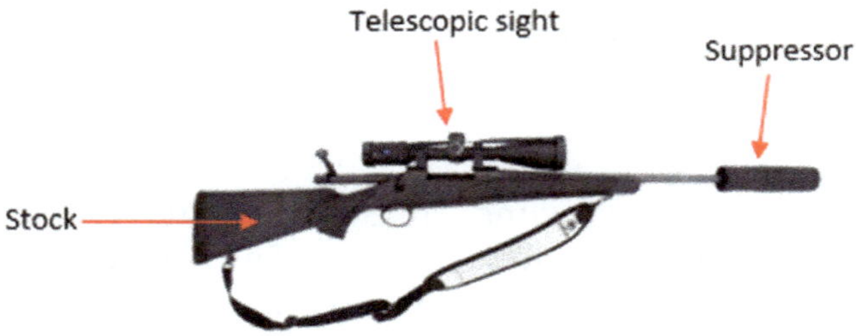

Figure 11.6 A modern rifle showing the stock which is positioned against the shoulder to provide stability during firing. This rifle is fitted with a telescopic sight and suppressor to reduce the noise and recoil when fired. Reproduced from ref. 2, https://commons.wikimedia.org/wiki/File:Hunting_rifle_02.png, under the terms of the CC BY-SA 4.0 license https://creativecommons.org/licenses/by-sa/4.0/

11.4.4 Assault Rifles

An assault rifle is a machine firearm that fires automatically but with an option to fire semiautomatically or in multiple shot burst modes (*e.g.*, 3 shot burst).

Two of the most well-known assault rifles are the Avtomat Kalashnikova (AK)-47 (Figure 11.7) and the M16. The AK-47 in particular is very durable and easy to use, contributing to its popularity since its introduction in 1947.

11.4.5 Machine Guns

Machine guns are rifles that automatically reload and fire as long as the trigger remains depressed, allowing for continual firing. It will continue to fire until the trigger is released, ammunition runs out or

Figure 11.7 An AK-47 assault (automatic) rifle showing the detachable magazine which, in this case, contains 30 rounds of ammunition (cartridges). Figure is in the Public Domain.

until the firearm jams. By contrast, a semi-automatic machine gun fires a projectile each time the trigger is pulled.

11.4.6 Shotguns

A shotgun does not usually have a rifled barrel. Shotgun cartridges (also called shells) can be loaded with projectiles of any size, ranging from either a single slug to hundreds of tiny pellets or even a super light load known as "dust shot". They can also be loaded with anything that fits into the cartridge, such as salt. If a cartridge containing multiple pellets is fired, as the pellets leave the barrel *via* the muzzle, they spread out in a cone shape and 'pepper' their target. Using knowledge of the type of ammunition and the particular shotgun in a case, the pattern of pellet spread at the time of impact can be used to assess the distance from the shot, which is a common question asked in cases where people have been injured or killed by shotgun fire or shots have been fired at property, vehicles or other items.

Because the energy from the propellant powder explosion is distributed between multiple pellets, each pellet does not usually travel as far as a single projectile (unlike from a rifle) because they have lower individual kinetic energies. For this reason, the effective range of a shotgun is usually less than that of a handgun or rifle.

11.5 Firearms Investigation

There are many potential aims when investigating criminal cases involving firearms.

11.5.1 Identifying Projectiles

In a case involving firearms, the projectile(s) and/or cartridge(s) might be recovered from the crime scene. For example, in a fatal shooting, the projectile might be lodged in the deceased's body and recovered during the *post-mortem* examination (see Chapter 8 *Forensic Pathology*) and a cartridge case might be recovered from the scene. The investigation would be significantly aided if the police could identify the type of firearm that fired the projectile. This, of course, can be possible because particular firearms use particular cartridge types, which fire particular projectiles. Therefore, comparing projectiles and cartridges can enable the identification of a potential firearm type.

This, in turn, might lead to a suspect. The question then is: did the suspect's firearm fire the cartridge and projectile in question? A closer investigation might show marks on the cartridge case and projectile; these marks can be caused by the loading, firing, extraction or ejection processes.

11.5.1.1 Rifling, Rifling Marks and Barrel Defects

When the propellant in the cartridge explodes, forcing the projectile down the firearm barrel at high speed, marks within the barrel will leave corresponding marks on the projectile. Assuming these marks survive the force of the impact, they can allow forensic scientists to compare a projectile to a firearm to determine if the former was fired from the latter.

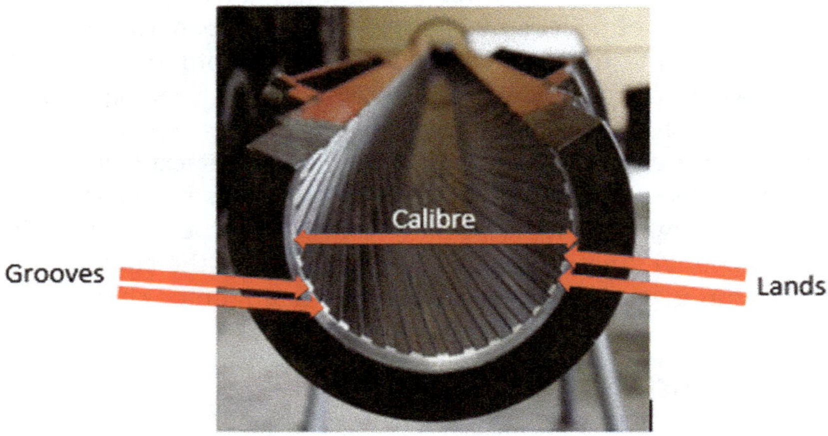

Figure 11.8 Left-hand twist firearm barrel rifling showing grooves, lands and calibre. This image is a cutaway model of a 105 mm calibre tank firearm on display at the Deutsches Panzermuseum in Münster, Germany – the principle of rifling in a smaller firearm (*e.g.*, a rifle) is the same. Reproduced from ref. 3, https://commons.wikimedia.org/wiki/File:105mm_tank_gun_Rifling.jpg, under the terms of the CC BY-SA 3.0 license https://creativecommons.org/licenses/by-sa/3.0/deed.en.

To understand these marks, it is necessary to understand a little about rifling. The purpose of rifling is to give the projectile spin, thereby improving stability and assisting accuracy. Rifling comprises spiral grooves cut into the inner surface of the barrel. The rifling grooves are called 'grooves' and the ridges are called 'lands' (Figure 11.8), and they can be either left-hand or right-hand twists.

To investigate the rifling and determine its parameters (*e.g.*, dimensions of lands), the barrel is illuminated *via* the muzzle and a photograph is taken – the measurements can be made from the photograph (Figure 11.9).

The projectile may show marks consistent with a firearm's rifling and any imperfections that are abrasive enough to mark the projectile as it travels down the barrel (Figure 11.10). The dimension of the marks on the projectile can be compared with a firearm's rifling pattern and dimensions, along with any other imperfections observed when examining the inner surface of the barrel.

11.5.1.2 Other Toolmarks

In addition to rifling marks, there are other processes occurring during loading and firing that create damage to, and marks on,

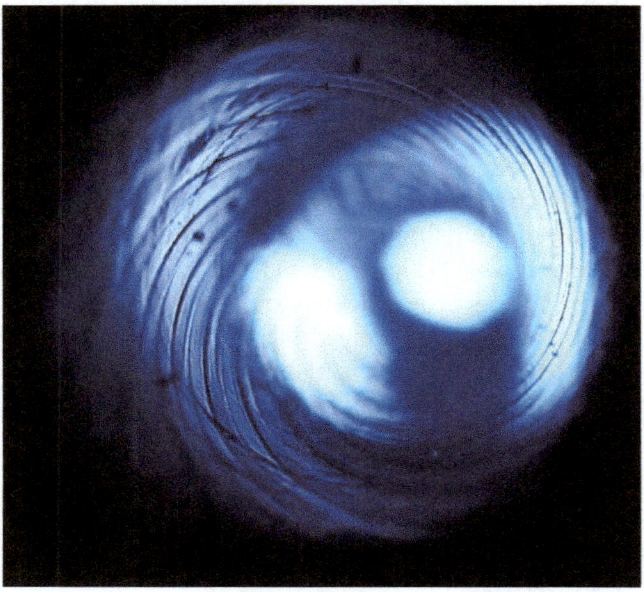

Figure 11.9 Photograph taken down the illuminated barrel of a Remington 35 mm calibre rifle showing right hand twist rifling with 20 lands and grooves. Figure is in the Public Domain.

Figure 11.10 Unfired (left) and fired (right) military bullets showing the rifling marks on the fired bullet. Reproduced from ref. 4, https://commons.wikimedia.org/wiki/File:762 × 39_bullets_-_unfired_and_fired.JPG, under the terms of the CC BY-SA 4.0 license https://creativecommons.org/licenses/by-sa/4.0/deed.en.

cartridge cases and projectiles. Firearms are complex, engineered items that have many moving parts. As a result of these movements, there are many contact areas and where surfaces come into contact, there is the opportunity for marks to be created. The most common areas of contact that firearm examiners consider include the movement of the ammunition into and through the chamber, the firing pin mark and extraction and ejection of the cartridge case.

Best practice in suitable cases where toolmarks are evident on cartridges and projectiles seized from a crime scene is to test-fire a selection of appropriate cartridges through the questioned firearm and then compare the marks on the test-fired projectiles and cartridge cases with the ones from the scene using high power comparison microscopy (see Chapter 5 *Forensic Biology*) (Figure 11.11). A Bayesian assessment of the likelihood of the scene projectile having been fired through the test firearm would usually accompany this interpretation (see Chapter 13 *Probability in Forensic Science*). It is also important to remember that in cases where a firearm is not recovered and the projectile has been extensively damaged at the point of impact, very little may remain for any meaningful examination and interpretation.

11.5.1.3 Gunshot Residues and Their Chemical Fingerprints

When a firearm is fired, the explosion (see Section 11.3.2 *Explosion* and Figure 11.12) creates a cloud of material composed of various

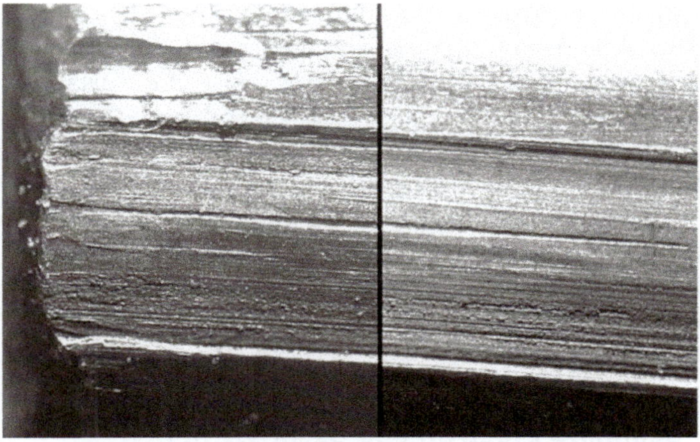

Figure 11.11 Marks on a recovered bullet and a test-fired bullet viewed under a comparison microscope. There are sufficient areas of common agreement to indicate that the recovered bullet was fired from the test firearm. Courtesy of Forensic Firearms Consultancy (FFC) Ltd.

Figure 11.12 When a firearm is fired, a cloud of particulates is produced that contains explosion combustion products, trace elements from the primer and powder mixtures, the firearm barrel, cartridge, cartridge case and the projectile(s), which can potentially settle on surrounding objects and surfaces, including the person who fired the firearm and anyone else nearby. Created with Bing Image.

gases, liquids and particulates originating from the primer, propellant, cartridge, projectile and the firearm itself. These settle on surrounding objects and surfaces, potentially including the person who fired the firearm, surfaces, objects and anyone else nearby. They can also be transferred through secondary contact, (*i.e.*, when one surface with gunshot residues on it comes in contact with another surface and transfer occurs from the original surface to the second surface – referred to as secondary transfer).

Collectively, the residues are referred to as gunshot residue or firearm discharge residue. They can be recovered and analysed, with the presence of trace elements being important. The combination (spectrum) of elements in the residues can provide a useful chemical fingerprint for assessing whether gunshot residues from an individual could have originated from a particular firearm that fired a particular type of ammunition.

11.5.1.4 Scanning Electron Microscopy/Energy Dispersive X-ray Spectrometry (SEM-EDX)

Scanning electron microscopy/energy dispersive X-ray spectrometry (SEM-EDX) is used to analyse elements in gunshot residues.

Samples can be collected either through gentle swabbing or direct transfer to an SEM stub and then analysed by SEM-EDX. Scanning electron microscopy (see Chapter 5 *Forensic Biology*) uses electrons to visualise the surface structure of a sample (Figure 11.13). When an element with an atomic weight greater than 23 (sodium) is bombarded with electrons, it emits X-rays: the wavelength of these X-rays

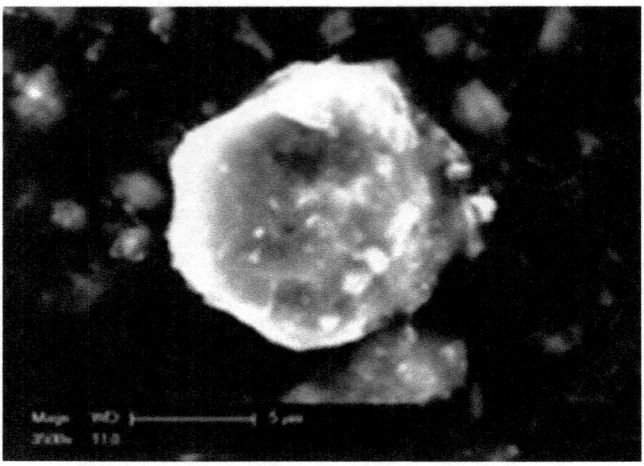

Figure 11.13 A gunshot residue particle visualised by scanning electron microscopy (approx. 8500×). The electrons bombarding the surface of the particle will cause the elements in the particle to emit a characteristic X-ray spectrum. Reproduced from ref. 5 with permission from the Royal Society of Chemistry.

is characteristic of the metal, and therefore the metal can be identified from its characteristic X-ray emission spectrum (Figures 11.14 and 11.15). The X-ray emission spectrum (elemental profile) of gunshot

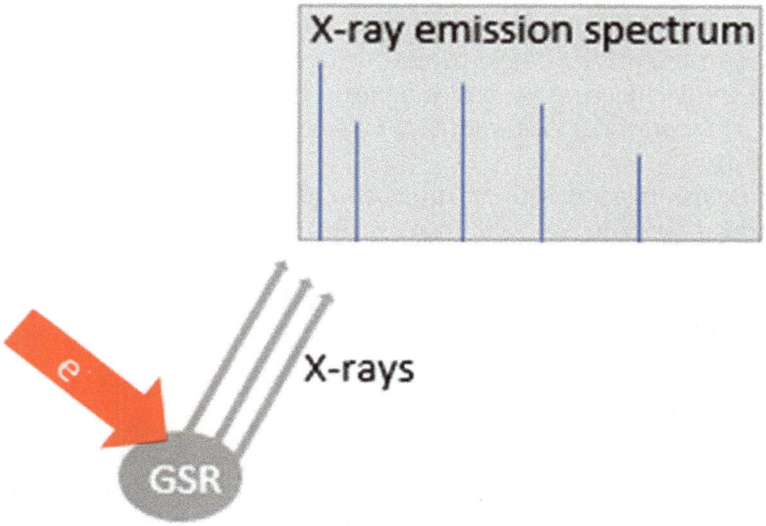

Figure 11.14 The principle of SEM-EDX: the gunshot residue particle is bombarded with electrons (e^-) in the electron microscope, resulting in emission of X-rays by its component elements, thus producing a characteristic X-ray emission spectrum (highly simplified here). Each line on the spectrum represents a different element.

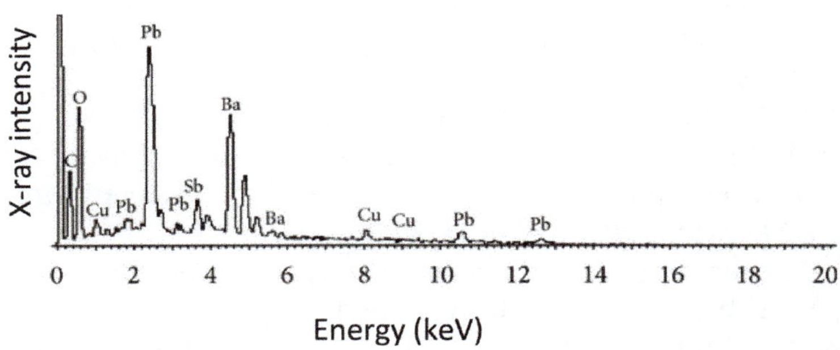

Figure 11.15 X-ray emission spectrum following scanning electron microscopy/X-ray microanalysis of a gunshot residue sample. The following metals are present: Cu = copper, Pb = lead, Sb = antimony, Ba = barium. The copper probably originated from the cartridge or the projectile. Reproduced from ref. 6, https://doi.org/10.1155/2014/428038, under the terms of the CC BY 3.0 license https://creativecommons.org/licenses/by/3.0/.

residues found, for example, on the hands of a suspect can be compared with residues recovered from a seized firearm or seized fired cartridge cases.

Very small samples can be analysed by SEM-EDX. If gunshot residue particles are visible under the scanning electron microscope, an X-ray emission spectrum can be produced. For example, hair samples might be examined under an electron microscope to reveal miniscule gunshot residue particles clinging to the hairs – even these tiny particles will emit X-rays and produce a characteristic spectrum.

Other spectroscopic techniques can also be used to determine the elemental profile of gunshot residue, including laser-induced breakdown spectroscopy (LIBS) and ion beam analysis (see *Further Reading*). Both methods produce spectra that enable the comparison of residues.

Further Reading

1. A. Jangra, S. Chandra and S. Mishra, Bullet fingerprinting analysis by digital comparison microscope, *J. Forensic Sci. Crim. Investig.*, 2019, **12**(5), 555847, DOI: 10.19080/JFSCI.2019.12.555847
2. M. E. Christopher, J.-W. Warmenhoeven, F. S. Romolo, M. Donghi, R. P. Webb, C. Jeynes, N. I. Ward, K. J. Kirkby and M. J. Bailey, A new quantitative method for gunshot residues analysis by ion beam analysis, *Analyst*, 2013, **138**, 4649–4655.
3. M. J. Silva, J. Cortez, C. Pasquini, R. S. Honorato, A. P. S. Paim and M. F. Pimentel, Gunshot residues: screening analysis by laser-induced breakdown spectroscopy, *J. Braz. Chem. Soc.*, 2009, **20**, 1887–1894.
4. Z. Brożek-Mucha, Scanning electron microscopy and X-ray microanalysis for chemical and morphological characterisation of the inorganic components of gunshot residue: selected problems, *BioMed Res. Int.*, 2014, **2014**, 428038.
5. F. C. Barnes, in *Cartridges of the World*, ed. K. Warner, DBI Books Inc, Northbrook, Illinois, USA, 6th edn, 1989, 448 p.
6. Firearms Protocol, Article 3(c), 55/255, Protocol against the Illicit Manufacturing of and Trafficking in Firearms, Their Parts and Components and Ammunition, supplementing the United Nations Convention against Transnational Organized Crime, United Nations General Assembly, 2001.

References

1. https://historiadelasarmasdefuego.blogspot.com/2011/11/pistola-browning-mod-1910-1922-y-1971.html.
2. File: Hunting rifle 02.png [Wiki], https://commons.wikimedia.org/wiki/File:Hunting_rifle_02.png [accessed 5 September 2023].
3. File:105 mm tank gun Rifling.jpg [Wiki], https://commons.wikimedia.org/wiki/File:105mm_tank_gun_Rifling.jpg [accessed 5 September 2023].
4. File:762 × 39 bullets – unfired and fired.JPG [Wiki], https://commons.wikimedia.org/wiki/File:762 × 39_bullets_-_unfired_and_fired.JPG [accessed 5 September 2023].
5. M. E. Christopher, J.-W. Warmenhoeven, F. S. Romolo, M. Donghi, R. P. Webb, C. Jeynes, N. I. Ward, K. J. Kirkby and M. J. Bailey, A new quantitative method for gunshot residues analysis by ion beam analysis, *Analyst*, 2013, **138**, 4649–4655.
6. Z. Brożek-Mucha, Scanning electron microscopy and X-ray microanalysis for chemical and morphological characterisation of the inorganic components of gunshot residue: selected problems, *BioMed Res. Int.*, 2014, **2014**, 428038.

12 Forensic Document Examination

Information can be transferred from one person to another and one place to another as words, numbers or symbols in a document. The concept of a document is very broad, ranging from a leg of lamb bearing a country-of-origin stamp to a sheaf of printed or typed pages or their electronic counterparts intended to convey information to its reader. This book is a document.

Alterations to the words or symbols in or on a document might have been carried out to deceive the reader – this is fraud. For example, New Zealand lamb is renowned worldwide for its quality: an inferior leg of lamb bearing a fraudulent New Zealand country-of-origin stamp might command the same price as the authentic product because the purchaser was fooled by the fraudulent stamp. Similarly, a bank cheque with an extra zero added to the amount might mean a handsome profit for a fraudster. The possibilities are endless and the ingenuity of fraudsters is endless too!

A forensic document examiner pits his or her wits against fraudsters by identifying modifications to documents. They might do this by showing differences between the make-up of inks used to write the document and an ink used to modify the document; they might scrutinize print or type characteristics or handwriting style – differences might indicate changes; on the other hand, they might look for deletions from a document that might change its meaning – imagine a will that reads 'My only son is not included in this final will and testament': simply removing 'not' conveys the opposite meaning and perhaps a great deal of money is inherited.

Forensic Science: The Science Behind the Truth
By Ian Shaw with Anna Sandiford

Published by the Royal Society of Chemistry, www.rsc.org

In this chapter, I will explore some of the methods used by forensic document examiners to identify fraud, counterfeiting and forgery. Forensic document examination is an extensive field and I can only attempt to give a flavour of the crimes and methods used to expose them – the subject deserves an entire book, not just a chapter – and indeed there are many such books for the interested reader.

12.1 What Is a Document?

The word 'document' comes from the Latin word, *documentum*, meaning lesson, instruction, warning, example, document or proof. It is clear from this etymology that a document carries authority. This authority might be conveyed *via* a series of facts, figures or images. In a legal setting, a 'document' is defined as 'a piece of written, printed, or electronic matter that provides information or evidence or that serves as an official record.' (*Oxford Languages* definition: https://languages.oup.com).

With the advent, rapid development and ubiquity of electronic transmission of information, document examiners now must consider not only physical documents, but also electronic documents, which exist as an array of electrons on a silicon chip.

A document can be anything that conveys information. It could be the stamp on a leg of lamb that denotes the country of origin (Figure 12.1), the label on a jar of mānuka honey that assures its authenticity (mānuka honey carries a huge price premium based on its methylglyoxal content), the numbers on a bank cheque that assures its value, the description of a piece of land on a title deed, the Dymo label messages that Donald Neilson left when he murdered Lesley Whittle (see Chapter 1 *The Whittle Case – 1976*); the possibilities are endless.

Not only are the forms that a document might take endless, but also the possibilities to alter a document to enhance its value or convey a different meaning are equally diverse. Using a counterfeit meat stamp to fool the purchaser of a leg of lamb that the meat is of high quality will affect the price the consumer pays and the profit the seller makes. Putting a mānuka honey label with a high methylgyoxal content onto mixed flower honey that does not contain methylglyoxal could increase its price from a few dollars a jar to tens of dollars a jar. Adding a zero to a bank cheque significantly increases its value. Altering the dimensions of land on a title deed could be highly beneficial to the purchaser. Careful scrutiny of the imperfections in

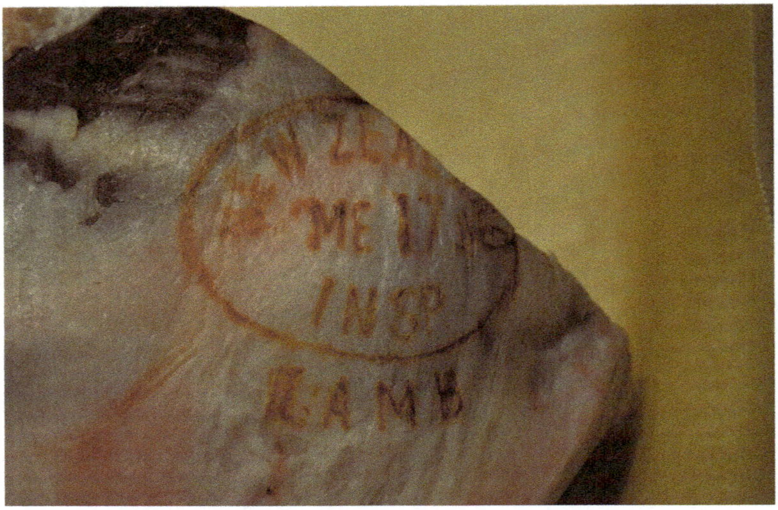

Figure 12.1 A leg of New Zealand lamb with its authentication stamp. New Zealand lamb is renowned worldwide for its quality: an inferior leg of lamb bearing a fraudulent New Zealand country-of-origin stamp might command the same price as the authentic product because the purchaser was fooled by the fraudulent stamp. In this case, the document is the leg of lamb.

Donald Nielson's Dymo label messages helped to trace them back to a particular Dymo label maker.

So, the simple answer to the question: what is a document? is anything that can be used to convey information.

12.2 The Tools of a Forensic Document Examiner

The document examiner looks for changes in documents and evidence of the chain of custody (*i.e.*, who had the document and when) and so the methods they use are either optical (*e.g.*, magnifying devices) to facilitate visualising changes, chemical to identify materials (*e.g.*, inks) that might have been used to change a document or physical, for example, to find writing impressions on paper under a document.

There is a cross-over between the work of a forensic document examiner and other forensic science professionals. For example, fingerprints (see Chapter 7 *Fingerprints*) might be important in establishing the chain of custody or ink analysis (see Chapter 4 *Forensic Chemistry*) might reveal a written change to a document.

12.2.1 Optical Devices

Usually only magnifying glasses (magnification ×3 or 4) and binocular microscopes (magnification up to ×25) are used by forensic document examiners to look for alterations in documents or perhaps impression marks on a sheet of paper beneath a written document – further details on magnifying glasses and binocular microscopes are covered in Chapter 5 *Forensic Biology*.

12.2.2 Electrostatic Detection Device

If pressure is applied (*e.g.*, writing) to a sheet of paper above another sheet of paper, the lower sheet will exhibit an indentation mark that corresponds to the impression on the upper piece of paper. So, if the impression is writing, the indentation marks on the lower sheet can be used to decipher the writing.

When an indentation mark is made on paper, it alters the surface charge on the paper; this changed surface charge can be picked up by an electrostatic detection device (EDD; Figure 12.2) using miniscule (approx. 5 μm diameter), coloured (usually black) charged particles which are attracted to the paper in different ways according to its surface charge. This highlights the indentation where the charge difference is greatest, effectively visualising the 'writing' (Figure 12.3).

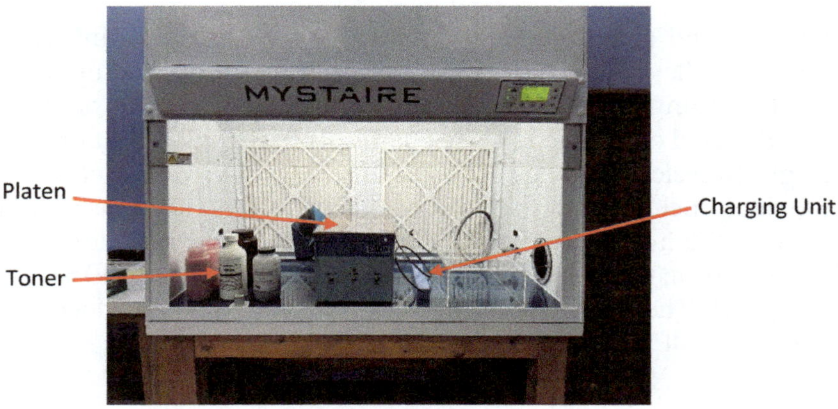

Figure 12.2 An electrostatic detection device: the document for investigation is placed on the platen and covered with a charging film, the charging unit is waved over the document/charging film which differentially introduces positive charges to the indentations according to their depth, *etc.* The toner is introduced and its negatively-charged particles adhere to the document showing up the writing indentations. Image courtesy of Dr Chris Palenik, Microtrace LLC, Elgin, IL USA; annotations by the author.

Figure 12.3 A writing impression visualised using an EDD. Reproduced from ref. 1 with permission from Elsevier, Copyright 1991.

The toner in a computer printer device works in the same way; in fact, the EDD used by forensic document examiners uses materials very similar to printer toners for visualising writing indentations.

12.2.3 Oblique Lighting

To visualise writing indentations on a sheet of paper beneath a written document, strong oblique lighting can be used. This technique casts shadows from the writing indentations on the paper, making the writing visible (Figure 12.4).

12.2.4 Infrared and Ultraviolet Light

Inks reflect and absorb light differently. The colour of an ink is determined by the visible light wavelengths that the ink components reflect; for example, if white light (*i.e.*, the entire spectrum of visible light) is directed onto blue ink, the chemicals in the ink reflect only the blue light (wavelength range 400–500 nm). The electromagnetic spectrum (which includes the visible region; Figure 12.5) also has regions invisible to the human eye, which lie beyond the extreme ends of the visible spectrum; these are the ultraviolet (UV) and infrared (IR) regions (Figure 12.5). The use of these spectral characteristics in document examination will be discussed later in this chapter.

12.3 Investigating Suspect Documents

As with many other aspects of this book, it must be remembered that our intention is to give an impression of the work conducted by various forensic science practitioners, rather than offering a comprehensive treatise on each. The above examples of the use of oblique

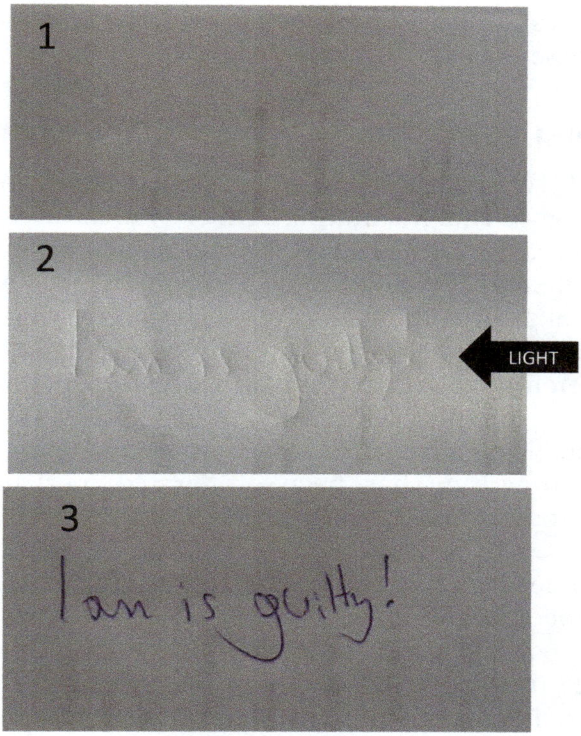

Figure 12.4 A sheet of paper (1) from beneath another piece of paper that was written on with a ballpoint pen and photographed under ambient light. You can clearly see printing on the reverse of the paper and, if you look very carefully, there is an inkling of writing indentation, but it is not possible to read it. When lit obliquely (2) with a strong light (direction shown by the arrow), the writing indentation becomes clear enough to read, revealing the writing on the top sheet of paper (3).

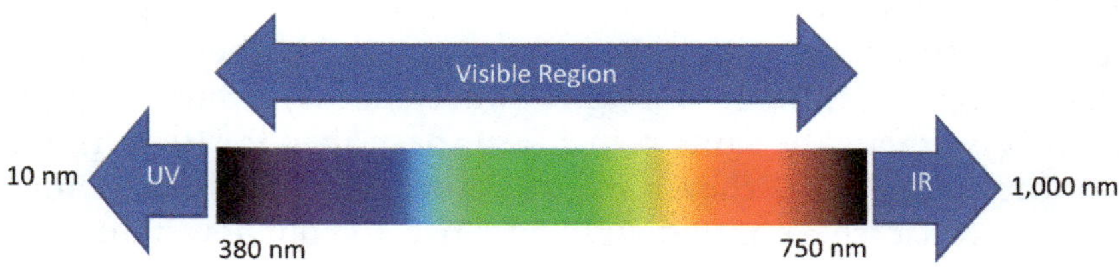

Figure 12.5 Part of the electromagnetic spectrum showing the visible, ultraviolet (UV) and infrared (IR) regions and wavelengths (nm).

lighting and EDD to show writing impressions, illustrate how these techniques are applied to document examination. More detailed examples of the application of the document examiner's tools to suspect document investigations, revealing any evidence of fraudulent activity, are discussed below.

12.3.1 Using IR to Reveal Changes in a Document

When IR is directed onto an ink (or any object), the ink might absorb some of the light's energy, emitting light of a lower wavelength (*i.e.*, in the visible spectrum). This means that an ink will exhibit a different colour (usually red) if it is observed under IR. This is an important property of inks (and other substances) because it can be characteristic. Therefore, if a document is modified (perhaps fraudulently) using a different, but same colour ink as the original document, it might appear a different colour when viewed under IR. This can be useful in visualising changes in a document.

Similarly, if writing has been obliterated, perhaps by being repeatedly crossed out, IR might reveal the original, obliterated writing.

The equipment used to examine emitted light from IR irradiated inks on a document can be as simple as an IR lamp with visual observation, but this is fast being replaced with spectral analysers with variable irradiation wavelengths. The document is placed in the analyser, irradiated and the reflected signal is analysed by computer software, which records the image showing any differences present in the inks on the document being investigated (Figure 12.6).

12.3.2 Using UV Light to Reveal Changes in a Document

Illuminating a document with a UV lamp might show up different inks because of the different degrees to which they absorb and transmit light of different energies (wavelengths) from the original UV – this is exactly the same principle as described for IR. In the case of IR-irradiated inks, they tend to appear red after absorbing IR energy because the emitted light is close to, but lower in energy compared to the original IR – *i.e.*, visible red light. UV tends to transmit visible blue light because this is close to UV in the spectrum but has lower energy. Similar to IR document examination, nowadays UV work often involves the use of spectral analysers with variable irradiation wavelengths; these instruments both irradiate the document and analyse the transmitted and/or reflected light.

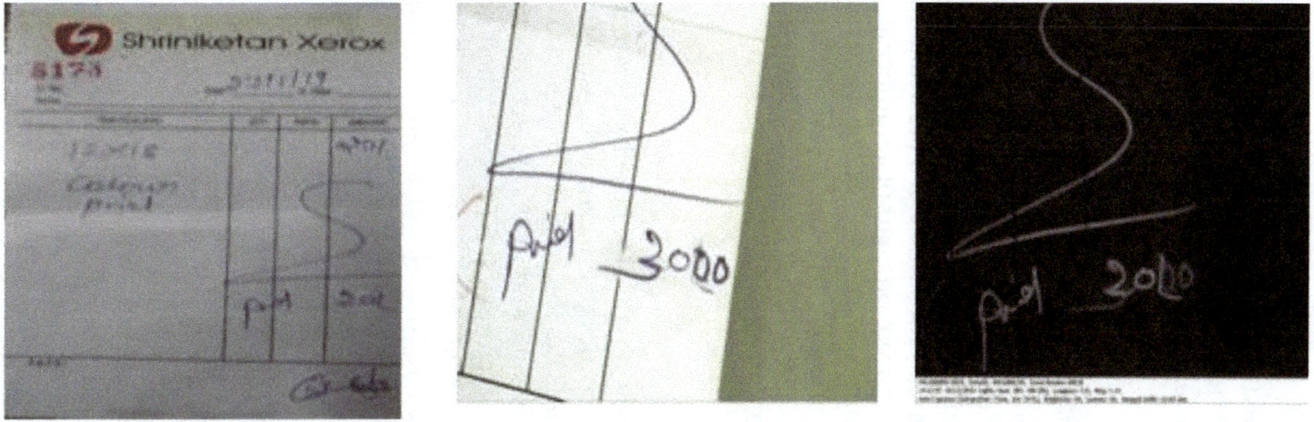

Figure 12.6 Left: A paid invoice showing the amount paid, 20 L. Centre: the same invoice with the amount paid adjusted to 3000. Right: multispectral image analysis using IR shows clearly the original 20 L under the adjusted (forged) 3000, and shows that the ink used to draw the original squiggle above the amount paid is the same as that used to write the original 20 L. L = Lakh, an Indian term meaning 10 000. Reproduced from ref. 2, https://www.ijsr.net/getabstract.php?paperid=ART20204054, under the terms of the CC BY 4.0 license https://creativecommons.org/licenses/by/4.0/.

12.3.3 Ink Extraction, Chromatography and Spectroscopy

A written, typed or printed section of a document, believed to have been modified (*e.g.*, from IR or UV irradiation studies), can be solvent extracted, and the extracted ink can then be analysed by chromatography or spectroscopy.

12.3.3.1 Chromatography

TLC or HPLC (see Chapter 4 *Forensic Chemistry*) can be used to separate the ink's components as distinct coloured bands on a TLC plate and peaks with specific retention times on an HPLC trace. The TLC band and HPLC peak patterns are characteristic of a particular ink; therefore, for example, if an ink extract from one part of a number (*e.g.*, a financial amount in a contract) has a different TLC or HPLC pattern from another part of the same number, this would strongly suggest that the number had been written with different pens (or printed or typed with different inks) (Figure 12.7).

12.3.3.2 Spectroscopy

The extracts described above might also be subjected to spectroscopic analysis. This involves determining the extract's absorption spectrum. This spectrum corresponds to the absorption spectra of the component dyes in a particular ink and is characteristic of the ink. The absorption spectra of different extracts (as described above) can be compared to decide whether any suspected modifications to a document might have been carried out with, for example, a different pen than that used to create the original wording or number (Figures 12.7 and 12.8).

12.4 Handwriting Analysis

Handwriting analysis is a very important aspect of the forensic document examiner's work. Examining handwriting by comparing a sample of, for example, a suspect's handwriting, with the handwriting on a suspect document might help decide whether the suspect forged a signature or changed the meaning of the document by adding words to it (Figure 12.9). Or, perhaps, handwriting analysis might be used to analyse an apparent suicide note: was the note written by the

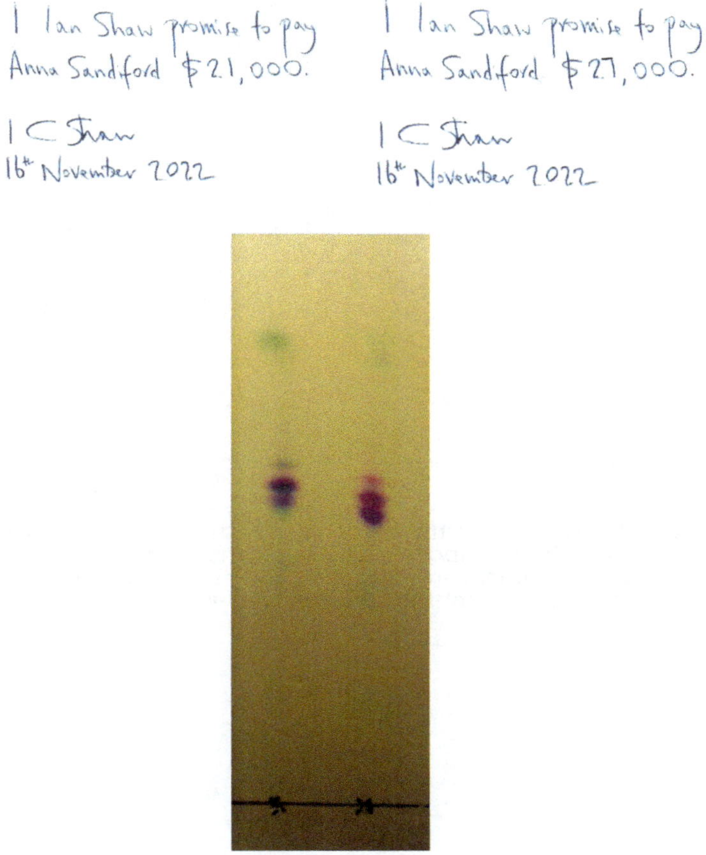

Figure 12.7 Spoofed documents to show how a small change can result in significant fraudulent financial gain to Anna Sandiford and a loss to Ian Shaw. The original document (top left) which promises $21 000 to Anna has been altered with a single bar on the number 1 in $21 000 to make $27 000 (top right). The ink from the added bar was extracted with trichloromethane (chloroform) and compared using TLC (see Chapter 4 *Forensic Chemistry*) (bottom). It is clear that the two inks are very different. The document was clearly adulterated, perhaps by Anna to defraud Ian of an additional $6000!.

deceased or was it written by somebody else? If it were the latter, this might suggest that the case was not a suicide, but that someone else was involved in the death and wanted it to look like a suicide. People who learn to write using a character-based language such as Japanese, and who then learn to write, say, English, often have a distinct writing style that document examiners can identify.

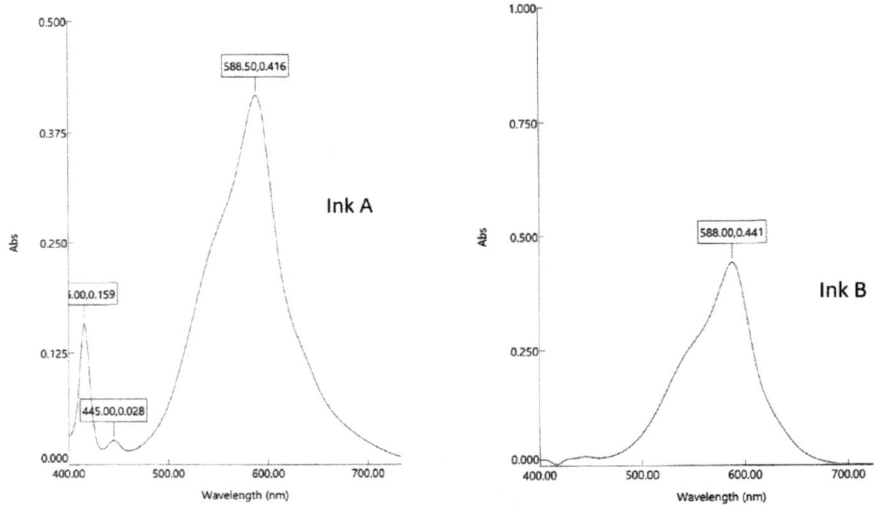

Figure 12.8 Absorption spectra of trichloromethane (chloroform) extracts of two blue ballpoint pen inks (A and B) from writing on paper (the document in Figure 12.7). The spectra of the two inks clearly show that the inks are different.

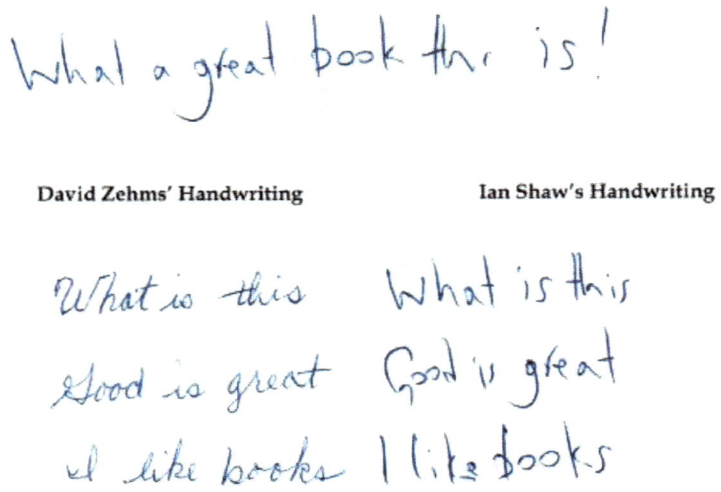

Figure 12.9 The author of the statement (top) can be determined when samples of handwriting from two 'suspects', David Zehms and Ian Shaw, are compared with the handwriting of the statement. For instance, compare the word *What* in the statement with the *What* written by David or Ian, and the word *book* in the statements with David's and Ian's *books*. It is clear that Ian wrote the statement (not surprisingly).

We can all look at samples of handwriting and decide whether or not they were written by the same person simply by looking at the writing style. It is obvious! Or is it? Could someone be mimicking another person's handwriting? Or could the writer be nervous, and could this change their handwriting style? Handwriting comparison is not as simple as it might initially appear.

The understanding that handwriting is a characteristic of an individual (this even applies to identical twins) is widely accepted (see *Further Reading* 5). This is why signatures have been used for centuries as a means of identification on business and legal documents. When I was a child and the post arrived, my mother might say, 'That's from your Auntie Betty', just by reading the address on the envelope – she recognised Auntie Betty's handwriting and identified her by its characteristic style...and she was always right!

Much of the process of comparing and identifying handwriting by forensic document examiners is empirical, indeed this is recognised by the Scientific Working Group for Forensic Document Examination (SWGDOC) and also in the American Society for Testing and Materials (ASTM) Standard Guide for Examination of Handwritten Items (E2290-07a):[3]

> *Among the features to be considered are elements of the writing such as abbreviation; alignment; arrangement, formatting, and positioning; capitalization; connectedness and disconnectedness; cross-strokes and dots; diacritics [accents] and punctuation; direction of strokes; disguise, embellishments; formation; freedom of execution; handedness; legibility; line quality; method of production; pen hold and pen position, overall pressure and patterns of pressure emphasis; proportion; simplification; size; skill; slant or slope; spacing; speed; initial, connecting, and terminal strokes; system; tremor; types of writing; and the range of variation.*

Usually, comparisons are made between the characteristics of a sample of handwriting and handwriting on the document being examined (*e.g.*, a suicide note; an altered legal document) as illustrated in Figure 12.9.

12.5 Graphology

Graphology is the art (a carefully chosen word) of attempting (another carefully chosen word) to determine personality and/or state of mind

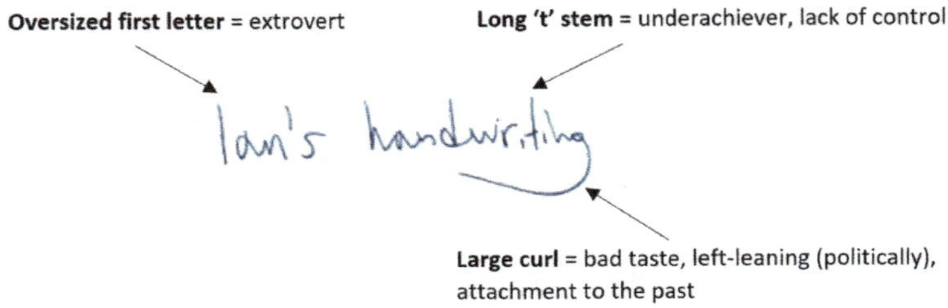

Figure 12.10 Ian Shaw's handwriting subjected to graphological analysis. According to graphology, I am an underachieving, politically left-leaning, extrovert, who is attached to the past and out of control! The question is: am I?

from someone's handwriting. It is no longer used by forensic document examiners because scientific studies have shown that it is of little or no value in a forensic context (see *Further Reading* 4). I have included graphology here because some popular forensic science TV shows and books seem to think otherwise!

Graphology is based on the belief that writing style can reveal the writer's character traits and state of mind. For example, for the letter 't', a looped stem apparently means that the writer is talkative and articulate, whereas a long beginning stroke denotes stubbornness. For fun, I have interpreted my own handwriting in Figure 12.10: I will leave any of my friends who might read this to assess the merits of graphology regarding my character, but I will tell you that I collect antiques!

12.6 Examining Printed Materials

In the 'old days' when handwritten documents were more common than they are in today's electronic world, handwriting variability between individuals and alterations using a different pen were both relatively easy to investigate as discussed above. Even typed documents had idiosyncrasies that reflected defects in individual typewriters, which made it possible to identify the typewriter used for a particular document and ink jet printers, fax machines and photocopiers produced (and still do) individualising characteristics in their printing. However, the advent of the laser jet computer printer put an end to most of this. Printers are remarkably consistent and use inks and toners that are produced by huge international companies, often varying little in their chemical compositions. In addition, fraudulent alterations to an electronic document are much more difficult to identify than handwritten or typed changes and require considerable information technology (IT) skills.

Crimes involving stealing, altering or infiltrating electronic documents in other ways are termed cybercrimes. These can range from breaches of copyright (*e.g.*, plagiarism) to international cybercrimes where electronic documents may be adulterated or stolen in order to harm an individual, company or even country's security or finances. Much of this is far outside the scope of this book, but it is important to recognise the importance of e-crimes or cybercrimes and that IT expertise and techniques are rapidly assuming great importance in document-related crimes.

In 2014, it was estimated that cybercrime costs the international economy approxiately US$45 billion a year (Technology News, 9th June 2014); this sets their importance in stark context and supports the huge effort devoted to developing new methodologies to investigate these crimes.

Cybercrimes can be divided into two categories:

1. The computer as the tool
2. The computer as the target

Using the computer as a tool can involve alteration of documents (*e.g.*, to increase financial gain), identity theft to facilitate access to someone's money, phishing where emails with computer links are sent to allow the sender's access to the recipient's personal details (*e.g.*, to allow access to a credit card), and much more. These basic crimes were carried out before the advent of computers, but now sophisticated IT is used to promulgate them.

The computer as the target involves the perpetrator sending information to a computer to corrupt it in some way (*e.g.*, using computer viruses or malware) or accessing information from the computer (hacking). This often does not result in document adulteration or fraud, but rather disables or incapacitates the receiving computer or IT network, preventing it from carrying out its normal tasks. If this was, for example, a payroll computer in a big company or government department, chaos would ensue.

12.6.1 Cybercrime Case Examples

One of the earliest cybercrimes was committed by Roswell Stefan, the chief teller at the Park Avenue, New York branch of Union Dime Savings Bank. Over a 3-year period beginning in 1970, he embezzled over US$1.5 million from hundreds of accounts. It was reported that Stefan had '*used the bank's computer to "shuffle" hundreds of individual accounts and then had fed fraudulent and inaccurate information into the computer so that those accounts always appeared up to date.*' By carefully covering his tracks and taking relatively small amounts of money from a huge number of accounts, his exploits went unnoticed for 3 years. He was charged with grand larceny and forgery; he pleaded not guilty in the Manhattan Criminal Court on 22nd March 1973, but he was later convicted.

In 1992, Kevin Poulson was found guilty of hacking into the telephone system of a Los Angeles radio station in order to win a competition. When the FBI began their pursuit of Poulson, he went on the run but was eventually found and convicted. He was sentenced to 5-years in prison. It was rather apropos that he was denied internet access throughout his time in prison!

By 2007, the incidence of hacking crimes had escalated to intergovernmental proportions when the Chinese government was accused of hacking into the USA government's legislative framework. Although this was never proved or disproved, it illustrates the potential magnitude of modern cybercrime. This was further illustrated in December 2019 by USA intelligence officials and New York Times investigations which revealed that the ToTok messaging system had been used by the United Arab Emirates government to track activities and communications of everyone who had installed the ToTok application (app) on their mobile phones. The espionage implications of this are huge.

Further Reading

1. H. K. Bamburde and M. P. Goutam, Questioned documents analysis for decipherment of obliterated writing, *Int. J. Sci. Res.*, 2018, **9**, 515–518.
2. C. D. Adam, S. L. Sherratt and V. L. Zholobenko, Classification and individualisation of black ballpoint pen inks using principal component analysis of UV-vis absorption spectra, *Forensic Sci. Int.*, 2008, **174**, 16–25.
3. J. A. Lewis, *Forensic Document Examination: Fundamentals and Current Trends*, Elsevier, 2014, ISBN 978-0-12-416693-6.
4. G. Ben-Shakhar, M. Bar-Hillel, Y. Bilu, E. Ben-Abba and A. Flug, Can graphology predict occupational success? Two empirical studies

and some methodological ruminations, *J. Appl. Psychol.*, 1986, **71**, 645–653.
5. S. N. Srihari, S-.H. Cha, H. Arora and S. Lee, Individuality of handwriting, *J. Forensic Sci.*, 2002, **47**, paper ID JFS2001227-474.

References

1. O. P. Jasuja and A. K. Singla, *Forensic Sci. Int.*, 1991, **52**, 21–23.
2. H. K. Bamburde and M. P. Goutam, Questioned documents analysis for decipherment of obliterated writing, *Int. J. Sci. Res.*, 2018, **9**, 515–518.
3. Standard Guide for Examination of Handwritten Items. https://www.astm.org/e2290-07a.html.

13 Probability in Forensic Science

Finding an understandable and easy-to-follow method of presenting scientific findings in court has been a challenge for forensic scientists for many years. The court seeks answers in black and white but science operates in shades of grey. Describing these shades of grey has been a challenge. We also need to incorporate the acknowledgement and management of human bias in interpretations.

If reports are balanced, robust, logical and transparent, then, in theory, bias risks can be reduced. There are various guidelines for crafting a well-written report, such as those provided by the European Network of Forensic Science Institutes, the England and Wales Criminal Procedures Rules for Expert Witnesses and the Australia and New Zealand Forensic Science Society. There are also numerous guidelines, scientific working groups and best practice documents that cover a wide range of forensic topics to help people understand what could and should be done in different disciplines.

The overall idea is to ensure that interpretations are reported as objectively as possible, thereby minimising subjectivity.

13.1 Probability

People use the word 'probably' all the time in everyday life. It is generally taken as a way of informally saying that you believe one outcome is more likely than another outcome: "Are you coming into work today?" "Probably." From this, I deduce the person I am asking

Forensic Science: The Science Behind the Truth
By Ian Shaw with Anna Sandiford

Published by the Royal Society of Chemistry, www.rsc.org

is more likely than not to come into work today but I do not know how likely it is. It will depend on how that person uses the word 'probably'; some people use it to mean extremely likely, almost yes, and some people might use it to mean only slightly more likely than not.

In the criminal justice system, the court (judge and jury) is tasked with determining whether the defendant is guilty or not guilty of the crime with which they are charged. The court listens to the evidence presented sequentially and puts together all the pieces of information and then makes a decision.

Some parts of the evidence can be proven as facts and do not need further assessment. For example, the incident took place at 10.00 pm. This is factual and can be proved by referencing a clock on a CCTV camera that has been proven to be accurate. No further assessment of the time is required and there is no vagueness involved.

However, when scientific findings are presented that pertain to an event that cannot be factually proven, the court needs to know how likely it is that the particular event occurred. This is where the concept of assigning a probability figure to the scientific findings given a particular proposition has developed. Furthermore, it is important that presenting such interpretations and probabilities maintains consistency, not only between different scientists in different laboratories and countries but also for the same scientist on different occasions.

The framework developed to assess probability in forensic science is a big topic and there are many publications that deal with it (see *Further Reading* for some suggested papers as a starting point).

As with anything involving humans, the issue of biased answers creeps in. In order to address bias, it first has to be acknowledged and then training must be provided so that the reporting scientist can mitigate it.

13.2 Bias

Objective interpretation is independent of individual thought, based on measurable and reproducible facts and an interpretation that any person looking at the same result would form the same conclusion. Subjective interpretation is an opinion influenced by life, memories, biases and prejudices, where two people looking at the same results may form different conclusions. The potential effect of bias is one of the key differences between subjective and objective interpretations.

One of the most important roles of an expert witness is to provide an objective interpretation of the scientific findings. Forensic science is about understanding what results mean in the context of a given case and explaining this to a lay audience; ideally, there should be no room for bias. However, it is inevitable that there will be some bias in interpretations because human beings are involved and human beings *are biased*.

There has been much discussion regarding the interpretation of experts' findings over the past few decades, reaching at least as far back as the investigations into bombings on the British mainland understood to have been carried out by the Provisional Irish Republican Army (IRA), particularly those in the 1970s. Those to which a woman called Judith Ward was initially linked were:

In September 1973, a bomb exploded at London's Euston Station, resulting in no fatalities but injuring a dozen people;
In February 1974, a bomb exploded on a coach that was carrying soldiers and family members along the M62 motorway in England; 12 people were killed and 38 were injured;
Also in February 1974, a bomb exploded at the National Defence College in England, injuring 12 people.

Scientific evidence presented at Ward's trial in November 1974 described that the bombs contained nitroglycerine, a powerful explosive. The scientific evidence also described the results of tests conducted on swabs taken from Ward and some of her items as "definitely positive" for nitroglycerine. It was explored at trial that there was no risk of accidental contamination because of the way the samples were collected and handled. The explanation for the results showing nitroglycerine under her fingernails was that she had been involved in making the bombs.

One of the key scientific findings in the cases of Judith Ward and others charged with bombing offences (such as the Birmingham Six) was reported by a prosecution-instructed expert, Dr Frank Skuse of the Home Office Forensic Science Laboratory, which was closely linked to the police teams. The test in question, the Griess test (Figure 13.1), should have only been a preliminary test or a screening test for the presence of nitrite in solution (nitrite is formed when nitroglycerine is treated with sodium hydroxide). The samples collected were not sufficient for *confirmation* testing and, in the case of Ward, were rejected by the laboratory. No findings were presented to the court about the probability of a false positive result (*i.e.*, other substances unrelated to explosives, such as cleaning products, components of some heart medication tablets and materials from which playing cards were made). Dr Skuse, along with other scientists, presented the findings at

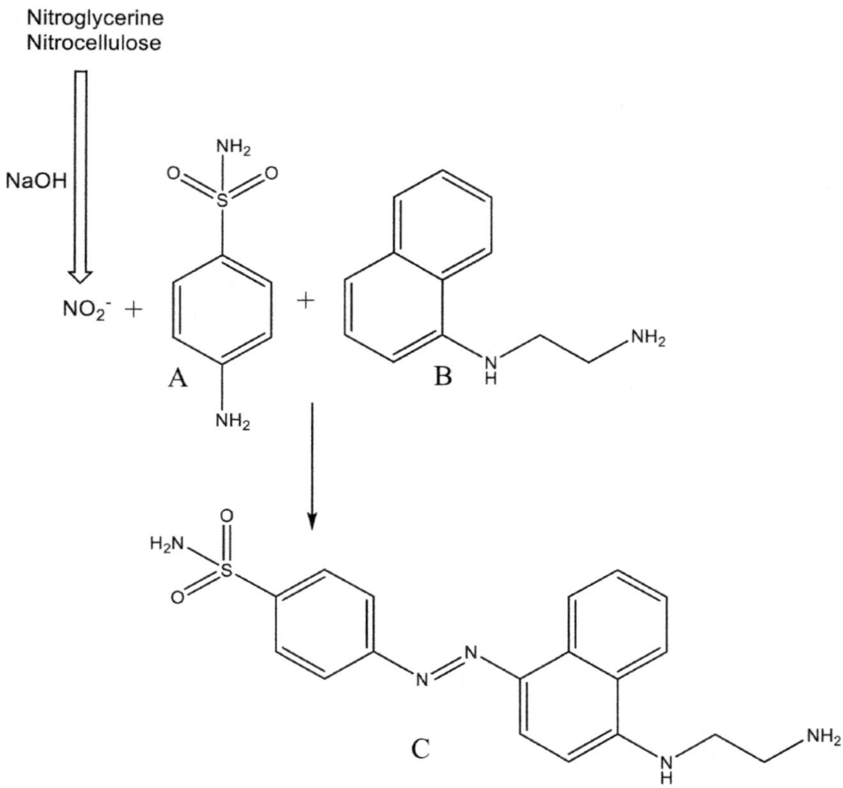

Figure 13.1 The Griess test for nitrite (NO_2^-): nitrite is reacted with sulfanilamide (A) to form a reactive intermediate (not shown), which reacts with *N*-(1-naphthyl)ethylenediamine (B) to form a red azo dye (C). Nitroglycerine and nitrocellulose when treated with sodium hydroxide (NaOH) release nitrite which the Griess test can detect.

trial in a way that overstated the interpretation and important information was withheld from the court, including the defence.

The test should be completed under specific conditions; otherwise, incorrect results can be generated. It was possible for the test to yield false positive results and its use was eventually discontinued because of its unreliability. The English Court of Appeal ruling in Regina v Ward indicated that: "*Dr Skuse's conclusion was wrong, and demonstrably wrong, judged even by the state of forensic science in 1974*".[4] The close association between the laboratory, the police and the potential for bias towards the instructing party was a contributing factor in this investigation. As a result, forensic science in England and Wales was then further separated from the police so as to increase confidence that expert opinions reached by forensic scientists were as independent as possible from the police investigation, thus reducing the potential effects of bias.

Human bias differs from errors that are wilful or made out of ignorance. Bias is a tendency or an inclination to think a certain way: to present or maintain a particular perspective at the expense of alternatives that may be equally or more likely. Bias creates a one-sided point of view and is therefore in direct opposition to impartiality and a neutral viewpoint, both of the latter being required by the expert witness.

Bias is a well-studied area and expert witnesses, no less than people in general, are susceptible to it. To psychologists' best understanding, biases tend to arise as a natural result of the normal operating procedures of the mind. Sometimes, when people are deciding between alternatives, problem-solving or reasoning, answers come quickly and without conscious deliberation. This mode of thinking is autonomous and heavily reliant on prior knowledge, memory and experience. It has its advantages, not least of all when a person has to make a decision rapidly or without considering all available information. However, it can lead to errors on occasion and there are professions (*e.g.*, forensic scientist), and life choices, such as choosing a partner or career, where such errors are less than ideal.

There is another way of thinking that is more deliberate but requires more effort and more conscious thought. This process leads to more rational decision-making, more effective problem-solving and less error-prone and biased thinking. It can override the influence of rapid production of answers that need to be reconsidered.

Human beings may never be able to free themselves entirely from bias, especially for people who refuse to accept that it affects them. As a balanced viewpoint is crucial for an expert witness, it is important for everyone to recognise that bias exists and to attempt to mitigate its effect as much as possible.

In England and Wales, more attention is currently being given to the existence of bias in forensic interpretation and there is a body of literature published on the subject as it affects expert witnesses. Experts are now being trained in the different types of bias, how they can affect their work and how to reduce the effects.

13.3 Contextual Bias

Contextual bias (stemming from specific context effects) refers to the influence of additional or extraneous information on a decision-making process. Either consciously or subconsciously, experts allow extraneous and external case information to affect their judgements such that a decision made under these circumstances would be different from a decision made strictly independently. It is not unusual for experts to receive detailed but unnecessary information from the police about the context of an allegation, such as *"the defendant has a history of violence towards women. He hit the victim multiple times with this baseball bat and so you should find his DNA on it"*. This context can prompt an expert to use an examination process that specifically 'looks' for DNA that matches the defendant without considering factors like transfer and persistence, the ownership of the baseball bat or the location of the alleged incident. If the bat belonged to the defendant then finding his DNA on the handle does not necessarily advance the investigation because his DNA would be expected to be on it – and the scientist should say that. The police could just as easily say *'please examine the defendant's baseball bat for DNA that may assist in identifying who last had contact with it.'* This is far more opened-ended and less biasing and should result in a more even-handed interpretation.

A great example illustrating the effects of contextual bias involves fingerprint experts, whose results were studied by cognitive neuroscientist Dr Itiel Dror. The objective of his research was to investigate *"whether experts can objectively focus on feature information in fingerprints without being misled by extraneous information, such as context"*.[1] He took fingerprints from crime scenes and known individuals, changed the context of the case and gave them to the same experts who had previously examined them. The difference was that he gave them a context that suggested the prints were a no-match, implying that the suspects could not be identified. He also did not tell the fingerprint experts that they had previously seen the same prints or disclose their original findings.

Given this second set of no-match contextual information, *"most of the fingerprint experts made different judgements, thus contradicting their own previous identification decisions"*.[1]

13.4 Confirmation and Expectation Bias

Confirmation bias is the human tendency to interpret information in a way that confirms their own already established suspicions, beliefs, hypotheses or explanations. This bias is evident when people seek information that confirms rather than challenges their explanations – they choose to ignore information that does not support their views.

Expectation bias is closely related to confirmation bias. It is a well-established phenomenon that occurs in scientific analysis. It occurs in the forensic setting when experts reach a premature conclusion without having examined or considered all of the relevant data in a logical and unbiased manner. Police are also affected; once an arrest is made, the police are usually committed to proving that the person they have arrested is guilty of the charges laid. From this point onwards, they may build a case for prosecution rather than continuing with wider enquiries that might exonerate a suspect or defendant. At this point, an investigation can become suspect-led instead of information-led.

13.5 Mitigating Bias

Potentially biasing factors exist in everyday life and in every scientific endeavour. To mitigate the effects of bias, experts should be trained to recognise biasing factors to avoid being influenced by them, for example, using a database during expert interpretation can reduce subjectivity. The more subjective the interpretation, the greater the importance of independent assessment.

13.6 Types of Interpretation

As indicated previously, interpretation of results and findings is one of the most significant roles of the expert witness, a responsibility that the international forensic scientific community has recognised for years. Experts need to find a way to describe the results they obtain fully and fairly, including any limitations. A forensic scientist should not perceive failure in a case if they cannot provide scientific findings to support the side of the person who instructed them. This means that an expert instructed by the police should not feel pressure to produce a result to support a guilty verdict if the scientific evidence is not there. Equally, a defence-instructed expert should not feel the need to undermine or pick holes in the prosecution scientist's findings just for the sake of trying to 'muddy the waters' and help a guilty person get off.

If an expert finds a footwear impression on a tiled floor at a scene, how sure can they be that it is related to the incident in question? If they *are* sure that it relates to the incident because the floor was cleaned that morning, is the mark good enough that they can make an assessment of whether the pattern is common or not?

When considering a footwear mark, the following things should be considered:

- Is the sole pattern rare? If not, the interpretation is likely to provide limited assistance to the investigation.
- Is the mark complete in length and, preferably, width (*i.e.*, can the shoe size be estimated?) If the shoe size can be estimated, the interpretation will be of more use for the investigation.
- Are there marks and defects in the impression at the scene that could reflect a short-lived feature that may affect several shoes at the time of manufacture, such as an air bubble in the mould used to produce the shoe sole? If so, this could narrow down the number of potential shoes and make the interpretation of greater use, although it would not necessarily be unique to one particular shoe.
- Are there marks and defects in the impression at the scene that could reflect a short-lived feature on a single shoe (*e.g.*, a glass cut?) If so, this could narrow down to a single shoe, which is potentially extremely useful for an investigation.

Recovering a shoe from a suspect that matches all the class characteristics and individualising characteristics found in a crime scene mark is a robust finding. This gives rise to the question of how the result should be worded in the report.

A lot of discussion has taken place about what phrase to use when reporting results because history has shown great variability in phrases used even by scientists within the same laboratory. For example, different scientists might report the same results using different phrases such as “highly probable that”, “more likely than not” and “I think that”. How does a jury perceive those words and does it affect or influence how they consider them when putting that information into the case they have heard?

As there have been so many different ways to describe the same results, the probability framework was established. This has helped standardise reporting within and between areas of expertise, countries, laboratories and individuals. This framework is an adapted mathematical equation into which information about two opposing propositions or hypotheses can be entered: that reflecting the prosecution’s account and that reflecting the defence account of events. This is the basis for the evaluative stage of interpretation, which is the most advanced stage of reporting, as outlined below.

13.6.1 Factual or Technical Reporting

Factual (also called technical) reporting simply describes the result of a test or an observation without any need for interpretation. If we go back to the footwear example, factual reporting would include the work of crime scene examiners who attended the scene, described the location from which a footwear mark was recovered, the surface type on or material in which it was made (*e.g.*, glass or wet sand) and the general appearance of that footwear mark, including its dimensions. No interpretation of this information is required.

13.6.2 Investigative Opinions and Interpretations

Investigative opinion is defined by the Association of Forensic Science Providers as "*the provision of an explanation for an observation*" and by the European Network of Forensic Science Institutes (ENFSI) in their *Guideline for Evaluative Reporting in Forensic Science* as "*explanations for technical/factual findings*" (see *Further Reading*). Scientists form this type of opinion when an explanation is required in order to understand technical results. Such opinions may involve estimating probability based on the experience and knowledge of the expert providing the opinion, especially in cases where there might be insufficient information available to construct two competing propositions/hypotheses. ENFSI provides the following as an example of an investigative opinion on page 25 of the Guidelines:[2]

> *(a) "The findings in relation to the blood on the hammer could be explained by the following:*
> - *the hammer was used in the assault of Mr X*
> - *the hammer was not used in the assault of Mr X, but came into contact with blood at the scene*
>
> *This is not an exhaustive list of possible explanations".*

13.6.3 Intelligence Reporting

Intelligence reporting provides information that can help inform the way an investigation develops and provide links between events and cases.

In New Zealand, for example, the most commonly-seen form of intelligence report is the DNA Databank Report. This is generated by the Institute of Environmental Science and Research (ESR, the forensic science laboratory that holds the national police contract for forensic services) when there is a "hit" between the DNA profile of a known individual that is already on the DNA Databank and a DNA profile recovered from a crime sample. A DNA Databank Report is generated that suggests the name of the person whose DNA was recovered from the crime sample. Police then need to acquire a reference DNA profile from the named individual for comparison with the crime sample for a specific case.

A DNA Databank Report cannot be used for evidential purposes under the current New Zealand Criminal Investigations (Bodily Samples) Act 1995. Therefore, its purpose is to provide intelligence that informs the police investigation. This means that once a DNA Databank Report has been issued, identifying a potential suspect, the police can then use that as the basis for developing further lines of investigation.

13.6.4 Evaluative Interpretations and Opinions

Evaluative opinions involve the establishment of two case-specific propositions and a clear framework of circumstances ("conditioning information") to evaluate what is termed a likelihood ratio (LR). As described by ENFSI, "*This framework for evaluative reporting applies to all forensic science disciplines. The likelihood ratio measures the strength of support the findings provide to discriminate between propositions of interest. It is scientifically accepted, providing a logically defensible way to deal with inferential reasoning*".[2]

The two propositions usually, between them, represent two opposing views: one based on the prosecution case and one based on the defence case. An example would be the likelihood of the findings if shoe A made the mark at a crime scene compared with whether another shoe made the mark at the crime scene – the more information from the sole pattern (size, manufacturing glitches, individualising scratches, and transit wear marks), the greater the strength of the scientific findings. This approach can help address the shades of grey in which forensic science operates and should provide the court with a verbal and understandable assessment of the findings. The LR approach is not about determining the likelihood of guilt or innocence – that determination is the role of the court.

This chapter is not designed to provide a detailed insight into the LR approach as it applies to expert witness evidence; there are plenty of textbooks and scientific papers that already cover this topic and some examples are given in *Further Reading*. However, it is important to be aware of how experts interpret scientific findings.

It is also important to understand that when the LR approach is used for interpreting findings, the wording in the expert's statement is likely to have been crafted in such a way as to avoid common problems and confusion. This is clear if we consider how a report should be set out, following the ENFSI guidelines:

Balance: demonstrated by the two propositions
Logic: the probability of the findings given the propositions (not the other way around – more on that later)
Robustness: the results should sustain scrutiny by other experts in the field, take into account up-to-date research and withstand cross-examination
Transparency: the interpretation written in the statement should be supported by the casefile and be written for a range of readers so that it is understandable. It should include any limitations on the findings or current state of research.

13.6.5 Likelihood Ratios and Verbal Equivalents

In forensic casework this approach is most widely used when comparing DNA results

$$\text{LR} = \frac{\text{probability of the scientific findings given the prosecution proposition}}{\text{probability of the scientific findings given the defence proposition}}$$

Normally, a numerical estimate is made for the numerator and the denominator, preferably with reference to a suitable database. The DNA database for known individuals in New Zealand holds the full reference DNA profiles of thousands of people. The results from a crime sample and a reference sample for a known person are compared. A statistical analysis is then conducted to assess the frequency of a given result in the population and compare the likelihood of the DNA profiling results if the DNA had originated from person X to the likelihood of the DNA profiling results if the DNA had originated from another individual selected at random from the population. A value, the LR, is determined from the statistical analysis.

A verbal scale is often used to transform the LR value into a verbal equivalent that can be spoken and hopefully provide meaning to the court (Table 13.1).

It can be seen from the numerical scale in Table 13.1 that the evidence must be at least 10 times more likely given the prosecution

Table 13.1 An example of a DNA LR scale.

Likelihood ratio	Verbal equivalent
10 to 100	Supports (the prosecution proposition)
100 to 1000	Strongly supports
1000 to 1 000 000	Very strongly supports
Greater than 1 000 000	Extremely strongly supports

proposition than given the defence proposition before it registers on the scale.

The same scale is used for negative values (*i.e.*, those less than zero). The negative values support the defence scenario over the prosecution scenario.

ENFSI has excellent guidelines on how to approach the likelihood ratio:

> *A likelihood ratio is a measure of the relative strength of support that particular findings give to one proposition against a stated alternative.... It is defined in terms of the ratio of two conditional probabilities: (i) the probability of the findings given that one proposition is true and given the conditioning information; and (ii) the probability of the findings given that the other proposition is true and given the conditioning information. The two conditional probabilities forming the likelihood ratio shall be assigned on the basis of published data ... or a body of data that can be made available for peer review. Additionally, and in the absence of such data, experience or knowledge may be used. All bases used should be disclosed.*

Even more important for the court to note from the same ENFSI guidelines is that:

> *The use of a likelihood ratio does not generally imply that one of the two propositions considered must be true. Though the considered propositions are those deemed most relevant, they do not need to be exhaustive, so both propositions could be false. The likelihood ratio says nothing about propositions other than the two that were considered.*

Experts in different countries use slightly different terminologies in their reports. However, the important issues are whether or not the verbal scale is described in any expert's report and whether there is an indication of how the expert chose the wording they used for a particular case.

Two areas of science that have databases, and therefore numerical figures to put into the LR equation, are DNA and certain parameters of glass analysis. Unless otherwise stipulated, interpretations without reference to a database and calculated figures are a subjective interpretation based on the scientist's knowledge and experience. Any scale of verbal equivalents has the potential to mislead and its use outside DNA and perhaps glass should be scrutinised by a suitably qualified independent expert to ensure that it is appropriate and that bias has been mitigated as much as possible.

Use of the LR approach should not be compulsory because the LR approach is not possible for all cases. However, when it is not possible, the relevance of the expert opinion may be significantly weakened, in a legal sense, and may be called into question because the opinion is not evaluative (*i.e.* not based on a full, statistical analysis).

13.6.6 The Pseudo-likelihood Ratio Approach

Some scientists quote a verbal scale and then choose their own level of interpretation for a given case, which implies that a logical and rational evaluation has been undertaken when it may not have been. This usually happens in a situation where there is insufficient research to establish numerical figures for the numerator and denominator and the interpretation is therefore subjective, not objective. If a scientist reports their findings by inadvertently using phrases from the LR scale without providing a verbal scale, it can imply a weight to the evidence that has not been objectively assessed (*i.e.*, it could be biased).

For these reasons, use of terms like "slightly supports", "supports", "strongly supports" and "extremely strongly supports" without reference to a verbal scale should be avoided. If there is not a consistent use of a phrase, the question arises as to whether or not the scientist understands and has assessed the LR and how they chose the terminology used. Otherwise, the scientist's word choice could be inadvertently biased, which can have an inaccurate but knock-on effect on the weight the court applies to it.

Employing a logical framework and formulating two propositions ensures that the expert

- remains within their field of expertise;
- does not usurp the role of the jury or judge;
- is protected from logical fallacy; and
- is protected from bias.

13.6.7 The Prosecutor's Fallacy and Attempted Rewording of Interpretations

As discussed, expert witness phraseology in evaluative reports, if presented in a LR framework, should have been chosen carefully. LRs can be confusing and it is common for lawyers to try to get experts to express interpretations more simply.

However, the LR and its verbal equivalent are obtained only by considering the specific result in light of two specially formulated and opposing propositions or hypotheses. If anyone tries to reword the findings, they run the risk of altering the meaning of the interpretation. A common example of this is called the Prosecutor's Fallacy.

The correct wording of a proposition in a forensic LR framework is "*the probability of the scientific findings given the prosecution/defence hypothesis*". The Prosecutor's Fallacy arises when this wording is swapped around to become "*the probability of the prosecution/defence hypothesis given the scientific findings*".

The difference between these two is not always obvious, so an easier explanation involves a cow: "*the probability that an animal has four legs (i.e., the scientific findings) if it is a cow (i.e., the proposition)*" is not the same as "*the probability that an animal is a cow (i.e., the proposition) if it has four legs (i.e., the scientific findings)*". Thinking along these lines is helpful when trying to work out whether the findings and hypothesis have been swapped around.

As the Prosecutor's Fallacy alters the meaning of an interpretation, pushing an expert to change his or her wording of an interpretation is fraught with danger. In fact, as a result the court can also get confused and misinterpret the findings of the expert. Particularly in cases where the LR is a calculated figure (such as in DNA cases), the calculation has been undertaken using specific data associated with the two propositions or hypotheses presented in the expert's statement: the calculated figure relates *only* to the calculation that was completed based on the two propositions/hypotheses as worded. It does not mean that other propositions are not possible.

13.7 The Misuse of Statistics

13.7.1 Winning the Grand National (or Other Famous Horse Race of Your Choice) Four Times in a Row

Probably one of the most infamous case examples of the misuse of statistics was the case of Sally Clark (also mentioned in Chapter 15

Assembling the Forensic Results – Case Histories in relation to altered medical findings). She was an English solicitor and also a mother of two boys, both of whom died at a very young age. Her first child died at 8 weeks, initially identified as Sudden Infant Death Syndrome (SIDS, also known as cot death). Her second child died at 11 weeks. The *post-mortem* of the second child was conducted by the same pathologist as the first child; the finding was that the second baby had been shaken and died as a result. As a result, the pathologist then reconsidered the cause of death of the first baby, reclassifying it as unnatural and indicating that the baby had been smothered. Clark was arrested and charged with the murder of both children.

There were issues with the *post-mortem* examinations (see Chapter 15 *Assembling the Forensic Results – Case Histories*) but one of the main pieces of prosecution evidence that resulted in a guilty verdict was statistical. It was presented at the trial by Professor Sir Roy Meadow, a paediatrician. His opinion on SIDS, as published in his book *ABC of Child Abuse*,[3] was that, "*'One sudden infant death is a tragedy, two is suspicious and three is murder until proved otherwise' is a crude aphorism but a sensible working rule for anyone encountering these tragedies*". At trial, he asserted that both babies had been murdered. During his evidence, he was asked about statistical data that was scheduled to be published in a government-funded report, for which he was writing a preface, regarding the occurrence of sudden unexplained deaths in infancy. The Court of Appeal judgment describes Professor Meadow's evidence as follows:[5]

> *96. It was explained to the jury that there were factors that were suggested as relevant to the chances of a SIDS death within a given family; namely the age of the mother, whether there was a smoker in the household and the absence of a wage earner in the family. None of these factors had relevance to the Clark family and Professor Meadow was asked if a figure of 1 in 8,543 [as published in the report] reflected the risk of there being a single SIDS within such a family. He agreed that it was. A table from the. . .report was placed before the jury. He was then asked if the report calculated the risk of two infants dying of SIDS in that family by chance. His reply was:*
>
> *"Yes, you have to multiply 1 in 8,543 times 1 in 8,543 and I think it gives that in the penultimate paragraph. It points out that it's approximately a chance of 1 in 73 million."*

97. It seems that at this point Professor Meadow's voice was dropping and so the figure was repeated and then Professor Meadow added:

"...in England, Wales and Scotland there are about say 700,000 live births a year, so it is saying by chance that happening will occur about once every hundred years."

98. Mr Spencer [for the prosecution] then pointed to the suspicious features alleged by the Crown in this present case and asked:

"So is this right, not only would the chance be 1 in 73 million but in addition in these two deaths there are features which would be regarded as suspicious in any event?"

He elicited the reply "I believe so."

99. All of this evidence was given without objection from the defence but Mr Bevan QC (who represented the appellant at trial and at the first appeal but not before us) cross-examined the doctor. He put to him figures from other research that suggested that the figure of 1 in 8,543 for a single cot death might be much too high. He then dealt with the chance of two cot deaths and Professor Meadow responded:

"This is why you take what's happened to all the children into account, and that is why you end up saying the chance of the children dying naturally in these circumstances is very, very long odds indeed one in 73 million."

He then added:

"... it's the chance of backing that long odds outsider at the Grand National, you know; let's say it's a 80 to 1 chance, you back the winner last year, then the next year there's another horse at 80 to 1 and it is still 80 to 1 and you back it again and it wins. Now here we're in a situation that, you know, to get to these odds of 73 million you've got to back that 1 in 80 chance four years running, so yes, you might be very, very lucky because each time it's just been a 1 in 80 chance and you know, you've happened to have won it, but the chance of it happening four years running we all know is extraordinarily unlikely. So it's the same with these deaths. You have to say two unlikely events have happened and together it's very, very, very unlikely."

100. The table that was produced to the jury gave just the figures for probability of a SIDS death in families where one or more factors thought to be relevant were present together with the figures when there was no such factor. In the...report the table was accompanied by explanatory text but although this was available to the prosecution and the defence, it was not before the jury. It made clear the purpose of the information saying:

"The identification of families at higher risk of SIDS is of importance in allowing the appropriate deployment of scarce health care resources and in attempting to achieve changes in lifestyle or patterns of child care that might reduce this risk."

It did not in any way suggest that it provided statistical information that would enable diagnosis of an unnatural death in an individual case.

101. The report also made clear that the figures did not "take account of possible familial incidence of factors other than those included" in the table. It ended with the warning: "When a second SIDS death occurs in the same family, in addition to careful search for inherited disorder, there must always be a very thorough investigation of the circumstances – though it would be inappropriate to assume maltreatment was always the cause".

102. None of these qualifications were referred to by Professor Meadow in his evidence to the jury and thus it was the headline figures of 1 in 73 million that would be uppermost in the jury's minds with the evidence equated to the chances of backing four 80 to 1 winners of the Grand National in successive years.

The Royal Society of Statistics was sufficiently disturbed about the evidence given by Professor Meadow that it published its concerns on 23 October 2001:[6]

In the recent highly-publicised case of R v. Sally Clark, a medical expert witness drew on published studies to obtain a figure for the frequency of sudden infant death syndrome (SIDS, or "cot death") in families having some of the characteristics of the defendant's family. He went on to square this figure to obtain a value of 1 in 73 million for the frequency of two cases of SIDS in such a family.

This approach is, in general, statistically invalid. It would only be valid if SIDS cases arose independently within families, an assumption that would need to be justified empirically. Not only was no such empirical justification provided in the case, but there are very strong a priori reasons for supposing that the assumption will be false. There may well be unknown genetic or environmental factors that predispose families to SIDS, so that a second case within the family becomes much more likely.

The well-publicised figure of 1 in 73 million thus has no statistical basis. Its use cannot reasonably be justified as a "ballpark" figure because the error involved is likely to be very large, and in one particular direction. The true frequency of families with two cases of SIDS may be very much less incriminating than the figure presented to the jury at trial.

Aside from its invalidity, figures such as the 1 in 73 million are very easily misinterpreted. Some press reports at the time stated that this was the chance that the deaths of Sally Clark's two children were accidental. This (mis-)interpretation is a serious error of logic known as the Prosecutor's Fallacy. The jury needs to weigh up two competing explanations for the babies' deaths: SIDS or murder. Two deaths by SIDS or two murders are each quite unlikely, but one has apparently happened in this case. What matters is the relative likelihood of the deaths under each explanation, not just how unlikely they are under one explanation (in this case SIDS, according to the evidence as presented).

The Court of Appeal has recognised these dangers (R v. Deen 1993, R v. Doheny/Adams 1996) in connection with probabilities used for DNA profile evidence, and has put in place clear guidelines for the presentation of such evidence. The dangers extend more widely, and there is a real possibility that without proper guidance, and well-informed presentation, frequency estimates presented in court could be misinterpreted by the jury in ways that are very prejudicial to defendants.

Society does not tolerate doctors making serious clinical errors because it is widely understood that such errors could mean the difference between life and death. The case of R v. Sally Clark is one example of a medical expert witness making a serious statistical error, one which may have had a profound effect on the outcome of the case.

This is an excellent, and tragic, case example of how catastrophic it can be for an expert to step outside their expertise and how statistics can be misrepresented in court. The Royal Society of Statistics finished off its press release:

Although many scientists have some familiarity with statistical methods, statistics remains a specialised area. The Society urges the

Courts to ensure that statistical evidence is presented only by appropriately qualified statistical experts, as would be the case for any other form of expert evidence.

Further Reading

1. T. Busey and M. Klutzke, Calibrating the perceived strength of evidence of forensic testimony statements, *Sci. Justice*, 2023, **63**, 38–53.
2. M. Liden and I. Dror, Expert Reliability in Legal Proceedings: Eeny, Meeny, Miny, Moe, With Which Expert Should We Go?, *Sci. Justice*, 2021, **61**, 37–46.
3. I. E. Dror, Cognitive and Human Factors in Expert Decision Making: Six Fallacies and the Eight Sources of Bias, *Anal. Chem.*, 2020, **92**, 7998–8004.
4. European Network of Forensic Science Institutes, ENFSI Guideline for Evaluative Reporting in Forensic Science: Strengthening the Evaluation of Forensic Results across Europe (STEOFRAE) (March 2015), Available at ENFSI <enfsi.eu>
5. Association of Forensic Science Providers, Standards for the formulation of evaluative forensic science expert opinion, *Sci. Justice*, 2009, **49**, 161.
6. D. Lucy, *Introduction to Statistics for Forensic Scientists*, John Wiley & Sons Ltd, Chichester, England, 2005, 251 p.
7. Regina v. Sally Clark [2003] EWCA Crim 1020.
8. *RSS Statement regarding statistical issues in the Sally Clark case*, October 23rd 2001, http://www.therss.org.uk/archive/evidence/sclark.html
9. Regina v. Judith Ward (1993) 96 Cr.App.R. 1 [M62 bombings], Court of Appeal (Criminal Division) 04 June 1991, Summary at http://netk.net.au/UK/Ward.asp and full text at http://netk.net.au/UK/Ward1.asp.
10. Regina v. McIlkenny, Hunter, Walker, Callaghan, Hill and Power (1991) 93 Crim. App. R. 287 [The Birmingham Six], Court of Appeal (Criminal Division) 27 March 1991.
11. S. Connor, The science that changed a minister's mind, *New Scientist*, 29 January 1987, 1545, p. 24, retrieved 06 February 2023.

References

1. I.E. Dror, Cognitive and Human Factors in Expert Decision Making: Six Fallacies and the Eight Sources of Bias, *Anal. Chem.*, 2020, **92**, 7998–8004.

2. European Network of Forensic Science Institutes, ENFSI Guideline for Evaluative Reporting in Forensic Science: Strengthening the Evaluation of Forensic Results across Europe (STEOFRAE) (March 2015), Available at ENFSI <enfsi.eu>.
3. R. Meadow, ABC of Child Abuse, 1989.
4. Regina v. Judith Ward, (1993) 96 Cr.App.R. 1 [M62 bombings], Court of Appeal (Criminal Division) 04 June 1991, Summary at http://netk.net.au/UK/Ward.asp and full text at http://netk.net.au/UK/Ward1.asp.
5. Regina v. Sally Clark [2003] EWCA Crim 1020, https://www.bailii.org/cgi-bin/format.cgi?doc=/ew/cases/EWCA/Crim/2003/1020.html&query=(EWCA)+AND+(Crim)+AND+(1020).
6. *RSS Statement regarding statistical issues in the Sally Clark case*, October 23rd 2001, http://www.inference.org.uk/sallyclark/RSS.html.

14 Presenting Forensic Evidence in Court

By its very nature, a criminal investigation seeks to establish the crime that occurred, gathering information to build a case and then presenting that case to the court, whether it is an adversarial court (such as in the USA) or an inquisitorial court (such as in France). Concerning expert witnesses, this usually involves two stages: report preparation and the presentation of that report as evidence in court. When considering a career as an expert witness, many people focus more on the investigation work than the courtroom. However, the courtroom experience is very challenging and should not be underestimated.

This chapter serves as an introduction to various aspects of the court experience, including report preparation, court preparation, attendance and giving evidence.

14.1 The Operations of a Criminal Court of Law

Upon being arrested and charged with an offence, an individual is often referred to as the accused or the defendant. The person initiating the prosecution is known as the prosecutor and the person assisting the defendant is the defence lawyer or defence counsel; the terms 'lawyer' and 'counsel' here are interchangeable and are terms often used in the USA and countries such as Australia and New Zealand. In the UK, terms such as 'solicitor' and 'barrister' are far more commonly used and represent different roles.

Forensic Science: The Science Behind the Truth
By Ian Shaw with Anna Sandiford

Published by the Royal Society of Chemistry, www.rsc.org

The two main types of criminal justice systems are the adversarial system and the inquisitorial system. The adversarial system involves advocates for each party presenting their version of events to the court, reflecting the information provided by their client. The court, which is usually composed of a judge, jury or both, then makes a decision based on which account they believe to be true.

In the adversarial system, expert witnesses can be instructed by either side. The risk is that experts get confused about their roles and try to be advocates for the person instructing them rather than being the impartial party who is there to assist the court to understand subjects generally considered beyond the realms of common knowledge of the jury or other trier of fact. The advantage is that every expert witness instructed by the prosecution faces the possibility that their expert opinion will be picked over by an expert instructed by the other side. In theory, the adversarial approach encourages transparency and up-to-date knowledge. The adversarial system operates in countries such as the United States of America, United Kingdom, New Zealand and Australia.

The inquisitorial or non-adversarial system is less combative and involves the active involvement of the court in the investigation of the case, undertaking an information gathering process for itself rather than functioning merely as an observer like it is in the adversarial system. Expert witnesses can be asked by the court to consider the case and provide an opinion for the court and the judge, rather than being instructed by one party or the other.

There is usually a hierarchy to court systems, with the lower courts (Magistrates Courts in England and Wales; District Courts in New Zealand) dealing with crimes that attract lesser sentences and the higher courts (Crown Courts in England and Wales; High Courts in New Zealand) dealing with the most serious crimes.

If a person is found guilty of a crime but they do not accept the court's decision then there are processes that allow them to appeal that decision to even higher courts. These appeal courts have different names in different countries, such as the Court of Appeal, Supreme Court and Privy Council. Presenting a case in the appeal courts requires a lot of effort to demonstrate to the court that the case deserves to be heard at the appeal level.

All countries have weaknesses in their criminal legal systems and mistakes do happen. Beyond the appeal courts, there exists an option to apply for a complete case reassessment although this is not available in every country. Some countries have a formal legal body to which convicted parties can apply. Scotland, Canada, England

and Wales and Norway have criminal cases review commissions; New Zealand has Te Kāhui Tātari Ture|Criminal Cases Review Commission. These organisations often have legal powers of investigation. In countries like the United States of America, The Innocence Project often undertakes cases that are particularly likely to benefit from post-conviction DNA analysis but it does not necessarily have legal investigative powers.

14.2 Layout of the Court

Courts are laid out in different ways depending on the country, the type of court and the buildings where they are located (Figure 14.1). Most courts are rectangular. However, Gloucester Crown Court in England, for example, is hemispherical, so the court rooms are semi-circular and set out a bit like a theatre with tiered seating into the public gallery. The design of the court can have a big impact on the atmosphere within the court: a grand, noble courtroom in an architecturally imposing building adds more *gravitas* to proceedings than a concrete tilt-slab, plain building.

In general, the court bench where the judge or judges sit is at the head of the court. Often a registrar or other court official will sit in front of that. The witness box is towards that same end of the court to

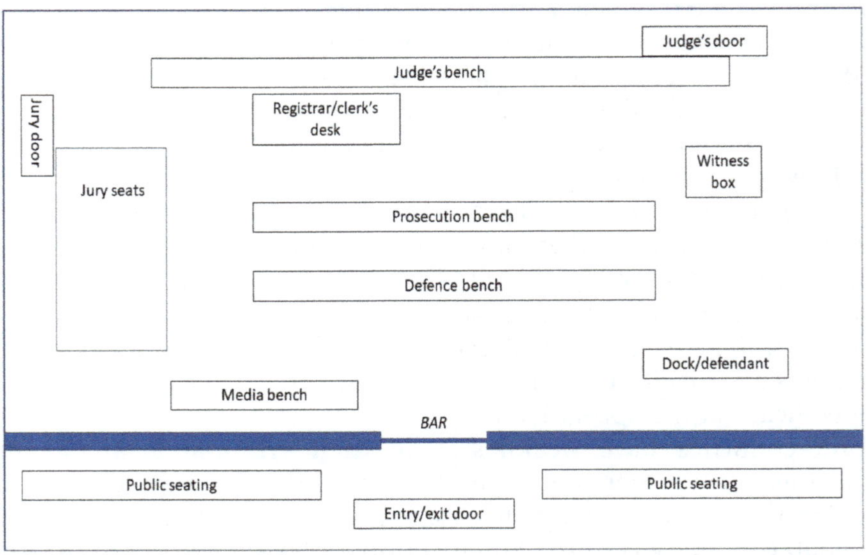

Figure 14.1 General layout of a court room in an adversarial jurisdiction.

one side. The prosecution will sit at the first main table facing the judge's bench and the defence will sit at a table behind the prosecution. The jury bench will be along one side of the court. The defendant will sit in a dock somewhere towards the back of the courtroom so that they can see proceedings; they could be behind a security screen and have security staff with them. There is a rail that crosses the whole room parallel to the judge's bench towards the back of the court. This is the 'bar' beyond which only lawyers, the defendant and court staff may go and representatives of the media and witnesses may enter only by invitation from the court. The general public sits at the back of the court and are there unless the judge says otherwise (for example, private legal discussions, sensitive witness evidence or if the case is not being heard publicly). Most trials are public so anyone can go and sit in on a trial.

Many movies show a court transcriber recording evidence on a transcribing device at the front of the courtroom – they are often asked to read back what someone just said in testimony. However, in developed countries, these days transcribers have largely been replaced by off-site, centrally located transcription agencies that work from recordings of court proceedings. The recordings are transcribed and sent back to the court electronically either to be printed and distributed or provided to counsel and the judge as an electronic document: the transcript of evidence (or equivalent term).

14.3 Roles of People in the Court

The judge or judges are arguably the most important people in the court. In the adversarial system, they control the trial process and set key dates by which things should happen, direct the jury on legal points throughout the trial, make legal decisions at relevant times throughout the trial process (such as whether or not particular witnesses will be allowed to give evidence) and, when applicable, issue sentences.

The court itself is often run by a clerk or a registrar. These are the people who help ensure that decisions of the judge are communicated to the relevant people, keep the court file updated, organise all the paperwork (electronic or old-school hard copy), liaise with the lawyers and police about witness appearances and basically keep the court running. For the Magistrates' Courts in England, the clerk of the court, who is legally trained, advises the magistrates on the law as the magistrates themselves are laypeople, not lawyers.

There will be a prosecutor representing the state or realm and they may also be a serving police officer trained in prosecuting cases. The prosecutor otherwise is a lawyer with a specialization in criminal law. The prosecution is required to prove its case beyond reasonable doubt and the defendant is presumed innocent until proven guilty.

There will also be a defence lawyer who acts on behalf of the defendant. The defence lawyer takes instructions from their client, acts on them in the court and is the legal representative for the defendant in court. In order to represent a defendant, the defence lawyer does not necessarily have to believe the defendant is not guilty; one of the most critical parts of the criminal justice system rests with the defence lawyer in ensuring equality of arms, representing the defendant fully and fairly in the system, arguing for the right for due process to be followed and that any sentence handed down by the court is appropriate within the law and case law.

Beyond the legally trained people and the court staff, cases are made up of witnesses. Witnesses are usually divided into two categories:

- lay witnesses who can give evidence about factual things, such as what they saw and
- expert witnesses who can give evidence based on their opinion, usually considered to be specialized knowledge beyond the normal or common knowledge of the general public.

The role and duty of the expert is to be impartial, assisting the court in understanding technical information and not to be an advocate for the party instructing them. While this may seem easy, bias can creep in because it is an inherent aspect of being human. The system therefore should incorporate checks and balances to make sure that experts stick to what they know and what they should be doing.

14.4 The Expert Witness

Expert witnesses face a variety of challenges when it comes to giving evidence in court. The court experience can be challenging and unpredictable. Anyone considering becoming an expert witness would benefit from watching a criminal trial, including expert witnesses, to see if they can imagine themselves in the witness box. It is a desolate and exposing place for the unwary and unprepared.

Coupled with this, expectations from the court, and juries in particular, require experts to manage those expectations while presenting their evidence.

A particular concern in the last decade or so has been what is called 'the CSI effect'. This developed after the widespread broadcast of television programs such as *CSI: Crime Scene Investigation*, which portray forensic science in an unrealistic manner, prompting courts and juries to have unrealistic expectations of what forensic science and expert witnesses can deduce from information gathered at a crime scene. Many papers have been published on the topic, succinctly summarised by the conclusion in Durnal's article:[1]

> *The CSI effect, as presented today is a far-reaching phenomenon in our criminal justice system. It both positively and negatively affects nearly all parts and pieces of the criminal process beginning with the time a crime is discovered to the day a suspect is sentenced. It is not the fault of the media that these types of effects occur. At no time has CBS or NBC, or ABC purported these television dramas to be a 100% accurate reflection of true life crime labs. It is in fact our society that likes to envision that sexy, smart, omniscient person that will be working on our side, should some false accusations arise. The CSI effect is very real. It may not influence every juror one way or another, but it most certainly has influence on the criminal justice system as a whole.*

Examples that have contributed to the CSI effect are episodes where, particularly in earlier series, forensic scientists enter a crime scene with little or no personal protective equipment and pronounce that they have found the single item within the scene that solves a major aspect of the crime, after having walked across the scene crushing underfoot many other potential items of evidence and destroying footwear marks on the floor in the process. They are also often involved in interviewing suspects or witnesses, which is often not the situation in reality. They also have an enormous range of cross-disciplinary skills that, in the real world, one person does not have. The analytical equipment and databases seemingly available on these programs are also highly unrealistic.

All of this presents to the audience a world of forensic science that simply does not exist except on television. Jurors are often unaware of this and are then surprised at the limitations of reality compared with the often science fiction-type laboratories, equipment and encyclopaedic knowledge of the 'experts' depicted on television.

To counter this effect, there have been instances where the prosecution or the defence needs to advise jurors that although the case being tried is an allegation of rape, a DNA expert will not make an appearance because the fact that sexual contact occurred between the two parties is not in dispute: DNA results generally cannot distinguish between consent and non-consent, so there is no need to test the results of the DNA interpretation in court in person with the reporting scientist.

In addition to the CSI effect, courts often expect expert witnesses to be able to answer questions in simple terms such as yes or no (*i.e.*, provide answers in black and white). In reality, many areas of forensic science do not provide answers in black and white terms but instead in shades of grey. The issue in any particular case will lie in determining the particular shade of grey to be described to the court and how it is taken into consideration with all other evidence that is heard by the court, much of which the expert witness will never hear and to which they will never be privy.

Another issue is that properly qualified and experienced expert witnesses who understand their overriding duty is to the court rather than the person instructing them may find themselves up against an expert who is either unsuitably qualified for the case in hand, who is a 'liar for hire' or a 'hired gun'. The perception is that this issue largely affects experts instructed by the defence but it also occurs in experts instructed by the prosecution.

A classic example of this was the case of an alleged expert witness in England called Gene Morrison who set himself up in 1977 as a forensic investigator. He acquired his qualifications from the internet because he said it "looked easier" than going to a real university, which might be true but does not make it the right thing to do. He was eventually investigated and prosecuted for a range of offences including perjury because although he had sham qualifications, including his doctorate, he refused to talk to the judge during his trial unless he was referred to as Dr Morrison. As he lied in court about holding a doctorate degree, he was subsequently charged with lying in court (perjury). It was revealed that the contents of many of his reports had been copied and pasted from the internet. The biggest issue was that he had worked on over 700 cases over more than 25 years before anyone realised he did not have the qualifications and training he said he had. By that time, he had undertaken a lot of his work for the Greater Manchester Police who then had to assess how many of the cases had to be re-investigated. To make matters worse, once he was on national television for his forensic science fraud, he was recognised and subsequently prosecuted and found guilty of rape and sexual assault offences against underage girls for which he was sentenced to an indeterminate term of imprisonment with a minimum sentence of 7.5 years. He was eligible for parole in 2017, by which time he would have been 58 or 59 years old; as of writing, he is still in prison. Not a poster boy for forensic science!

Although the court in theory is the gatekeeper to ensure that an expert is appropriate, it often takes an expert to recognise an expert, which is why inappropriate experts have ended up giving evidence in court. Once a person has been accepted by a court as an expert in a particular area, it is very hard to challenge them in subsequent cases on the grounds of not being an expert. The best option therefore is to ensure that only qualified experts give evidence in court.

A further issue for experts, particularly those who are relatively new to the forensic arena, is the unpredictability of a trial. Any experienced expert will say that although they can be as prepared as it is possible to be prior to giving evidence, the chances are that something obscure or unforeseen will be asked when they are giving evidence. This is because expert witnesses do not generally know everything about a case; they are instructed to address a relatively restricted issue and the dynamics of lay witness evidence can change the course of evidence as a trial progresses. An expert witness therefore has to be mentally prepared for the possibility that the experience of giving evidence in a particular case may take an unexpected turn.

14.5 Before the Evidence: The Report

Expert evidence is based on reports and statements prepared by the expert and provided to their instructing party in advance of a trial or other substantive hearing.

As expert evidence can be so significant in a trial outcome, setting and complying with standards for written reports is a critical step in ensuring that the court can trust and rely on an expert's opinion.

There are many forensic professional organisations around the world that set appropriate standards for approaching, recording, interpreting and reporting scientific investigations. The European Network of Forensic Science Institutes sets standards and there are many Scientific Working Groups based in North America that outline guidelines on good practice in specific areas of forensic science. Some forensic professional societies and associations require their members to go through an independent vetting process before being granted membership (*e.g.*, the Chartered Society of Forensic Sciences or the Faculty of Forensic & Legal Medicine of the Royal College of Physicians) and such organisations often have codes of conduct with which members musts comply or risk exclusion from the organisation. In some countries, there is legislation with which experts must comply, such as Part 19 Expert Evidence of the England and Wales Criminal Procedure Rules and the Forensic Science Regulators Act 2021. Other countries have guidelines that, although not compulsory in the criminal arena, experts are expected to follow, such as the New Zealand High Court Code of Conduct for Expert Witnesses. Overall, there are various ways in which an expert can demonstrate recognition of their understanding of the duty to the court. It is then the responsibility of the court to scrutinize and ensure that it is a genuine understanding, not just lip service.

Reports are meant to be clearly written, balanced (*i.e.*, consider both sides of the indicated events), logical, robust (including citing the relevant literature) and transparent (not hiding findings that may be considered unhelpful nor hiding mistakes) and shouldn't present interpretations that only support one side of the case. Any limitations of results and interpretations should be clearly communicated and reports should follow a sensible format, not unlike the method taught to many high school students, incorporating introduction, methods, results and conclusions. Technical information is often included, perhaps as an appendix, as any other expert should be able to assess how the report author reached the reported interpretations and conclusions. The report is a stand-alone document so that if it is picked up and read without any other context or documentation available to the reader, the reader can understand what the author was asked to do and how they accomplished it. Reports should be written with the final audience in mind, which could be people who have little or no knowledge of the topic being covered by the expert. There is little point including long, technical phrases if the reader does not know what they mean: there is no point in a doctor talking about a mandible if they do not immediately explain that this is basically the lower jaw.

14.6 Assistance Pre-trial and at Court: Evidence and Cross-examination

Suitably experienced experts often provide advice and assistance to lawyers prior to and during a trial.

14.6.1 Pre-trial

Communication with the expert witness prior to trial is never a waste of time. Many experts say that there is insufficient communication or inadequate briefing prior to report preparation and also pre-trial. Pre-trial discussions help the lawyer to understand not only the content of the expert's report but also the limits of what the expert and the area of expertise are saying. It is quite common, for example, for photographs in assault cases to be taken by police officers and the photographs often lack a size scale or colour scale. If the photographs are of really poor quality and better photographs are provided during the trial, the expert's opinion might change – better quality photographs often appear just prior to or during a trial.

Experts need to be sufficiently prepared for attendance at court including re-reading any statements and re-familiarising with the contents of the case file and the work completed. Particularly in extensive cases, where the expert can have several hundred pages of case notes, it helps the court and the jury if the expert can easily and quickly reach the relevant pages. Remembering the basic work completed helps the expert's confidence in giving evidence with the benefit of keeping the court engaged; long pauses while the expert reads his or her own file to find out what was done is boring.

14.6.2 At Court

Courts constitute an unusual working environment and they are all different. Practical issues that experts have to think about in advance are transport links to and from court, proximity of the court to train stations and parking because all of these can delay an expert's arrival at court; no court wants to be delayed and no expert wants to be the cause for a court's delay.

Additionally, many courts now have security procedures to follow before people are allowed into the building including bag searches and metal detectors or X-ray scanners. At peak times, these procedures are slow, in much the same way as airport security measures.

In addition to getting into the court, experts have to appreciate the practicalities of giving evidence, some of the most common of which are covered below.

14.6.2.1 The Amount of Waiting Time at Court

Trials and hearings are unpredictable and even the most efficiently prepared court proceedings are derailed or postponed by unpredictable events (changes or developments in the evidence of witnesses called before the expert; illness (the Covid19 global pandemic created a whole new set of problems); other more urgent court matters that delay current proceedings, such as a murder suspect needing to appear before the court the day after arrest). Many experts and jurors say they underestimated the tediousness of court proceedings.

Although giving evidence can be difficult, waiting around to give evidence can be just as arduous; five hours waiting at court is mentally exhausting before even getting into the witness box. Reading a book or a newspaper does not necessarily help because it helps to keep the case information foremost in the mind. Waiting can be a particular problem for young or inexperienced experts, so everyone manages this situation in their own way.

14.6.2.2 How to Address the Judge

Although lawyers are taught how to address members of the court and the roles of court staff, expert witnesses aren't always aware of the same protocols. "My Lord" is specifically an English term that witnesses and defendants sometimes pick up from television and use incorrectly. Judges are usually safely addressed as "Sir", "Ma'am" or "Your Honour".

14.6.2.3 The Written Report as Oral Evidence

When an expert prepares a report for court, they do not know to what use the report will ultimately be put. Experts therefore have to assume that any letter, formal report, brief of evidence or similar document will eventually be given to an opposing expert and to opposing counsel. Reports are written on the basis that they may be read out in court, either by the expert or by someone else so an expert learns early on never to document anything on paper that they do not want read out in open court.

Every report prepared by an expert has a casefile to go with it. In really big cases, such as reinvestigations of historical cases, the expert's casefile can be hundreds of pages long; in a recent New Zealand case, the total number of pages for the lawyer to review was 500 000, among which were several thousand pages of expert-related material that had to be hand-plucked and separated out – it's a massive task.

Many courts now allow experts to use a laptop computer to access electronic casefiles, but the efficacy of this method depends on the internet quality in the court, which is highly variable.

Some cases involve dozens of ring binder folders that are taken into the courtroom on a trolley, which presents a challenge when it comes to accessing them from the witness box and storing them in the court while the expert gives evidence. In one high profile case being filmed by television news stations for the 6.00 pm broadcast, the police officer in charge of the case was so focussed on getting his stacked-high trolley of files to the witness box without spilling them and embarrassing himself that when he got there he realised he had left his glasses in his car; court had to adjourn so that he could go and find them.

14.6.2.4 Oral Evidence

The skill of writing a good and clear report is different from being able to convert that into understandable, engaging information for the court to hear and understand. There is little point if an expert discusses an interesting topic in a monotone voice that makes the jury or other members of the court either falls asleep or want to. When giving evidence, a crash investigator once stopped partway through his evidence and said to the judge "your honour, Juror number 2 has fallen asleep." The judge's answer was, "you put him to sleep; you wake him up again."

Generally, experts are not allowed in the court to listen to other evidence before they give their own evidence unless in special circumstances. An example is in a drink driving case where an expert based their report on what the driver and other witnesses said in their statements. Sometimes, witnesses change their minds about their evidence between the time they signed a written statement and the time they give evidence in court (which can be weeks, months or years later). As the expert's calculations are based on what the witnesses said in their statements, the expert can, with the permission of the judge, listen to the oral evidence of witnesses who talk about the alcohol that was consumed. The expert then compares that information with the information originally provided and confirms whether or not the new evidence changes the expert's calculations. It is surprisingly common for witness evidence to change over time: in one drink-driving case, the written statements indicated that the driver had consumed one litre of beer prior to driving – based on this information, over the time period in question, the driver would not have exceeded the legal limit for driving at the time of the crash.

By the time of the trial, the oral witness evidence included a full bottle of wine in addition to the beer, which changed the calculations. When that happens, the court decides which account of drinking it believes. In that case, it believed both wine and beer had been drunk and the driver was found guilty.

When entering the witness box, the expert takes either a religious oath or makes an affirmation they will tell the truth. After that, the expert states their full name and occupation, and they might be asked to specify their geographical location. They will be asked to describe their qualifications and experience, which allows the court to assess whether or not the expert is an expert in the area relevant to the case at hand.

After that, the lawyer who called the expert leads the initial evidence from the expert; if an expert was instructed by the prosecution, the prosecutor leads the initial evidence after which the defence will be able to cross-examine the witness. Following that, the prosecution is allowed to re-examine the expert. The judge and jury might have questions after that. Once all the questions have been asked and answered, the expert is released by the judge and they can leave the court.

14.6.2.5 Cross-examination

Expert witnesses sometimes help lawyers conducting cross-examination of other experts to narrow down contentious issues and save court time. They can also help by assessing whether or not another expert is appropriately qualified, experienced or both to be providing comment in any given case – it takes an expert to know an expert.

An expert witness has to be prepared to account for their whole written or verbal opinions; if there is any doubt, the questionable material has to be omitted. Experts should be trained to use neutral language because bias creeps in everywhere. Even using a word such as "trauma" in forensic medical cases can cause an emotive association by lay witnesses. Medics often refer to blunt force trauma because they are talking about the skin and body receiving trauma as a result of something hitting it but for other people, trauma has emotional connotations. Blunt force impact is more neutral and means the same thing.

Sometimes, an expert feels a question cannot be answered with a simple yes or no. The way they manage this is by not answering with a yes or no but starting the answer with the *proviso* to the answer and expanding on their reasoning. Counsel does not like this approach but it allows the expert to get into evidence all the information the expert thinks the court needs to know. It is important for an expert to know when to stop talking because too much detail confuses matters for the court and the judge can intervene and halt the expert in mid-flow.

Expert witnesses are in the witness box to convey to the court often complex information in easily understood words and speech. They are not there to show off their personalities or make friends. They serve as information exchange and translators: complex scientific issues translated into understandable and accurate oral delivery. It can be tempting for experts to be condescending or give a smart answer but this carries risks because good counsel takes the opportunity to make a point for their own case. This is an example of an exchange during cross-examination between counsel and an expert witness:

> Counsel: "*You wouldn't take the silencer off before you shot yourself, would you?*"
> Expert: "*I probably wouldn't shoot myself*".

At this point, there was laughter within the court including from the expert and the jury, but not from the judge, court staff or counsel for either side.

> Counsel: "*I would hope you wouldn't but the tragedy is...that a lot of people who are seriously depressed do...commit suicide, that's the tragedy, isn't it?*"
> Expert: "*That's correct*".
> Counsel: "*It's not a laughing matter, is it?*"
> Expert: "*No, it is not*".

Although the expert had given valuable evidence up until this point, credibility was lost. Counsel chose this point to say "no further questions" and end the expert's cross-examination on a bad note for the expert.

Further Reading

1. E. W. Durnal, Crime scene investigation (as seen on TV), *Forensic Sci. Int.*, 2010, **199**, 1–5.
2. European Network of Forensic Science Institutes [ENFSI], ENFSI Guideline for Evaluative Reporting in Forensic Science:

Strengthening the Evaluation of Forensic Results across Europe (STEOFRAE), 2015, Available at ENFSI <enfsi.eu>.
3. England & Wales Criminal Procedure Rules 2020 Part 19, Expert Evidence, https://www.legislation.gov.uk/uksi/2020/759/part/19/made.
4. Faculty of Forensic & Legal Medicine of the Royal College of Physicians, Frequently asked questions, https://fflm.ac.uk/about/frequently-asked-questions-faqs/.

Reference

1. E. W. Durnal, Crime scene investigation (as seen on TV), *Forensic Sci. Int.*, 2010, **199**, 1–5.

15 Assembling the Forensic Results – Case Histories

Investigating criminal offences requires a team with varied skills employing a multidisciplinary approach. In many cases, this team identifies the right person who is ultimately convicted of the crime. Sometimes though, a case remains unsolved or the wrong person is investigated and convicted. This chapter compiles some examples of cases where changes over time have resulted in changes in the ultimate outcome. Every case can teach us something.

15.1 Introduction

Forensic science has advanced significantly from the early days when a blood stain 20 mm in diameter was required for blood grouping; we now only need a few cells from a crime scene to be able to generate a complete DNA profile of the individual from whom the blood originated. Our ability to detect microscopic amounts of biological material such as cells and obtain a DNA reference profile now far outstrips our ability to interpret those results in a particular case. For example, in a murder case where swabs are taken from a crime scene for DNA analysis to determine who was in the room at the time of the murder, detecting someone's DNA profile from a few cells that were picked up on the swab using a highly sensitive DNA technique does not mean that person was present in the room at the material time. Interpreting a result in that situation means that scientists have to consider whether the person whose DNA has been detected had

Forensic Science: The Science Behind the Truth
By Ian Shaw with Anna Sandiford

Published by the Royal Society of Chemistry, www.rsc.org

legitimate access to the room prior to the murder (could be hours, days or weeks before, depending on from where the DNA profile was recovered). If so, the presence of their DNA might not be related to the murder. The question these days generally revolves more around how the DNA was transferred to the place from where it was recovered, how long it could have been there and therefore whether it is relevant to the murder investigation or it may be just a remnant from a previous transfer of cells from that person (either directly from that person being there or through another transfer mechanism such as someone else's hands or clothing).

There are many examples of cases that have been revisited after several or many years where new scientific techniques have been applied to reveal important new information. This approach is particularly helpful in so-called 'cold cases', where police revisit a case and its exhibits, conducting testing of samples using new analytical techniques. These efforts have led to the identification, charging and conviction of new suspects. The converse is also true: some people who were convicted have been demonstrated to be innocent and released from prison for crimes they did not commit. This chapter compiles examples of both of these extremes.

Most criminal cases are complicated. Although advances in scientific techniques may add new pieces of evidence to a prosecution case or remove them, new scientific results may not always be enough to change the ultimate outcome of a case. Having said that, once the door to reinvestigation has been opened (*e.g.*, by applying new methods), it is not uncommon for other issues, that were not evident in the original investigation, to be identified. In some areas of expertise, there have been changes in the way of thinking and of interpretations of results rather than any significant advancements in analytical techniques – these changes are as important as the development of new methodology.

15.2 Standard DNA Analysis

One of the greatest advances in cold case reviews and historical case reinvestigations has been the developments in DNA techniques and the expanded possibilities they offer for analyzing results in old and cold cases. The Innocence Project in the USA was established in 1992 by two lawyers to address a gap in the system concerning the testing of samples for the presence of DNA (see Chapter 6 *DNA*). The Innocence Project is specifically focussed on undertaking DNA analyses on case samples after a person has been convicted; in the USA, just because samples exist does not mean that the court will allow them to be processed even if the defence believes that they may assist in the investigation and potentially prove a defendant's innocence of charges they face. As summarised by the Innocence Project,

> *"Because shifts in scientific understanding often take decades to emerge, individuals whose wrongful convictions were based on misapplied science might face difficulties in proving their innocence due to time limitations and high evidentiary standards. In addition, some state courts do not recognize discredited scientific evidence as new evidence of a wrongful conviction."*

The Innocence Project has advocated for post-conviction DNA testing and there are many cases of post-conviction DNA analysis demonstrating that the convicted person did not commit the crime. One of the most famous is the Kenny Waters case.

15.2.1 The Kenny Waters Case, Massachusetts, USA, 1983

In 1983, Kenny Waters was convicted of murdering his next-door neighbour in 1980. The scene was messy and bloody with the deceased having been stabbed multiple times, hairs including some associated with the hand of the deceased were recovered, and fingerprints were recovered from various items. Betty Anne Waters, sister of Kenny, put herself through law school so that she could be her brother's lawyer. Ultimately, she got access to exhibits held at the court and, in conjunction with the Innocence Project, in 2001 she had items from the crime scene examined and tested for DNA. The DNA results demonstrated that her brother had not been the perpetrator; he was released after 18 years in prison.

As a result, the case was reinvestigated and evidence was released from old police files that showed that Waters had been excluded as a likely suspect very early on in the original investigation but that information had been withheld from prosecutors and the defence. The story was made into a 2010 movie *Conviction* starring Hilary Swank. There was much more to the case than just DNA, but because it is such a powerful discriminating tool, good DNA results can provide the strong leverage needed to either overturn a conviction, restart a police investigation, or both - this case illustrates that very well indeed.

15.2.2 The Bain Family Murders, Dunedin, New Zealand, 1994

The Bain family consisted of Margaret and Robin Bain who had four children aged as follows at the time of the deaths: David the eldest (22), then Arawa (19), Laniet (18) and Stephen (15), the youngest.

In 1994, the children and Margaret lived in a large, two-storey, run-down wooden house in Dunedin, New Zealand. As Robin and Margaret were estranged, Robin spent his weekends living in a small caravan in the back garden of the property and, during the week, in a van at the school where he was the headteacher.

On the morning of 20th June 1994, four of the five members of the family died of gunshot wounds from a 0.22 rifle fitted with a suppressor, a rifle that was known to belong to David. Only David survived; he was uninjured. The scene was initially examined by a forensic pathologist who concluded it was a murder-suicide and that Robin had killed Margaret, Arawa, Laniet and Stephen and then killed himself. This was later reappraised and changed to murder by David of Robin, Margaret, Arawa, Laniet and Stephen.

It was clear that an altercation or interaction had taken place between Stephen and his assailant in Stephen's bedroom. He had been shot through the hand and had a scalp injury, both of which generated a lot of blood.

A focus of the first trial in 1995 was a graze to one of David's knees. A fragment of skin was recovered from Stephen's bedroom and the Crown argument was that this piece of skin had been removed from David's knee during the struggle, thus placing David in Stephen's room at the time and suggesting he was the person who shot Stephen. A photograph of David's grazed knee was included in the police photograph booklet that was presented to the jury.

Under New Zealand law, cases can be reviewed by the Judicial Committee of the Privy Council in London if the case pre-dates 2004 (due to a law change in NZ, recourse to the UK's Privy Council ceased from 1st January, 2004). After a review by the Privy Council in 2007, a retrial was ordered. Both prosecution and defence worked towards the 2009 new trial date. Part of that work involved the defence assessing which exhibits still existed. This revealed that although most exhibits in New Zealand had been destroyed by that time, the piece of skin from Stephen's room was still held at an Australian police laboratory where it had been sent by the police. The defence team instructed the laboratory to analyse the skin for DNA. The result showed it originated from Stephen, not David. This changed a significant part of the case and the prosecution was no longer able to establish David's presence in Stephen's room at the time of the struggle and ultimately Stephen's death. The first trial in 1995 lasted around three weeks and as a result of his conviction, David Bain spent around 11 years in prison before the retrial process started. The second trial in 2009 lasted over three months with David Bain being found not guilty of all four murders.

15.3 Familial DNA

Familial DNA (see Chapter 6 *DNA*) searching is a technique whereby DNA recovered from a scene is compared with a DNA database (see Chapter 6 *DNA*). Specific statistical analyses are applied to the DNA results in order to identify a possible relative of the person who left DNA at the crime scene. Police will consider the use of familial DNA when they have little information to go on and need to consider other avenues of investigation (see also Chapter 6 *DNA*: The Bobby Dunbar case). Examples of two such cases that have been successfully prosecuted are the Golden State Killer in the USA and Joseph Reekers in New Zealand.

15.3.1 The Golden State Killer, California, USA, 1974 to at Least 1986

The Golden State Killer is now infamous because he was caught using a nonstandard statistical approach to DNA databases. His criminal activities started in 1974 and continued until 1986 when they seem to have stopped abruptly. The offences to which he eventually confessed include at least 13 murders, more than 50 rapes and 120 burglaries across California, USA. His offences were such that they contributed to the establishment of California's DNA database, which feeds into the FBI's Combined DNA Index System (CODIS) that was established in 1998. The database now holds more than 20 million DNA profiles of convicted individuals from across the whole country (see also Chapter 6 *DNA*).

In 2017, with numerous cases linked by *modus operandi* but no identified suspect, the police decided to upload the DNA profile obtained from swabs collected from murder or rape victims to an open-source genealogy website. This led to the identification of up to twenty people who shared the same great-great-great grandparents as the offender. From this, and with the assistance of a genealogist, they created a very large family tree. Each member of the family tree was eventually excluded as a potential suspect with the exception of one.

Police needed a reference DNA sample from their suspect in order to compare his DNA profile directly with the crime scene DNA samples. This they did by collecting covert samples, one from the driver's door of his vehicle and one using a facial tissue from his rubbish bin left at the roadside for collection. Following confirmation that his DNA profile could not be excluded as being the source of the crime scene DNA, he was arrested. In June 2020, Joseph DeAngelo (Figure 15.1) entered guilty pleas to 13 counts of murder and 13 counts of kidnapping with which the police were able to charge him (the statute of limitations meant he could not be charged with many of the other crimes he had actually committed). This was part of a plea deal in which he would not face the death penalty; instead, he was sentenced to multiple terms of life imprisonment.

The case has since raised the ethics of using ancestry DNA databases for familial DNA searches in criminal investigations, given that a search warrant is not required as the databases hold publicly available data albeit not provided for use in criminal investigations. It has also raised the issue of ethical considerations surrounding the collection of covert samples. Covert samples are those that are collected without the knowledge of the person from whom the sample is

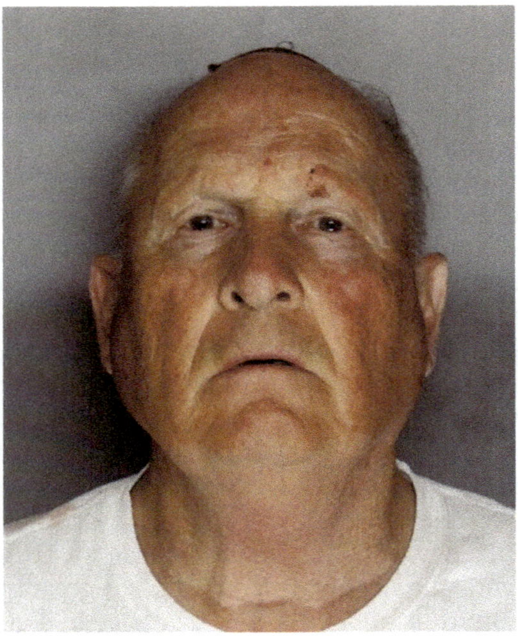

Figure 15.1 The Golden State Killer, Joseph DeAngelo. Figure is in the Public Domain.

being collected. An example would be a used facial tissue from a rubbish bin or a cup from which someone has drunk. Not all countries approve either of these approaches and therefore the approach used by police differs from country to country.

15.3.2 Murder of Marie Jamieson, Auckland, New Zealand, 2001

Marie Jamieson was a 23-year-old who disappeared on her way home from a night out in central Auckland, New Zealand, on 10th February, 2001. Her body was found 9 days later, on 19th February. February corresponds to summer in the southern hemisphere and her body had started to decompose, which made it difficult for the pathologist to determine the exact cause of death. However, he was able to determine that she had received at least three stab wounds. Swabs were taken from her body and clothing and tested for DNA but there were no matches with the DNA profiles from the national DNA database.

It took another seven years for the police investigation to progress: the police requested a familial DNA search of the national DNA databases from the state scientific laboratory, the Institute of Environmental Science and Research (ESR); the search identified (because of DNA similarities) the name of a close relative of a man called Joseph Reekers.

Reekers himself was not on the DNA database and although the police suspected him of being involved in the death of Marie Jamieson, he could not be forced to provide a DNA sample – unless he committed another crime requiring a DNA sample to be taken. He was later convicted for shoplifting, which allowed police to collect a reference DNA sample from him. This was compared with the DNA from the crime scene and the DNA result indicated that he could not be excluded as the source of male DNA from the swabs taken from Jamieson's body and/or clothing. Reekers entered a guilty plea in December 2009; this was the first case in New Zealand in which familial DNA was used to secure a conviction.

15.4 Pathology

Opinions on certain topics in forensic pathology have changed over time, such as time since death based on stomach contents or the interpretation of biomechanics of injuries caused to babies and infants. Additionally, not all pathologists are forensic pathologists and not all *post-mortems* adhere to forensic standards; forensic *post-mortems* are generally far more intricate than non-forensic *post-mortems* (see Chapter 8 *Forensic Pathology*).

A facet of forensic pathology often depicted in television programmes is the determination of whether an offender is left- or right-handed based on the position of stab wounds on a deceased individual or another surface like a cavity wall (which can happen if the person doing the stabbing misses their human target and strikes the wall behind them). In reality, determining handedness is extremely difficult as there are too many variables that cannot be determined because of constant changes not only in people's relative orientations during an incident but also the positions of their arms, legs and hands; people can move a weapon from their dominant hand to a less dominant hand; weapons can slip and move at the moment an injury is inflicted; clothing and surfaces can influence the way an injury occurs to a person or how damage can be caused to property. Overall, assaults are dynamic where participants do not remain stationary in a face-to-face stance while events unfold.

A prime example of the change in this evidence was the case of Alan Hall.

15.4.1 The Alan Hall Case, Auckland, New Zealand, 1986

Alan Hall (Figure 15.2) was convicted of murder in 1986. He was charged with stabbing to death Arthur Easton, who was attacked in his own home in 1985. Hall was also charged with the injury of Easton's two sons who tried to stop the attack on their father. Part of the prosecution's original case was that the offender was left-handed based on the forensic pathologist's opinion; Hall was left-handed, so the forensic pathologist's report stating the offender was left-handed proved to be immensely helpful for the prosecution case. Reinvestigation of Hall's conviction included assessment of the forensic pathology. One issue explored through a contemporaneous lens was the handedness issue; it was concluded it is not possible in this case to determine the handedness of the murderer. There are many scenarios that could have led to the injury, eloquently put by a forensic pathologist in a similar case where he could not exclude the stab wound being caused by a "*monkey hanging upside down on a trapeze*" – this humorous comment makes the point that there was, and often is, simply not enough information to support one scenario over any other. The case reinvestigation also revealed that police had

Figure 15.2 Alan Hall. Convicted of murder in 1986, partly due to unfounded science. His conviction was quashed by the New Zealand Court of Appeal in 2022. Image courtesy of Alan Hall, December 2022.

altered witness evidence indicating that Hall was not the offender and withheld other information from the defence team that supported Hall's innocence. Hall's conviction was eventually overturned 36 years later, in 2022; he spent 19 years in prison in what is currently considered New Zealand's worst miscarriage of justice.

15.5 Origin and Cause

'Origin and cause' (sometimes incorrectly referred to as 'cause and origin') is the process by which fire experts assess fire scenes: the first step is to identify the origin of the fire and then progress onto what caused it. Origin and cause work can be critical in criminal investigations and a great example of testing a theory was undertaken by the television program, *Mythbusters*, in 2005. Although it had been known in the forensic science arena for a long time that it was not – and is not – possible, *Mythbusters* tested the myth that a lit cigarette thrown into a pool of gasoline will ignite the fuel or the vapour, causing a fire. The *Mythbusters* crew made several attempts to light a pool of gasoline using a lit cigarette, including rolling a lit cigarette into a pool of gasoline on the floor – there was no joy, and no fire. They concluded it was highly unlikely for a fire to be started by that method.

15.5.1 Arson and Murder: The Case of John Galvan, Chicago, Illinois, USA, 1986

A 1986 case example illustrates the importance of 'origin and cause': John Galvan was accused of starting a fire by throwing a bottle filled with gasoline against the side of a building, and then he threw a lit cigarette into the pool of fuel, causing it to ignite and resulting in the deaths of two men. Galvan was found guilty in 1989, with the fire-starting evidence playing a significant role in the prosecution's case.

In 2007, 21 years into his life-without-parole sentence, Galvan watched a re-run of the *Mythbusters* episode. He raised the issue with his lawyer, who had coincidentally seen the same re-run; this initiated a chain of events which led to a review of the fire findings from the first trial. Also in 2007, the US Bureau of Alcohol, Tobacco and Explosives undertook experiments to ignite gasoline with cigarettes in different scenarios. After over 2000 attempts, they had still not managed to ignite the fuel. In 2012, during a case hearing, a forensic scientist gave evidence that the method for starting the fire was scientifically impossible and it is agreed in modern fire science that a burning cigarette cannot ignite a flammable vapour.

It was not until 2022 that the charges against Galvan were dismissed; by this time, he had spent more than 35 years in custody and prison based on scientifically impossible evidence.

15.6 Medical Evidence Changed – The Sally Clark Case, England, 1998

The Sally Clark case was mentioned in Chapter 13 *Probability in Forensic Science* in a discussion on the incorrect use of statistics by a paediatrician. Clark was charged with and convicted of the murders of her two infant sons. As is often the situation in these sorts of cases, more than one issue contributed to the outcome of the trial. In Clark's case, the issues were compounded by changes in expert opinion by the pathologist, Dr Alan Williams, who undertook the *post-mortems* on her two children, Christopher (born in September 1996 and died in December the same year) and Harry (born in November 1997 and died in January 1998).

He also failed to disclose to anybody, including the police and prosecution, that microbiology tests were completed on multiple samples from Harry, the results of which, in light of other findings, may have suggested a death from natural causes. Queries were raised by the jury during the trial about what blood samples were taken, but Dr Williams did not take the opportunity to explain that, as is often done in *post-mortems*, samples had been taken from multiple sites around the body (see Chapter 8 *Forensic Pathology*). When the test results were uncovered and the impact of his failure to disclose them in the case was exposed, he responded

"It is not my practice to refer to additional results in my post-mortems unless they are relevant to the cause of death, as the specimens were referred to another consultant".[1]

The Court of Appeal summarised this response in its judgment.[1]

"We find that explanation wholly unacceptable. If it does correctly state Dr Williams' practice, then on the evidence available to us his practice is completely out of line with the practice accepted by other pathologists to be the standard. It is likely to mislead others, who may work on the same case and who will be denied the opportunity of considering the material in the way that Dr Williams explained that he found necessary, in reaching their own properly informed conclusions. It runs a significant risk of a miscarriage of justice."

Significantly,

"It is tantamount to saying "If I can discount it, nobody else need consider it". As an approach it only has to be voiced for the inherent danger to be obvious."

The matter was reviewed by a specialized forensic pathologist in Australia with expertise in sudden infant deaths, who helpfully summarised the issues relating to Dr Williams' work as follows:[1]

"Unfortunately there appears to have been significant and ongoing problems in the investigation of these deaths. Standard protocols were not followed and essential steps such as routine dissection and histology were omitted which prevented verification of alleged autopsy findings. As well, a number of potentially important diagnoses and conclusions were altered over time. For example, Christopher's initial cause of death of lower respiratory tract infection was withdrawn, observations of no significant haemorrhage within his lungs were changed to marked haemorrhage...brain lacerations were found to represent post-mortem *artefact, swelling of the spinal cord was not present and bruising of paraspinal tissues was also not able to be substantiated. This is not a*

unique situation with statements in the literature in recent years that 'investigations into the pathology and circumstances of sudden infant death are often scanty and inexpert' with significant omissions being documented when cases were audited. The Clark brothers demonstrate difficulties that may arise if cases are not fully investigated with all of the results being clearly summarised and discussed in the autopsy report. Trying to clarify findings, diagnoses and circumstances of death at a later stage may simply not be feasible due to a wide variety of possibilities other than inflicted injury".

Clark's convictions were overturned in 2003 on various grounds including the incomplete and changed pathological evidence provided by Dr Williams and the erroneous statistical evidence presented by Professor Meadow, five years after she was first arrested. Her family said that she never recovered from the ordeal of the death of her children, the subsequent trial and convictions. She died in 2007 at the age of 42.

Further Reading

1. Decision of the Judicial Committee of the Privy Council: Bain v. The Queen (New Zealand) [2007] UKPC 33 (10 May 2007), http://www.bailii.org/uk/cases/UKPC/2007/33.html.
2. P. Saukko and B. Knight, *Knight's Forensic Pathology*, CRC Press, London, 4th edn, 2015, p. 666.
3. D. Selby, How Discovery Channel's 'MythBusters' Helped a Wrongly Convicted Man Prove His Innocence, *Innocence Project News*, 10 May 2022, https://innocenceproject.org/discovery-channel-mythbuster-john-galvan-wrongful-conviction-innocence/, retrieved 16 December 2022.
4. England and Wales Court of Appeal Judgment: R v Sally Clark [2003] EWCA Crim 1020.
5. C. Dyer, Pathologist in Sally Clark case suspended from court work, *Br. Med. J.*, 2005, **330**, 7504.
6. https://innocenceproject.org/overturning-wrongful-convictions-involving-flawed-forensics/.

Reference

1. England and Wales Court of Appeal Judgment: R v Sally Clark [2003] EWCA Crim 1020.

Epilogue Forensic Science: Where Next?

On that day in Lyon, France, in 1912, when Dr Edmund Locard persuaded the French police to fund a lab to allow him to apply science to crime investigation, he could never have dreamt that the seeds of forensic science that he was just about to sow would germinate and grow into an international discipline utilising sophisticated technology underpinned by key, fundamental scientific principles. He could not have imagined that forensic science would constitute the important and often indisputable scientific evidence in some of the world's most notorious crimes, and that it would provide the evidence that led to acquittals and convictions.

Locard's basic principles coupled with the introduction of DNA technology into forensic casework have rendered DNA results almost uncontestable. As a result of this, DNA findings are now so important that they often displace most other forensic investigations. This is an incredible speed of evolution of a scientific discipline: in just over a century, forensic science has moved from focussing on the transfer of fibres identified by their colour and form using light microscopy to the sophistication of unique DNA sequences transferred during a crime. Amazing!

In many countries, the importance of forensic science in crime investigations prompted their police forces to establish forensic labs, which were subsequently grown and developed by the countries' relevant government ministries (*e.g.*, the Home Office in the UK). To some extent, this arrangement introduced a bias: the government department responsible for crime investigation also controlled the

Forensic Science: The Science Behind the Truth
By Ian Shaw with Anna Sandiford

Published by the Royal Society of Chemistry, www.rsc.org

laboratories that provided forensic evidence as part of the same crime investigations. Indeed, I remember in my days as a forensic science student in the 1970s, when the police officer in charge of a case brought the material for investigation to the forensic lab, he or she would tell you what they were looking for. For example, 'We're looking for fibres from the accused on the murder victim's clothing.' This might have influenced the forensic investigation and so introduced bias. I hope the scientists eliminated this bias, but we cannot know. I introduce this here because this issue may underpin some of the changes that occurred in the organisation of forensic science in the 1990s and might be an important focus for future forensic science research. But we will delve into more of this later.

Because of the need for highly sophisticated technology, forensic science investigations are expensive. This meant that the 1970s approach, where the police requested the lab work they wanted (rather than necessarily needed) for their case investigations, incurred huge costs. In the 1980s and 1990s, there was a much greater awareness of government department costs, which meant that the police had to become accountable for the money they spent. They had to adhere to a budget and the concept of 'pay for service' was introduced. From this point on, forensic science labs charged the police for the investigations they carried out. This, to some extent, was futile because it was in reality simply cross-charging between divisions in the same government department. Nevertheless, it made the 'bean counters' happy, but more importantly it precipitated further change in the provision of forensic science services to the police.

At this point, it is important to appreciate that similar changes were happening in many countries; the speed of change and exact nature of the changes were different, but the overall outcome was the same: payment for forensic services.

The inclusion of monetary costs as a factor in deciding which forensic investigations to carry out in a particular case significantly changed the face of forensic investigations. There was no longer carte blanche. Forensic scientists could no longer operate freely as a multidisciplinary team until they found something interesting. Instead the police told the scientist what they wanted based, in part, on the cost of the particular service and their budget for the case.

As forensic methodology developed, particularly the introduction of DNA technology to forensic casework in the mid-1980s, the police requested the techniques that they thought might provide the best evidence in their case. After 1985, DNA was the method of choice because of its evidential strength in court. This was the beginning of the most significant change in forensic science to date. Forensic labs shifted a significant portion of their operational laboratory work to DNA extraction and analysis because this is what their client, the police, wanted. This also marked the change from police and forensic science colleagues working together to solve a crime, to a provider/client relationship. The police purchased the investigations they wanted and the forensic science labs developed their portfolios of provision to suit their client's needs. DNA ruled! The forensic science service had been engulfed by the market economy. This was a significant change in the way forensic science expertise was applied to crime investigations.

Regardless of whether finances determined which forensic investigations were applied to a particular case, forensic scientists continued, driven by their innate desire (and perhaps need) as scientists, to advance their science. So, forensic science moved on apace as new scientific techniques were developed and applied to forensic investigations, and as our understanding of the basic sciences that underpin forensic investigations evolved. This arguably had the most significant impact on the application of DNA technology in forensic investigations. Back in the 1980s when DNA technology was first applied to forensic casework, the process required 'large' amounts of extracted DNA that was separated by electrophoresis to produce bands according to the array of DNA fragments in the extract. The electrophoresis pattern was compared between samples. Thirty years later, Kary Mullis (1944–2019) and Michael Smith (1932–2000) shared the Nobel Prize in chemistry for inventing PCR (see Chapter 6 *DNA*), which 20 years later revolutionised the use of DNA in forensic casework because it significantly reduced the amount of DNA needed for analysis – our ability to detect DNA now far outstrips our ability to interpret its meaning and relevance in a particular case. This is an excellent example of the scientific evolution of a key forensic method. But is it possible to make more revolutionary changes in forensic methodology in the future?

In 2015, the American Academy of Forensic Sciences (AAFS) Interdisciplinary Symposium brought together 'preeminent visionaries' in their respective forensic fields in Orlando, Florida, USA. The objective of the meeting was to 'provide a diverse interdisciplinary view of the direction forensics will take in the years to come.' They focused on one decade from the year of the meeting (*i.e.*, 2025). And rather than attempting to predict what might happen in forensic science as a whole, the meeting focused on some individual disciplines (*e.g.*, forensic toxicology and document examination) that comprise forensic science. Four years after the AAFS meeting, a book entitled *The Future of Forensic Science* was published (see *Further Reading*). The book's title is a little grandiose, since many aspects of forensic science are not covered; nevertheless, it is worth taking a look at a few of the experts' predictions with the benefit of hindsight (I am sitting in my study writing this epilogue in 2023).

The definition of a document has evolved over the past decade or so, with electronic documents now taking precedence over 'traditional' handwritten or typed documents. The AAFS symposium interestingly focused on the need to hone document examination techniques, rather than predicting a sea change in the approach to document examination. The application of neuroscience to disguised handwriting analysis, using IT to increase the objectivity (*i.e.*, removing analyst bias) of handwriting comparisons, and automating IT approaches for handwriting comparisons were predicted to be the futuristic developments in the document examination field. There was no mention of e-documents or computer printers and cybercrimes. Perhaps this was not on the horizon in 2015. The AAFS symposium document examination predictions seem rather pedestrian to me. I perceive advances in the investigation of cybercrimes as of paramount importance as the electronic world takes over so many of our transactions.

The evolution of toxicology over the past few decades has been driven by significant advances in methodology, particularly mass spectrometry coupled with liquid chromatographic techniques (see Chapter 6 *Forensic Chemistry*). These changes have revolutionised the identification and quantitation of drugs and toxins in biological samples. Only 30 years ago, forensic toxicologists had to extract grams of *post-mortem* liver tissue from poisoning cases to solvent extract enough drug from the tissues to reliably identify and quantify it. Nowadays, small (μL) quantities of blood can be analysed directly simply because the mass spectrometry technology has improved tremendously. I have no doubt that this will continue. Indeed, we are fast moving into the era of small-scale high resolution mass spectral devices that can identify and quantify minute quantities of drugs or toxins in tiny samples. We are almost at the point when a small blood sample in a forensic case could be reliably analysed directly using such technology. The role of the forensic toxicologist would then evolve from reliance on their analytical skills to simply interpreting the results from sophisticated, small-scale analysers and deciding whether the drug or toxin identified was responsible for the victim's death.

It is impossible to know what is around the corner for forensic science, but reading the current scientific literature, it seems to me that we may not witness a DNA-like revolution that our predecessors experienced in the 1980s, but rather there seems to be a more gentle (but nevertheless significant) improvement in methodology. In addition, I see financial constraints as a significant determinant of which methods are applied to forensic investigations; this in turn, is likely to drive funding towards method development rather than researcher-led curiosity. I worked for a short time at the UK's Home Office Central Research Establishment (HOCRE) many years ago; this was a vibrant focus of forensic science research and employed some of the world's top forensic scientists. It no longer exists: I suspect that someone in a government department deemed it too costly...

Finally, my reading of the forensic science literature shows an increasing interest in assuring the validity of the results of lab studies, in particular, removing operator bias. This is crucially important, but not particularly alluring in a research context. I think it is likely that researchers will continue to scrutinise forensic science decision making to increase the robustness of the decisions we make. I see this as a very important focus for the future of forensic science.

Further Reading

1. D. A. Martell, *The Future of Forensic Science*, John Wiley & Sons Ltd., Chichester, UK, 2019.
2. A. F. Moussa, Electronic evidence and its authenticity in forensic evidence, *Egypt. J. Forensic Sci.*, 2021, **11**, 20.
3. G. S. Cooper and V. Meterko, Cognitive bias research in forensic science: a systematic review, *Forensic Sci. Int.*, 2019, **297**, 35–46.
4. I. E. Dror, Cognitive and Human Factors in Expert Decision Making: Six Fallacies and the Eight Sources of Bias, *Anal. Chem.*, 2020, **92**, 7998–8004.

Subject Index